T0295473

Fundamentals of Electroweak Theory

Second Edition

Jiri Horejsi
Charles University, Czech Republic

NEW JERSEY • LONDON • SINGAPORE • BEIJING • SHANGHAI • HONG KONG • TAIPEI • CHENNAI • TOKYO

Published by

World Scientific Publishing Co. Pte. Ltd.
5 Toh Tuck Link, Singapore 596224
USA office: 27 Warren Street, Suite 401-402, Hackensack, NJ 07601
UK office: 57 Shelton Street, Covent Garden, London WC2H 9HE

Library of Congress Control Number: 2024938771

British Library Cataloguing-in-Publication Data
A catalogue record for this book is available from the British Library.

FUNDAMENTALS OF ELECTROWEAK THEORY
Second Edition

ISBN 978-981-12-9166-1 (hardcover)
ISBN 978-981-12-9167-8 (ebook for institutions)
ISBN 978-981-12-9168-5 (ebook for individuals)

For any available supplementary material, please visit
https://www.worldscientific.com/worldscibooks/10.1142/13798#t=suppl

Desk Editor: Carmen Teo Bin Jie

Typeset by Stallion Press
Email: enquiries@stallionpress.com

Printed in Singapore

Preface to Second Edition

In 2012, 10 years after the first edition of my book *Fundamentals of Electroweak Theory*, the long-awaited Higgs boson was discovered, and the edifice of the standard model (SM) was thereby completed. The discovery came despite the skepticism and doubts of many prominent physicists (in this context, see, e.g., the essay in [1]), and it certainly represented one of the most important milestones in the development of particle physics. On the other hand, during the past two decades, there was no other breakthrough discovery that would reveal clearly a new physics beyond SM (though some well-known open problems persist, which apparently cannot be solved within SM). Taking all this into account, I have found it appropriate to refurbish the relatively old original text of my book, since it is still commonly in use [however, the first edition is currently out of stock]. Apart from incorporating the Higgs boson as a physical reality, I have also corrected some misprints and other minor mistakes found in the original version. Furthermore, as a bonus for true enthusiasts, I have added some new problems to be solved (or at least contemplated); these extend the corresponding sections of Chapters 2, 5 and 7. As in the first edition of this work, the main emphasis is placed here on fundamentals rather than the current phenomenology of electroweak physics. Anyway, the phenomenological aspects of SM are extremely important, and so, it is very helpful that several other new books by other authors appeared during the past 20 years, where one may find, among other things, a detailed treatment of various aspects of electroweak physics that are not covered by the present text. At least some of them are duly included in the updated bibliography; they are marked as [Alt], [Lan], [Pal] and [Pas], respectively. Some other new items have also been added explicitly to the list of references. Needless to say, the relevant literature on SM physics (and beyond) is vast; a rather comprehensive list can be found, e.g., in [Lan].

I am indebted to Karol Kampf for preparing the current LaTeX file of the manuscript. I would also like to thank Jonáš Dujava, one of the students who attended my classes in the spring semester of 2023, for carefully reading the whole text and finding several misprints and other minor errors.

Prague, January 2024 Jiří Hořejší

Preface to First Edition

This work is an extended version of a one-semester course of lectures on the theory of electroweak interactions that I taught regularly at the Faculty of Mathematics and Physics of the Charles University in Prague during the 1990s. I have also included here some selected material from my earlier courses, delivered at the same school in the second half of the 1980s. Throughout those years, I could benefit from the feedback provided by the students who attended my lectures; thus, I believe that the contents of the present text is properly tuned to the needs of an uninitiated reader who wants to understand the basic principles of the electroweak Standard Model (SM) as well as their origin and meaning. For pedagogical reasons, I have adopted partly a historical approach — starting from a Fermi-type theory of weak interactions, explaining subsequently the theoretical motivation for intermediate vector bosons and only then proceeding to the basic concepts of the gauge theory of electroweak interactions.

The main point of the discussion of the old weak interaction theory (Chapters 1 and 2) is to demonstrate that an effective Fermi-type Lagrangian can in fact be "measured" in a series of appropriate experiments and one is thus led to the universal $V-A$ theory, providing a key input for the construction of SM. In many other treatments of the electroweak theory, it has become quite common to start directly with an *a priori* knowledge of the $V-A$ theory; nevertheless, I believe that one should not take lightly the fact that establishing the $V-A$ structure of charged weak currents took more than 20 years (between 1934 and 1958), and this development was a rather dramatic interplay between experiment and theory. Besides that, the first two chapters can help the students of *nuclear* physics — who usually do not exploit the full SM — to understand the origins of our present-day knowledge concerning "ordinary" weak interactions.

In Chapter 3 and also later, in connection with the successive construction of the electroweak SM, the issue of "good high-energy behaviour" of scattering amplitudes, or — in common technical parlance — the "tree unitarity", is often emphasized (note that this is a necessary condition for perturbative renormalizability). In my experience, such a kind of argumentation is very helpful and natural when explaining the construction of SM and an "inevitability" of its essential ingredients. In this regard, the reader may find it useful to consult occasionally a companion work, namely my earlier book *Introduction to Electroweak Unification: SM from Tree Unitarity* published some years ago (and quoted here as [Hor]) — this supplies a lot of technical details omitted in the present text. Of course, an impatient reader, who wishes to arrive at a formulation of the electroweak theory as soon as possible, can start immediately with Chapter 4 devoted to the basics of non-Abelian gauge theories. Finally, the most pragmatic student may jump immediately into Section 7.10 (that contains a succinct overview of the SM), skipping the rest of the book contents altogether.

One more remark concerning the contents is in order here. In accordance with the book title, the emphasis is placed on the "fundamentals", which means that a discussion of applications of the electroweak theory is rather suppressed in favour of a thorough elucidation of the basic principles and their origin. This may be disappointing for a more phenomenologically oriented reader, but I am deeply convinced that for mastering a theory, the understanding of its genesis is as important as a precise formulation of the theory itself. Moreover, there are many other textbooks, specialized monographs and review articles devoted to the practical applications and phenomenology, where the interested reader can find the required information; some of these sources are quoted in our bibliography.

The bulk of the present text is devoted to the electroweak Standard Model, but after going through its exposition, one should keep in mind the remark made at the end of the synoptic Section 7.10: despite the stunning phenomenological success of the SM, the prevailing opinion now is that it cannot be the whole story. The SM should be viewed as an effective theory valid (with remarkable accuracy) within a limited energy domain, explored till the end of 20th century. The contours of a deeper electroweak theory are to be unveiled in the forthcoming decades — to this end, experimental input provided by the new accelerator facilities (such as the Large Hadron Collider at CERN, etc.) will be of crucial importance.

For understanding of the presented material with all technical details, a preliminary knowledge of quantum field theory at the level of Feynman diagrams is necessary. In order to make the reader's life easier, several appendices have been included, which contain a lot of important formulae and/or describe some special techniques employed in the main body of the text. The list of quoted or recommended literature is divided into two parts: "References" represent mostly (with several exceptions) the original articles that are particularly important in the considered context, while "Bibliography" contains books and review articles. Needless to say, the list is far from complete, and I apologize in advance to all authors whose important work was not mentioned here.

Each chapter of the main text is supplemented with a set of problems, or exercises, to be solved. Some of them are not entirely trivial and may require long and tedious calculations. In any case, a diligent reader should not be discouraged by finding out that an appropriate answer cannot be obtained within less than half an hour or so. On the other hand, some exercises should stimulate the student's appetite for further reading; in fact, we thus also partly make up for a broader discussion of applications of the electroweak theory — in particular, this can be said about the computation of the Z boson production cross-section (see Problem 7.7 in Chapter 7).

In the course of writing this book, I was helped, in various ways, by many people. I would like to thank all students who read the preliminary versions of the manuscript when preparing for their exams and pointed out to me errors and numerous misprints; in this respect, I am particularly grateful to Jaromír Kašpar, who came through a substantial portion of the text and also checked most of the formulae. For technical assistance in the early stages of the whole process, I am indebted to Marie Navrátilová. My special thanks are due to Karol Kampf for preparing the final LaTeX version of the manuscript as well as for the ultimate proofreading of the complete text. Last but not least, let me add that this work was partially supported by the Centre for Particle Physics, the Czech Ministry of Education Project No. LN00A006.

Prague Jiří Hořejší
December 2002

Conventions and Notation

Some of the conventions employed in this book are given in the main text and, in particular, in Appendix A. For reader's convenience, and to avoid any misunderstanding, we summarize the most important items here.

Unless stated otherwise, we always use the natural system of units in which $\hbar = c = 1$. Numerical values of observable quantities (such as decay rates or scattering cross-sections) are converted into ordinary units by setting

$$1\,\text{MeV}^{-1} \doteq 6.58 \times 10^{-22}\,\text{s}$$

or

$$1\,\text{MeV}^{-1} \doteq 197\,\text{fm}$$

where $1\,\text{fm} = 10^{-13}\,\text{cm}$ (fm stands for "fermi" or "femtometer").

Most of the other conventions correspond to [BjD]. The indices of any Lorentz four-vector take on values 0, 1, 2, 3. The metric is defined by

$$g_{\mu\nu} = g^{\mu\nu} = \text{diag}(+1,\ -1,\ -1,\ -1)$$

so that, e.g., the scalar product $k \cdot p$ is

$$k \cdot p = k_0 p_0 - \vec{k} \cdot \vec{p}$$

Dirac matrices γ^μ, $\mu = 0, 1, 2, 3$ are defined by means of the standard representation [BjD]. We also employ the usual symbol $\not{p} = p_\mu \gamma^\mu$ for an arbitrary four-vector p. We should particularly stress the definition of the γ_5 matrix

$$\gamma_5 = i\gamma^0\gamma^1\gamma^2\gamma^3$$

that coincides with [BjD] (see also [ItZ], [PeS], and [Ryd]), but differs, e.g., from [Wei]. Further, the fully antisymmetric Levi-Civita tensor is fixed by

$$\epsilon_{0123} = +1$$

(let us remark that this convention differs in sign, e.g., from that used in [ItZ] and [PeS]).

Our conventions for Dirac spinors are described in Appendix A. Let us emphasize that the normalization employed here differs from [BjD] (it coincides, e.g., with [LaL], [PeS]).

Finally, the Lorentz invariant transition (scattering) amplitude, or simply "matrix element" $\mathcal{M}_{fi}$, has an opposite sign with respect to [BjD]; the convention adopted here coincides, e.g., with that of [LaL].

Contents

Chapter 1

Beta Decay

1.1 Kinematics

The oldest and best known example of a process caused by weak interaction is the nuclear beta decay, i.e. the spontaneous emission of electrons (or positrons) from an atomic nucleus. Experimentally observed for the first time at the end of the 19th century, it subsequently played a very important role in establishing the present-day theory of weak interactions. Thus, we will start our road towards the famous $V - A$ theory by discussing some essential features of the beta-decay processes.

As we know now, a process of that kind can be described as the decay of a neutron into proton, electron and an electrically neutral particle called (anti)neutrino:

$$n \to p + e^- + \bar{\nu}_e \tag{1.1}$$

(throughout this chapter, we will write simply ν instead of ν_e). Let us remind the reader that a third particle in the final state of (1.1) is necessary (though very difficult to detect) to explain the observed continuous spectrum of the beta-electron energies; in fact, the existence of such a neutral elusive particle was postulated by Pauli in the early days of the beta-decay theory to save — in a natural way — the fundamental law of energy conservation. It is also well known that the neutrino must be very light — the current upper bound for its mass is a few electronvolts, i.e. five orders of magnitude less than the electron mass. While m_ν can be safely neglected for most practical purposes, an ultimate resolution of the puzzle of neutrino mass (and of possible related phenomena) constitutes one of the most challenging experimental goals of particle physics.[1]

[1] In fact, until 1998, even the possibility of a strictly massless neutrino was acceptable. Since then, the accumulating experimental evidence concerning the so-called oscillation phenomena (cf., e.g., [Bil]) made it clear that neutrinos must have non-vanishing masses.

At least a rough estimate of the m_ν value can be obtained from simple kinematical characteristics of the decay process (1.1). In particular, one can calculate the maximum energy of the emitted electron (which of course depends on the masses of the particles involved) and compare it with a measured value; this can in principle give a desired bound. The endpoint value for the electron energy spectrum will occur frequently in our future considerations, so let us now calculate it explicitly. Denoting the four-momenta of the neutron, proton, electron and antineutrino in (1.1) by P, p, k and k', respectively, the energy–momentum conservation requires that

$$P = p + k + k' \tag{1.2}$$

For our purpose, it is most helpful to utilize a simple consequence of (1.2) for suitable kinematical invariants, namely

$$(P - k)^2 = (p + k')^2 \tag{1.3}$$

One may observe that the quantity on the right-hand side of Eq. (1.3) cannot be less than $(m_p + m_\nu)^2$: indeed, taking into account Lorentz invariance, it can be calculated in the proton — antineutrino c.m. system, where one obviously gets

$$(p + k')^2 = (E_p^{c.m.} + E_\nu^{c.m.})^2 \geq (m_p + m_\nu)^2 \tag{1.4}$$

Of course, because of the Lorentz invariance, the lower bound (1.4) is actually independent of the reference frame. Thus, expressing the invariant $(P - k)^2$ in terms of the variables corresponding to the *laboratory system* (the neutron rest frame), one gets an inequality

$$m_n^2 - 2m_n E_e + m_e^2 \geq (m_p + m_\nu)^2 \tag{1.5}$$

which immediately yields the desired upper bound for the electron energy; the endpoint of the electron energy spectrum obviously corresponds to the value

$$E_e^{\max} = \frac{m_n^2 - (m_p + m_\nu)^2 + m_e^2}{2m_n} \tag{1.6}$$

Analogous bounds for the proton and antineutrino energies can be obtained by means of the same method. These are

$$E_p^{\max} = \frac{m_n^2 + m_p^2 - (m_e + m_\nu)^2}{2m_n} \tag{1.7}$$

$$E_\nu^{\max} = \frac{m_n^2 - (m_p + m_e)^2 + m_\nu^2}{2m_n} \tag{1.8}$$

The formula (1.6) makes it obvious that from the position of the endpoint of the electron energy spectrum one may infer information about the value of the neutrino rest mass. Since the relevant experimental data are still compatible with zero, we will set $m_\nu = 0$ for the time being; we shall comment on some effects of $m_\nu \neq 0$ later on. Taking now into account the known values of the relevant masses

$$m_e = 0.51 \text{ MeV}, \quad m_p = 938.27 \text{ MeV}, \quad m_n = 939.56 \text{ MeV} \tag{1.9}$$

the kinematical bound (1.6) can be approximately written as

$$\begin{aligned} E_e^{\max} &= \frac{m_n^2 - m_p^2 + m_e^2}{2m_n} \doteq \frac{m_n^2 - m_p^2}{2m_n} \\ &\doteq m_n - m_p = 1.29 \text{ MeV} \end{aligned} \tag{1.10}$$

The value (1.10) indicates that an electron emitted in the beta decay can be — at least near the endpoint of the spectrum — highly relativistic. To see this clearly, let us calculate the maximum electron velocity; using (1.10) and the familiar kinematical formulae, one gets

$$\begin{aligned} \beta_e^{\max} &= \sqrt{1 - \frac{m_e^2}{(E_e^{\max})^2}} \doteq \sqrt{1 - \frac{m_e^2}{(\Delta m)^2}} \\ &\doteq 1 - \frac{1}{2}\frac{m_e^2}{(\Delta m)^2} \doteq 0.92 \end{aligned} \tag{1.11}$$

where we have denoted $\Delta m = m_n - m_p$. On the other hand, the maximum proton velocity is of the order $\Delta m/m_p$, as one can deduce easily from (1.7): indeed, neglecting m_e, one has

$$\begin{aligned} \beta_p^{\max} &\doteq \sqrt{1 - m_p^2 / \left(\frac{m_n^2 + m_p^2}{2m_n}\right)^2} = \frac{m_n^2 - m_p^2}{m_n^2 + m_p^2} \\ &= \frac{\Delta m}{m_p} + O\left(\left(\frac{\Delta m}{m_p}\right)^2\right) \doteq 1.37 \times 10^{-3} \end{aligned} \tag{1.12}$$

which means that the recoil proton is certainly non-relativistic over the whole kinematical range.[2]

The lesson one can learn from these simple considerations is essentially twofold. First, in the beta-decay process (1.1), a particle (neutron) is

[2]However, the reader should keep in mind that the maximum proton velocity (1.12) amounts to about 400 km s^{-1}, i.e. such a particle is pretty fast by everyday standards.

annihilated and three new particles are created — this obviously calls for employing the framework of quantum field theory, which is able to incorporate such processes in a very natural way (in other words, the ordinary quantum mechanics is not quite adequate for such a purpose). Second, the final-state electron (to say nothing of the quasi-massless neutrino) is relativistic, at least for a certain part of its energy spectrum — it means that one should use a *relativistic quantum field theory model* to achieve a satisfactory treatment of the dynamics of this decay process. At the same time, one may expect some technical simplifications in connection with the non-relativistic nature of the recoil proton.

1.2 Fermi theory

The first quantitative theory of beta decay was formulated in 1934 by Fermi. In his pioneering work [2], he suggested a direct interaction of four spin-$\frac{1}{2}$ quantum fields corresponding to the particles involved in the process (1.1). The interaction Hamiltonian density proposed by Fermi can be written as

$$\mathscr{H}_{\text{int}}^{(\text{Fermi})} = G(\bar{\psi}_p \gamma^\mu \psi_n)(\bar{\psi}_e \gamma_\mu \psi_\nu) + \text{h.c.} \tag{1.13}$$

where the ψ's stand for the relevant four-component spinor (i.e. fermionic) fields, the γ^μ are standard Dirac matrices and G is a coupling constant. Alternatively, one may write Lagrangian density, which in this case corresponds simply to changing the sign in (1.13), i.e.

$$\mathscr{L}_{\text{int}}^{(\text{Fermi})} = -G(\bar{\psi}_p \gamma^\mu \psi_n)(\bar{\psi}_e \gamma_\mu \psi_\nu) + \text{h.c.} \tag{1.14}$$

It is easy to see that the first term of the interaction Lagrangian describes, in a straightforward way (in the first order of perturbation theory), the neutron decay (1.1) (and related processes like, e.g., $e^+ + n \to p + \bar{\nu}$, etc.), while its hermitian conjugate incorporates the nuclear transition of a proton into neutron, positron and neutrino (and related reactions). Now, it should also be clear why the third particle produced in the neutron beta decay is called *anti*neutrino. First of all, to describe the process in question, one needs an annihilation operator for the neutron and creation operators for proton and electron — this in turn means that $\bar{\psi}_p$ and ψ_e must occur in the interaction Lagrangian. The Lorentz invariance then requires that $\bar{\psi}_e$ be paired with ψ_ν into a bilinear covariant form; however, according to the conventional terminology, ψ_ν contains a creation operator for antiparticle (along with an annihilation operator for particle).

The form (1.14) reflects the original Fermi idea that the "weak nuclear force" responsible for beta decay has essentially zero range, i.e. that — unlike, e.g., the electromagnetism — there is no relevant bosonic particle mediating the weak interaction. Although we know now that an intermediary does exist (this is the famous W boson), the original "conservative" assumption of the contact character of the weak force is, in fact, a very good approximation to reality *at sufficiently low energies* (the reason is, of course, that the W boson is very heavy). Thus, in this chapter, we will stick to the framework of a direct four-fermion interaction using the paradigm of (1.14) and its subsequent generalizations.

A note of historical character is perhaps in order here. In spite of some clear differences between the weak and electromagnetic forces, the original Fermi form (1.14) has certainly been inspired by electrodynamics — the bilinear combinations ("currents") of the fermion fields appearing in (1.14) are Lorentz *four-vectors* similar to the electromagnetic current (coupled to vector four-potential) familiar from QED. (Needless to say, in the early 1930s, there was no previous empirical evidence that would support such a theoretical construction for beta decay.) In this sense, the original Fermi ideas clearly constitute the first step towards an electroweak unification and represent thus a rather fortunate conjecture indeed.

Before examining the phenomenological consequences of the Fermi theory, let us add one more remark concerning general structural aspects of the Lagrangian (1.14). A generic feature of any four-fermion Lagrangian of the type (1.14) is a specific dimensionality of the corresponding coupling constant G; in our natural system of units, this is

$$[G] = M^{-2} \tag{1.15}$$

where M is an arbitrary mass scale. As we shall see later, such a fact plays an important technical role in the development of weak interaction theory, so let us now show how the result (1.15) can be inferred directly from the structure of the relevant Lagrangian. To this end, one should first realize that the action integral of any Lagrangian density $\mathscr{L}$

$$\mathscr{I} = \int \mathscr{L} d^4x \tag{1.16}$$

is dimensionless in natural units (remember that an action has dimension of $\hbar$ in general); this in turn implies

$$[\mathscr{L}] = M^4 \tag{1.17}$$

since the four-dimensional volume element in (1.16) has dimension of M^{-4} (a length unit is M^{-1}). Using the general result (1.17), one can determine easily the dimensionality of a fermion field: the kinetic term of Dirac Lagrangian density has the usual form $i\bar{\psi}\partial\!\!\!/\psi$ and the derivative has dimension of M (namely that of an inverse length), so one immediately gets

$$[\psi] = M^{3/2} \tag{1.18}$$

From (1.14), (1.17) and (1.18), the result (1.15) is then obvious.[3] A note on terminology is in order here. For an interaction Lagrangian density, a corresponding "scale dimension" is usually introduced, which can be defined as the dimension of the corresponding field monomial alone (i.e. of the considered interaction term with the coupling constant removed); more precisely, it is taken to be the relevant exponent of the arbitrary mass referred to above. Thus, we will write, e.g.

$$\dim \mathscr{L}_{\text{int}}^{(\text{Fermi})} = 6 \quad \left(= 4 \cdot \frac{3}{2}\right) \tag{1.19}$$

(for the individual fields, we write similarly $\dim \psi = \frac{3}{2}$ and $\dim B = 1$, respectively). Taking into account (1.17), it is obvious that the scale dimension of an interaction Lagrangian fixes uniquely the dimensionality of the corresponding coupling constant and vice versa.

Let us now discuss some specific physical implications of the original Fermi Lagrangian (1.14). In the first order of perturbation expansion of the S-matrix, the neutron beta decay can be represented by a simple Feynman diagram shown in Fig. 1.1. The corresponding matrix element reads

$$\mathcal{M}_{fi} = -G[\bar{u}_p(p)\gamma^\mu u_n(P)][\bar{u}_e(k)\gamma_\mu v_\nu(k')] \tag{1.20}$$

where u and v are momentum-space wave functions (Dirac spinors) for the particles involved. From (1.20), one can easily calculate the corresponding decay rate and fit the value of the coupling constant G to the measured neutron lifetime; one thus gets roughly $G \doteq 10^{-5}$ GeV^{-2}. We will discuss the decay rate calculation and the determination of the Fermi constant later in this chapter within the framework of a more general (improved) four-fermion interaction Lagrangian. Now, we are going to examine more

[3] In a similar way, it is easy to see that for a bosonic field B (e.g., scalar or vector), one has $[B] = M$. Proving this is left to the reader as a simple but instructive exercise. Then it is also obvious that, e.g., in QED, the relevant coupling constant is dimensionless.

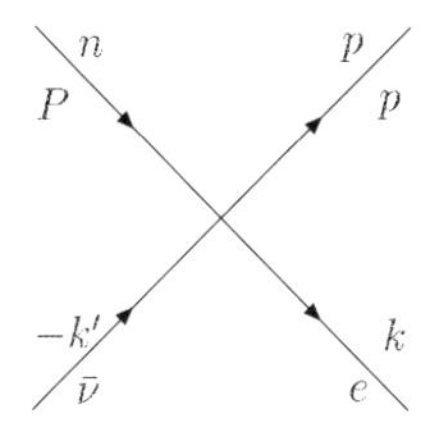

Figure 1.1. Lowest-order Feynman graph for the neutron beta decay in a model of direct four-fermion interaction.

closely the structure of the matrix element (1.20) to get information about possible limitations of the original Fermi model (1.14).

We consider the neutron decay in its rest system. As we have seen in the preceding section, the final-state proton is then safely non-relativistic; in the first approximation, we will therefore neglect the proton momentum altogether. Using the well-known general formulae for solutions of the Dirac equation in the standard representation (see Appendix A), the nucleon spinors can then be approximately written as

$$u_n = \begin{pmatrix} U_n^{(r)} \\ 0 \end{pmatrix}, \quad u_p \doteq \begin{pmatrix} U_p^{(r)} \\ 0 \end{pmatrix} \tag{1.21}$$

where $U_{p,n}^{(r)}$, $r = 1, 2$ are two-component objects

$$U_{p,n}^{(1)} = \sqrt{2M} \begin{pmatrix} 1 \\ 0 \end{pmatrix}, \quad U_{p,n}^{(2)} = \sqrt{2M} \begin{pmatrix} 0 \\ 1 \end{pmatrix} \tag{1.22}$$

and we have denoted by M the nucleon average mass,

$$M = \frac{1}{2}(m_p + m_n) \tag{1.23}$$

Note that the neglected proton momentum is actually of the order of the nucleon mass difference (cf. (1.12)), so writing the M instead of a nucleon mass (whenever it does not lead to an inconsistency) fits precisely into our "quasi-static" approximation scheme. Of course, the lower components of the u_n are exactly zero, since the decaying neutron is at rest by definition. Taking now into account the explicit form of the standard Dirac gamma matrices, it is easy to see that in the non-relativistic approximation (1.21), the nucleon part of the matrix element (1.20) becomes

$$\begin{aligned} \bar{u}_p \gamma^0 u_n &= u_p^\dagger u_n \doteq U_p^\dagger U_n \\ \bar{u}_p \gamma^j u_n &= u_p^\dagger \alpha^j u_n \doteq 0 \quad \text{for } j = 1, 2, 3 \end{aligned} \tag{1.24}$$

Note that the last implication obviously holds as the matrices α^j

$$\alpha^j = \begin{pmatrix} 0 & \sigma_j \\ \sigma_j & 0 \end{pmatrix} \tag{1.25}$$

(with σ_j being the Pauli matrices) only connect "large" and "small" components of Dirac spinors. The matrix element (1.20) thus can be written, in our quasi-static approximation for the recoil proton, as

$$M_{fi} \doteq -G(U_p^\dagger U_n)(\bar{u}_e \gamma_0 v_\nu) \tag{1.26}$$

The result (1.26) shows that within the Fermi model, an "effective transition operator" for nucleons is actually the *unit matrix* and thereby a nucleon spin flip is not possible. At the level of atomic nuclei, this means that the original Fermi Lagrangian can only account for beta decay processes with no change of the nucleonic spin. However, with the development of nuclear spectroscopy, it has become clear that beta transitions involving a spin change ($\Delta J = 1$ in particular) do occur, with intensity comparable to the $\Delta J = 0$ case. Examples include $\text{He}^6(0^+) \to \text{Li}^6(1^+) + e^- + \bar{\nu}$ or $\text{B}^{12}(1^+) \to \text{C}^{12}(0^+) + e^- + \bar{\nu}$. (For historical reasons, the spin-changing beta-decay processes are called Gamow–Teller transitions, while those caused by the effective unit operator as in (1.26) are Fermi transitions.[4])

One may thus conclude that the original Fermi model, though conceptually correct (and applicable at least in a limited sense), is certainly incomplete and must therefore be generalized if one wants to get a realistic effective theory of weak nuclear force.

1.3 Generalization of Fermi theory and parity violation

In fact, there is a straightforward way how to generalize the simple Fermi model. Along with the vector-like currents appearing in (1.14), other possible covariant bilinear combinations of the relevant spinor fields may be included as well, i.e. one can construct a four-fermion interaction Lagrangian using the whole set of scalar (S), vector (V), tensor (T), axial vector (A) and pseudoscalar (P) bilinear Dirac forms (cf. Appendix A). Such an extension of the original Fermi model was suggested first by Gamow and Teller [3], and it is not difficult to realize that one is thus indeed capable

[4]It should be emphasized that here and in what follows we always have in mind only the so-called **allowed** transitions — these occur in the lowest order even when the quasi-static approximation for nucleons is adopted. For a more detailed discussion of allowed and forbidden beta decay processes, see, e.g., [CoB], Chapter 5.

of describing both the spin-conserving and the spin-changing nuclear beta transitions (as we shall see later in this section, interaction terms of the types A and T are those which can account for the Gamow–Teller $\Delta J = 1$ transitions).

While the above-mentioned construction represents a rather straightforward and natural step in building a realistic theoretical framework for the description of weak nuclear force, a real breakthrough came in 1956, when Lee and Yang in their fundamental paper [4] suggested that one could abandon the traditional assumption of parity symmetry (i.e. the invariance under spatial inversion) of the weak interaction Lagrangian. They observed that such a "mirror symmetry", though naively taken for granted (e.g., in analogy with electrodynamics), actually had no support in the available data and proposed therefore a set of experiments which could truly test this fundamental issue.[5] A series of subsequent experiments revealed clearly the envisaged parity-breaking phenomena (some of these effects will be discussed explicitly later on) and the parity violation in weak interactions has thus become one of the most dramatic discoveries of the 20th century physics (Lee and Yang received the Nobel Prize in 1957).

Let us now see how these aspects of the weak nuclear force can be described formally. A general four-fermion interaction Lagrangian for beta-decay processes, including all the algebraic structures mentioned above and taking into account a possible parity violation, can be written as

$$\mathscr{L}_{\text{int}}^{(\beta)} = \sum_{j=S,V,A,T,P} C_j(\bar{\psi}_p\Gamma_j\psi_n)[\bar{\psi}_e(1+\alpha_j\gamma_5)\Gamma^j\psi_\nu] \tag{1.27}$$

(tacitly assuming the presence of the h.c. term), where

$$\begin{aligned} \Gamma_j &= 1, \gamma_\mu, \gamma_5\gamma_\mu, \sigma_{\mu\nu}, \gamma_5 \\ \Gamma^j &= 1, \gamma^\mu, \gamma_5\gamma^\mu, \sigma^{\mu\nu}, \gamma_5 \end{aligned} \tag{1.28}$$

for $j = S, V, A, T, P$ consecutively; the symbol $\sigma_{\mu\nu}$ means

$$\sigma_{\mu\nu} = \frac{i}{2}[\gamma_\mu, \gamma_\nu] \tag{1.29}$$

and 1 denotes the 4×4 unit matrix. The parameters C_j in (1.27) have dimension of M^{-2} in analogy with the original Fermi coupling constant G.

[5] Note that Lee and Yang came up with their radical idea in order to solve a conundrum concerning the strange meson (K^+) decays into pions — a problem that is usually quoted as the "$\tau - \theta$ puzzle" in the literature. For a detailed discussion of this important piece of particle physics history, see, e.g., [Adv] or [CaG].

The α_j are dimensionless and provide a measure of parity violation, as it should be clear from the familiar transformation properties of the fermion bilinear forms under space inversion $\mathcal{P}$. Indeed, for each $j = S, V, A, T, P$, there are two terms in the Lagrangian, descending from the factor of $1 + \alpha_j\gamma_5$: the term corresponding to the unity is $\mathcal{P}$-even (i.e. true Lorentz scalar) while that involving $\alpha_j\gamma_5$ is $\mathcal{P}$-odd (Lorentz pseudoscalar). The point is that the presence of an extra γ_5 always changes the parity of a Lorentz-covariant bilinear form under $\mathcal{P}$ (cf. Appendix A). Note, however, that both terms coming from $1 + \alpha_j\gamma_5$ must be present if the Lagrangian is designed to describe parity-violating effects: if one drops the $\mathcal{P}$-even terms in (1.27) and keeps only those involving $\alpha_j\gamma_5$, then the remaining Lagrangian is in fact parity-conserving (though naively $\mathcal{P}$-odd) since one can redefine the neutrino field by means of a unitary transformation $\psi'_\nu = \gamma_5\psi_\nu$ without changing the physical contents of the theory.

The generalized four-fermion interaction (1.27) is described in terms of 10 arbitrary parameters α_j, C_j. For the sake of simplicity, we take all these parameters to be *real*, which in fact means that invariance of (1.27) under time reversal is tacitly assumed (we will discuss this issue in more detail within the framework of the standard model of electroweak interactions). Note also that in principle one could add to (1.27) infinitely many other terms involving derivatives of the fermion fields (i.e. interaction terms of dimension higher than six). The form (1.27) represents, in this sense, a minimal model involving non-derivative (i.e. lowest-dimensional) four-fermion couplings only. Even so, introducing as many as 10 arbitrary parameters into our "realistic" beta-decay Lagrangian certainly makes it much less elegant than the original Fermi model. In subsequent sections, we will see that the number of relevant parameters can in fact be significantly reduced when the Ansatz (1.27) is confronted with experimental data. At the end of the day, a rather simple and elegant interaction Lagrangian emerges, which in certain sense is quite similar to the old Fermi model. In other words, it turns out that Fermi was "almost right" when writing his provisional theory of weak nuclear force *a priori* in terms of vectorial currents.

Next, we discuss the properties of the relevant transition amplitude. Obviously, the lowest-order matrix element for neutron beta decay (corresponding to the Feynman graph in Fig. 1.1) now becomes

$$\mathcal{M}_{fi}^{(\beta)} = \sum_{j=S,V,A,T,P} C_j(\bar{u}_p\Gamma_j u_n)[\bar{u}_e(1+\alpha_j\gamma_5)\Gamma^j v_\nu] \tag{1.30}$$

Let us examine how the last expression is simplified if one employs the non-relativistic approximation for nucleons. Using the standard representation

$$\gamma^0 = \begin{pmatrix} \mathbb{1} & 0 \\ 0 & -\mathbb{1} \end{pmatrix}, \quad \gamma^j = \begin{pmatrix} 0 & \sigma_j \\ -\sigma_j & 0 \end{pmatrix}, \quad \gamma_5 = \begin{pmatrix} 0 & \mathbb{1} \\ \mathbb{1} & 0 \end{pmatrix} \tag{1.31}$$

it is not difficult to find that the non-relativistic (static) reduction of the nucleon part of the matrix element (1.30) follows the pattern shown in Table 1.1 (the result for the V term has already been discussed in the preceding section).

More precisely, such a scheme means that the matrix element (1.30) can be recast as

$$\mathcal{M}_{fi}^{(\beta)} = \mathcal{M}_S + \mathcal{M}_V + \mathcal{M}_A + \mathcal{M}_T + \mathcal{M}_P \tag{1.32}$$

where

$$\begin{aligned} \mathcal{M}_S &\doteq C_S(U_p^\dagger U_n)[\bar{u}_e(1+\alpha_S\gamma_5)v_\nu] \\ \mathcal{M}_V &\doteq C_V(U_p^\dagger U_n)[\bar{u}_e(1+\alpha_V\gamma_5)\gamma_0 v_\nu] \\ \mathcal{M}_A &\doteq C_A(U_p^\dagger \sigma_j U_n)[\bar{u}_e(1+\alpha_A\gamma_5)\gamma_5\gamma^j v_\nu] \\ \mathcal{M}_T &\doteq 2C_T(U_p^\dagger \sigma_j U_n)[\bar{u}_e(1+\alpha_T\gamma_5)\Sigma^j v_\nu] \\ \mathcal{M}_P &\doteq 0 \end{aligned} \tag{1.33}$$

with

$$\Sigma^j = \frac{1}{2}\epsilon^{jkl}\sigma^{kl} = \begin{pmatrix} \sigma_j & 0 \\ 0 & \sigma_j \end{pmatrix} \tag{1.34}$$

We thus see that the pseudoscalar (P) term does not contribute at all in the considered approximation and the remaining algebraic structures

Table 1.1. Scheme of the reduction of nucleonic part of a beta-decay matrix element in non-relativistic (static) approximation.

Algebraic type of coupling	Nucleon matrix elements	
	Covariant form	**Static approximation**
S	$\bar{u}_p u_n$	$U_p^\dagger U_n$
V	$\bar{u}_p \gamma_\mu u_n$	$U_p^\dagger U_n$
A	$\bar{u}_p \gamma_5 \gamma_\mu u_n$	$U_p^\dagger \vec{\sigma} U_n$
T	$\bar{u}_p \sigma_{\mu\nu} u_n$	$U_p^\dagger \vec{\sigma} U_n$
P	$\bar{u}_p \gamma_5 u_n$	0

come in pairs with similar properties: the S and V couplings are effectively represented by the 2×2 unit matrix (i.e. a spin-zero transition operator), while the A and T couplings both lead to Pauli matrices and constitute thereby an effective transition operator carrying spin 1; these can therefore account for spin-changing Gamow–Teller processes (remember the good old Wigner–Eckart theorem). Let us remark that the traditional terminology, mentioned briefly in previous section, can now be made more precise: processes due to S and/or V couplings (i.e. effectively mediated by unit matrix) are called Fermi (F) transitions and those caused by A, T couplings (i.e. effectively mediated by Pauli matrices) are Gamow–Teller (GT) transitions. This rather technical definition can be translated into a more physical language as follows. With regard to the spin of the initial and final nucleon system $J_{i,f}$, there are essentially three types of beta-decay processes. If $J_i = J_f = 0$, only the S and/or V couplings can contribute and such a process is therefore **pure F transition**. If $\Delta J = |J_f - J_i| = 1$, then only the A and/or T terms contribute and we have a **pure GT transition**. For $J_i = J_f \neq 0$, one can in principle get a contribution from both types of couplings and such a process may be naturally called **mixed transition**. We have already given examples of pure GT transitions in the preceding section. A well-known case of a pure F process is the $0^+ \to 0^+$ transition $\mathrm{O}^{14} \to \mathrm{N}^{14^*} + e^+ + \nu$, while the free neutron decay or the tritium decay $\mathrm{H}^3 \to \mathrm{He}^3 + e^- + \bar{\nu}$ can serve as examples of mixed transitions.

Now, we have the necessary technical prerequisites at hand and we can employ the matrix elements (1.33) to calculate some observable dynamical characteristics of beta-decay processes that will help us to determine the values of the free parameters in the Lagrangian (1.27). This will be the subject of subsequent sections.

1.4 The electron energy spectrum

Observable quantities for the considered processes are expressed in terms of appropriate decay rates. The starting point of our calculations will be the differential decay rate for a free neutron in its rest frame, involving the element of the corresponding three-particle phase space

$$\begin{aligned} dw = {} & \frac{1}{2m_n} |\mathcal{M}|^2 \frac{d^3k}{(2\pi)^3 2E(k)} \frac{d^3k'}{(2\pi)^3 2E(k')} \\ & \times \frac{d^3p}{(2\pi)^3 2E(p)} (2\pi)^4 \delta^4(P - k - k' - p) \end{aligned} \tag{1.35}$$

where we have denoted the relevant momenta in accordance with (1.2). Various interesting quantities can then be obtained by integrating (1.35) over some kinematical variables (in other words, over the phase-space volume elements). We will discuss the phase-space integrations later on and calculate first the matrix element squared, which is the object of central importance, reflecting the weak interaction dynamics.

To begin with, let us consider the situation where the particles are unpolarized, i.e. one does not care about a particular spin (projection) of a decay product and the decaying neutron is supposed to have spin "up" or "down" with equal probability. In such a case, $\overline{|\mathcal{M}|^2}$ is to be summed over the spins of final-state particles and averaged over the initial neutron spin, i.e. the relevant quantity is

$$\overline{|\mathcal{M}|^2} = \frac{1}{2} \sum_{\substack{\text{spins} \\ n,p,e,\bar{\nu}}} |\mathcal{M}|^2 \tag{1.36}$$

This brings about some simplifications, especially within our quasi-static approximation for nucleons. In particular, as a result of the summation over nucleon spins, there is no interference between the Fermi (S, V) and Gamow–Teller (A, T) parts of the matrix element (1.32), (1.33). Let us prove this simple statement for the reader's convenience. Obviously, such an interference term in $|\mathcal{M}|^2 = \mathcal{M}\,\mathcal{M}^*$ would certainly contain a nucleonic factor

$$U_p^\dagger U_n \; U_n^\dagger \sigma_j U_p \tag{1.37}$$

that can be identically recast as the trace

$$\mathrm{Tr}(U_p^{(r)} U_p^{(r)\dagger} U_n^{(r')} U_n^{(r')\dagger} \sigma_j) \tag{1.38}$$

Spin sums for the two-component Pauli spinors (1.22) are proportional to the unit matrix:

$$\begin{aligned}\sum_{r=1}^{2} U_{p,n}^{(r)} U_{p,n}^{(r)\dagger} &= (\sqrt{2M})^2 \left[\begin{pmatrix} 1 \\ 0 \end{pmatrix} (1 \quad 0) + \begin{pmatrix} 0 \\ 1 \end{pmatrix} (0 \quad 1) \right] \\ &= 2M \left[\begin{pmatrix} 1 & 0 \\ 0 & 0 \end{pmatrix} + \begin{pmatrix} 0 & 0 \\ 0 & 1 \end{pmatrix} \right] = 2M \begin{pmatrix} 1 & 0 \\ 0 & 1 \end{pmatrix}\end{aligned} \tag{1.39}$$

The expression (1.38) summed over r, r' thus becomes proportional to $\mathrm{Tr}\,\sigma_j$, which of course vanishes for any $j = 1, 2, 3$.

The quantity (1.36) can thus be written (within the non-relativistic approximation (1.33)) as

$$\overline{|\mathcal{M}|^2} = \overline{|\mathcal{M}_S + \mathcal{M}_V|^2} + \overline{|\mathcal{M}_A + \mathcal{M}_T|^2} = \overline{|\mathcal{M}_F|^2} + \overline{|\mathcal{M}_{GT}|^2} \tag{1.40}$$

Let us now work out the Fermi part of the last expression. Using the explicit form of $\mathcal{M}_S$ and $\mathcal{M}_V$ as given in (1.33), some familiar properties of gamma matrices and the usual trick of introducing traces of matrix products, one first gets

$$\begin{aligned}
|\mathcal{M}_F|^2 &= |\mathcal{M}_S|^2 + |\mathcal{M}_V|^2 + \mathcal{M}_S\mathcal{M}_V^* + \mathcal{M}_S^*\mathcal{M}_V \\
&= C_S^2 \mathrm{Tr}(U_p U_p^\dagger U_n U_n^\dagger)\mathrm{Tr}[u_e \bar{u}_e (1+\alpha_S\gamma_5) v_\nu \bar{v}_\nu (1-\alpha_S\gamma_5)] \\
&\quad + C_V^2 \mathrm{Tr}(U_p U_p^\dagger U_n U_n^\dagger)\mathrm{Tr}[u_e \bar{u}_e (1+\alpha_V\gamma_5)\gamma_0 v_\nu \bar{v}_\nu (1+\alpha_V\gamma_5)\gamma_0] \\
&\quad + C_S C_V \mathrm{Tr}(U_p U_p^\dagger U_n U_n^\dagger)\mathrm{Tr}[u_e \bar{u}_e (1+\alpha_S\gamma_5) v_\nu \bar{v}_\nu (1+\alpha_V\gamma_5)\gamma_0] \\
&\quad + \text{c.c.}
\end{aligned} \tag{1.41}$$

(of course, the complex conjugation refers only to the last line in (1.41)). After the summation over spins (cf. (1.39) and (A.67)), this becomes

$$\begin{aligned}
\overline{|\mathcal{M}_F|^2} &= 4C_S^2 M^2 \mathrm{Tr}[(\not{k} + m_e)(1+\alpha_S\gamma_5)\not{k}'(1-\alpha_S\gamma_5)] \\
&\quad + 4C_V^2 M^2 \mathrm{Tr}[(\not{k} + m_e)(1+\alpha_V\gamma_5)\gamma_0\not{k}'(1+\alpha_V\gamma_5)\gamma_0] \\
&\quad + 4C_S C_V M^2 \mathrm{Tr}[(\not{k} + m_e)(1+\alpha_S\gamma_5)\not{k}'(1+\alpha_V\gamma_5)\gamma_0] + \text{c.c.}
\end{aligned} \tag{1.42}$$

The leptonic traces can be simplified to

$$\begin{aligned}
\overline{|\mathcal{M}_F|^2} &= 4C_S^2(1+\alpha_S^2)M^2 \mathrm{Tr}(\not{k}\not{k}') \\
&\quad + 4C_V^2(1+\alpha_V^2)M^2 \mathrm{Tr}(\not{k}\gamma_0\not{k}'\gamma_0) \\
&\quad + 4C_S C_V(1-\alpha_S\alpha_V)m_e M^2 \mathrm{Tr}(\not{k}'\gamma_0) + \text{c.c.}
\end{aligned}$$

and a straightforward calculation then yields

$$\begin{aligned}
\overline{|\mathcal{M}_F|^2} &= 16C_S^2(1+\alpha_S^2)M^2(k\cdot k') \\
&\quad + 16C_V^2(1+\alpha_V^2)M^2(2E_e E_{\bar{\nu}} - k\cdot k') \\
&\quad + 32C_S C_V(1-\alpha_S\alpha_V)m_e M^2 E_{\bar{\nu}}
\end{aligned} \tag{1.43}$$

The Lorentz scalar product $k\cdot k'$ can be expressed as

$$k\cdot k' = E_e E_{\bar{\nu}} - |\vec{k}||\vec{k}'|\cos\vartheta = E_e E_{\bar{\nu}}(1-\beta_e\cos\vartheta) \tag{1.44}$$

and (1.43) thus finally becomes

$$\begin{aligned}
\overline{|\mathcal{M}_F|^2} &= 16M^2 E_e E_{\bar{\nu}}\Big[C_S^2(1+\alpha_S^2)(1-\beta_e\cos\vartheta) \\
&\quad + C_V^2(1+\alpha_V^2)(1+\beta_e\cos\vartheta) + 2C_S C_V(1-\alpha_S\alpha_V)\frac{m_e}{E_e}\Big]
\end{aligned} \tag{1.45}$$

The Gamow–Teller part of (1.40) can be evaluated in a similar way using the standard trace techniques. We defer the calculation to the next section in order not to clutter the present section with too many technicalities. The result has a form analogous to (1.45); it reads

$$\begin{aligned}\overline{|\mathcal{M}_{GT}|^2} &= \overline{|\mathcal{M}_A + \mathcal{M}_T|^2} \\ &= 16M^2 E_e E_{\bar{\nu}} \left[3C_A^2(1+\alpha_A^2)\left(1 - \frac{1}{3}\beta_e \cos\vartheta\right) \right. \\ &\quad \left. + 12C_T^2(1+\alpha_T^2)\left(1 + \frac{1}{3}\beta_e \cos\vartheta\right) - 12C_A C_T(1-\alpha_A\alpha_T)\frac{m_e}{E_e}\right]\end{aligned} \tag{1.46}$$

Note that while there is no interference between $\mathcal{M}_F$ and $\mathcal{M}_{GT}$, the $S-V$ or $A-T$ interference terms can in principle occur, depending on the values of the relevant parameters C_j, α_j.

Let us now turn to the calculation of some interesting differential decay rates defined with respect to the leptonic kinematical variables. Our ultimate goal will be the electron energy spectrum (we still have in mind the case of unpolarized particles). To this end, we start from the basic formula (1.35) and integrate first over the proton momentum. Such an integration is essentially trivial — we simply use up the three-dimensional delta function corresponding to the momentum conservation, i.e. replace the $\vec{p}$ by $-(\vec{k}+\vec{k}')$. This immediately yields

$$\begin{aligned} dw &\doteq \frac{1}{2m_n}\overline{|\mathcal{M}|^2}\frac{d^3k}{(2\pi)^3 2E_e}\frac{d^3k'}{(2\pi)^3 2E_{\bar{\nu}}}\frac{1}{2m_p} \\ &\quad \times 2\pi\delta\left(m_n - E_e - E_{\bar{\nu}} - \sqrt{(\vec{k}+\vec{k}')^2 + m_p^2}\right) \end{aligned} \tag{1.47}$$

Note that for the sake of brevity we denote the once integrated differential decay rate by the same symbol as the original quantity. In (1.47), we have already neglected the proton momentum (setting $E_p \doteq m_p$) in the normalization coefficient at the corresponding phase-space factor. We will neglect quantities of the order of $\Delta m/M$ in the course of our calculation, whenever such an approximation is safely under control and does not lead to an inconsistency, e.g. in energy–momentum balance. The next step is an integration over d^3k'. For a fixed direction of the electron momentum $\vec{k}$ and fixed angle ϑ between $\vec{k}'$ and $\vec{k}$, we will integrate first over the modulus $|\vec{k}'| = E_{\bar{\nu}}$. Let us denote $x = |\vec{k}'|$ for brevity; the energy-conservation delta function in (1.47) can be written as $\delta(f(x))$, with

$$f(x) = m_n - E_e - x - \sqrt{|\vec{k}|^2 + 2x|\vec{k}|\cos\vartheta + x^2 + m_p^2} \tag{1.48}$$

The energy-conservation condition $f(x_0) = 0$ implies

$$x_0 = E_{\bar{\nu}} = \frac{m_n^2 - m_p^2 + m_e^2 - 2m_n E_e}{2m_n - 2E_e + 2|\vec{k}|\cos\vartheta} \tag{1.49}$$

(note that from (1.49), one can recover the kinematical upper bound (1.6) for electron energy). To carry out the x-integration, one can now use the well-known relation

$$\delta(f(x)) = \frac{1}{|f'(x_0)|}\delta(x - x_0) \tag{1.50}$$

The evaluation of the $f'(x_0)$ is straightforward; differentiating plainly the expression (1.48) and utilizing the condition $f(x_0) = 0$, i.e.

$$\sqrt{|\vec{k}|^2 + 2x_0|\vec{k}|\cos\vartheta + x_0^2 + m_p^2} = m_n - E_e - x_0$$

one readily gets

$$|f'(x_0)| = \frac{m_n - E_e + |\vec{k}|\cos\vartheta}{m_n - E_e - x_0} \tag{1.51}$$

Now, we can make our usual approximations, neglecting the terms of relative order $\Delta m/M$. The expression (1.49) then becomes

$$\begin{aligned} x_0 = E_{\bar{\nu}} &\doteq \frac{1}{2m_n}(m_n^2 - m_p^2 + m_e^2 - 2m_n E_e) \\ &= E_e^{\max} - E_e \end{aligned} \tag{1.52}$$

where we have taken into account (1.6). We thus see that for a given value of E_e, the antineutrino energy can approximately be written as

$$E_{\bar{\nu}} \doteq E_e^{\max} - E_e \doteq \Delta m - E_e \tag{1.53}$$

(of course, this is an expected result — it follows simply from energy conservation if the proton motion is neglected). Similarly, from (1.51), we obtain

$$|f'(x_0)| \doteq 1 \tag{1.54}$$

The integration of (1.47) over the $|\vec{k}'|$ can now be done easily — it essentially consists in dropping the delta function and replacing everywhere the antineutrino energy by its approximate physical value (1.53). Since the

d^3k' can be written in spherical coordinates as $x^2 dx d\Omega_{\bar{\nu}}$ (where $d\Omega_{\bar{\nu}}$ is an element of a corresponding solid angle), the result of the x-integration can be written as

$$dw \doteq \frac{1}{8M^2}\overline{|\mathcal{M}|^2} E_{\bar{\nu}} \frac{d^3\vec{k}}{(2\pi)^3 2E_e}\frac{d\Omega_{\bar{\nu}}}{4\pi^2} \tag{1.55}$$

where we have also replaced m_n and m_p by the average nucleon mass M. Note that $d\Omega_{\bar{\nu}} = \sin\vartheta d\vartheta d\varphi$ with ϑ being the angle between the directions of $\vec{k}'$ and $\vec{k}$. The decay rate (1.55) thus describes the angular correlation between electron and antineutrino — we will discuss this experimentally interesting quantity in more detail in the next section.

To arrive at the electron energy spectrum, we have to integrate over the angular variables in (1.55). For the moment, let us consider, e.g., only the Fermi part of the $\overline{|\mathcal{M}|^2}$, i.e. the expression (1.45). Obviously, the terms proportional to $\cos\vartheta$ vanish upon integration over $d\vartheta$ (and the constant terms are simply multiplied by 4π). Further, $d^3k = |\vec{k}|^2 d|\vec{k}| d\Omega_e$, where $d\Omega_e$ is the element of solid angle corresponding to the electron momentum direction related to an arbitrarily (conventionally) chosen coordinate system. Of course, there is no non-trivial angular dependence that would survive after the preceding integration over $d\Omega_{\bar{\nu}}$ (as there is no natural preferred spatial direction in the considered problem), so the remaining angular integration over $d\Omega_e$ is trivial — it amounts to a multiplication by 4π. As the last step, one should pass from the differential $d|\vec{k}|$ to dE_e; this is done easily, since the relation $E_e = (|\vec{k}|^2 + m_e^2)^{1/2}$ implies immediately $|\vec{k}|\ d|\vec{k}| = E_e dE_e$. The form of the electron energy spectrum is thus given by

$$dw(E_e) \doteq \frac{1}{2\pi^3}\left[C_S^2(1+\alpha_S^2) + C_V^2(1+\alpha_V^2) + 2C_S C_V(1-\alpha_S\alpha_V)\frac{m_e}{E_e}\right]$$
$$\times |\vec{k}| E_e (\Delta - E_e)^2 dE_e \tag{1.56}$$

where we have employed the explicit expression for $\overline{|\mathcal{M}_F|^2}$ given by (1.45) and we have also introduced the usual shorthand notation Δ for the endpoint of the spectrum ($\Delta = E_e^{\max} \doteq \Delta m$).

Let us now discuss the obtained result. We have performed the calculation for a free neutron, but it is not difficult to realize that the energy dependence shown in (1.56) should be valid in the case of an allowed nuclear beta decay as well. Indeed, when dealing with atomic nuclei, the usual quasi-static approximation for nucleons can be employed; the relevant nuclear matrix element is then of course independent of the electron energy and the form of $dw(E_e)$ comes out to be the same as in (1.56). In other words, for an allowed nuclear beta decay, only a constant factor in (1.56) may get modified, but not the functional dependence on the E_e. The result (1.56) is thus appropriate for the description of the electron energy spectrum corresponding to a pure F transition. Similarly, one could use the form (1.46) for the spin-averaged matrix element squared and thus obtain a straightforward analogy of the relation (1.56) for pure GT transitions (of course, for a free neutron decay, one has to include both types of matrix elements). The contribution proportional to m_e/E_e in the generic formula (1.56) is called, for historical reasons, a Fierz interference term. The existing experimental data show that the value of the corresponding coefficient of such a term is consistent with zero for both F and GT transitions (for typical numbers, see, e.g., [CoB] or [Ren]). Obviously, this empirical fact represents a certain constraint on the parameters C_j and α_j. If taken at face value (i.e. assuming that the Fierz interference term is exactly zero), this would mean that for F transitions, one has

$$C_S C_V (1 - \alpha_S \alpha_V) = 0 \tag{1.57}$$

and this in turn implies that either $C_S = 0$, or $C_V = 0$, or $1 - \alpha_S \alpha_V = 0$. We will obtain further constraints on the parameters later on (by utilizing other relevant experimental data), and for the time being, we will simply keep in mind the condition (1.57). Similarly, we will interpret the empirical evidence for the absence of Fierz interference terms in GT transitions as a constraint

$$C_A C_T (1 - \alpha_A \alpha_T) = 0 \tag{1.58}$$

(cf. (1.46)). Let us emphasize that we do not attempt to accomplish a "best fit" of the free parameters of the general four-fermion Lagrangian (1.27) to the available experimental data — rather, we will try to show that the wealth of empirical data clearly point towards a very particular and remarkably simple theoretical scheme for weak interactions.

From the preceding considerations, it is clear that — in the absence of the Fierz interference terms — the electron energy spectra do not provide any further information about the properties of the weak interaction (for example, on the basis of the spectrum alone, one cannot distinguish between S and V couplings in Fermi transitions, etc.). The characteristic functional dependence

$$\frac{dw(E_e)}{dE_e} = \text{const.} \times \sqrt{E_e^2 - m_e^2}\, E_e (\Delta - E_e)^2 \tag{1.59}$$

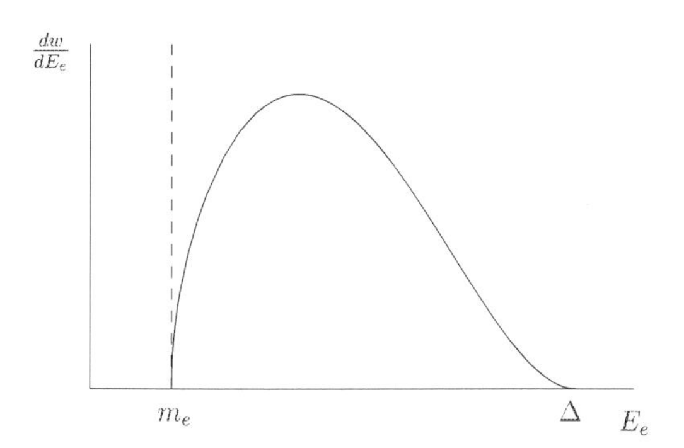

Figure 1.2. The typical form of the electron energy spectrum for an allowed beta decay.

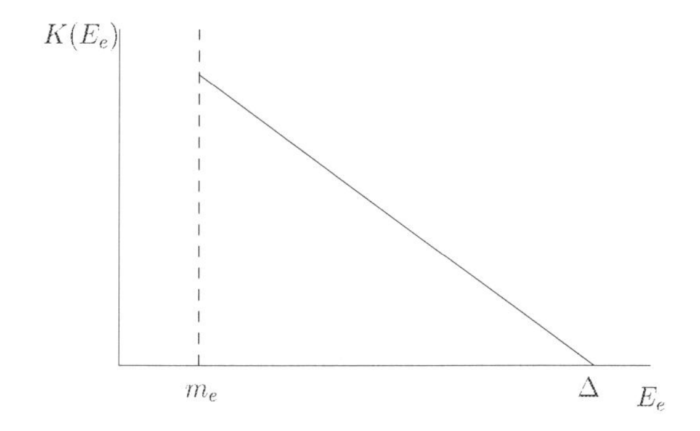

Figure 1.3. The straight-line Kurie plot for a beta decay spectrum, which exhibits the absence of a Fierz interference term.

is sometimes called the "statistical form" of the energy spectrum, since this is essentially determined by the phase-space factors (it is instructive to trace the origin of the individual factors in (1.59) back to our starting point (1.35) and to the normalization of the matrix element $\mathcal{M}$). The function (1.59) is depicted in Fig. 1.2.

Note that beta-decay spectra are usually represented in the form of the so-called Kurie (or Fermi–Kurie) plot, which displays the energy dependence of the quantity

$$K(E_e) = \left(\frac{1}{|\vec{k}|E_e} \frac{dw}{dE_e} \right)^{1/2} \tag{1.60}$$

For $dw(E_e)$ given by (1.59), the Kurie plot is a falling straight line, $K(E_e) =$ const. $\times (\Delta - E_e)$, as shown in Fig. 1.3. It is obvious that the presence of a

Fierz interference term would manifest itself as a distortion of the straight-line Kurie plot that would be most pronounced in the low-energy part of the spectrum, i.e. near its beginning at $E_e = m_e$.

Last but not least, one should note that the linear dependence (1.60) also relies on the assumption of vanishing rest mass of the neutrino. Indeed, from the antineutrino phase-space volume element in (1.35), one gets, in general, a factor of $|\vec{k}'|E_{\bar{\nu}}$ in (1.59) (which of course coincides with $E_{\bar{\nu}}^2$ for $m_\nu = 0$), with $|\vec{k}'| = \sqrt{E_{\bar{\nu}}^2 - m_\nu^2} < E_{\bar{\nu}}$ for $m_\nu \neq 0$. This could then also yield a deviation from the straight-line Kurie plot (namely, a downward deflection), in particular near its endpoint: if the electron energy is close to its maximum value, the antineutrino energy is small and a relative difference between $|\vec{k}'|$ and $E_{\bar{\nu}}$ due to $m_\nu \neq 0$ then becomes largest. This simple observation in fact provides a conceptual basis for several experiments that play an important role in the present-day quest for a neutrino mass. For details, the reader is referred, e.g., to [Vog], [Kay] or [CoB] (see also [Gre]).

1.5 The e–$\bar{\nu}$ angular correlation: Dominance of V and A couplings

We are now going to examine the angular distribution of leptons produced in a beta-decay process. Such an observable quantity is rather sensitive to the type of coupling responsible for a given process and its analysis hence provides some powerful constraints on the parameters of the effective Lagrangian (1.27). Before discussing this issue in detail, let us derive — as promised in the preceding section — the formula (1.46) for the spin-averaged GT matrix element squared.

Looking back at (1.33), we have

$$\begin{aligned}\mathcal{M}_{GT} &= \mathcal{M}_A + \mathcal{M}_T \\ &= C_A(U_p^\dagger \sigma_j U_n)[\bar{u}_e(1+\alpha_A\gamma_5)\gamma_5\gamma^j v_\nu] \\ &\quad + 2C_T(U_p^\dagger \sigma_j U_n)[\bar{u}_e(1+\alpha_T\gamma_5)\Sigma^j v_\nu] \end{aligned} \tag{1.61}$$

Squaring (1.61), using the familiar trace techniques and summing over spins, one first gets

$$\begin{aligned}\overline{|\mathcal{M}_{GT}|^2} &= \frac{1}{2}\sum_{\text{spins}} |\mathcal{M}_{GT}|^2 \\ &= 2C_A^2 M^2 \mathrm{Tr}(\sigma_j\sigma_k)\mathrm{Tr}[(\not{k} + m_e)(1+\alpha_A\gamma_5)\gamma_5\gamma^j \not{k}'(1+\alpha_A\gamma_5)\gamma_5\gamma^k] \\ &\quad + 8C_T^2 M^2 \mathrm{Tr}(\sigma_j\sigma_k)\mathrm{Tr}[(\not{k} + m_e)(1+\alpha_T\gamma_5)\Sigma^j \not{k}'(1-\alpha_T\gamma_5)\Sigma^k]\end{aligned}$$

$$+ 4C_A C_T M^2 \mathrm{Tr}(\sigma_j \sigma_k) \mathrm{Tr}[(\not{k} + m_e)(1 + \alpha_A \gamma_5)\gamma_5 \gamma^j \not{k}'$$
$$\times (1 - \alpha_T \gamma_5)\Sigma^k] + \text{c.c.} \tag{1.62}$$

(where we have employed, among other things, the completeness relation (1.39) for the two-component nucleon spinors); the complex conjugation of course refers only to the last line in (1.62). To work out the traces involving the spin matrix Σ^j (see (1.34)), it is useful to remember the identity

$$\Sigma^j = \gamma_5 \alpha^j = \gamma_5 \gamma_0 \gamma^j \tag{1.63}$$

(cf. (A.72)). The expression (1.62) then becomes, after some manipulations,

$$\begin{aligned}
\overline{|\mathcal{M}_{GT}|^2} &= 2C_A^2 M^2 \mathrm{Tr}(\sigma_j \sigma_k) \mathrm{Tr}[\not{k}\gamma^j \not{k}' \gamma^k (1 + \alpha_A^2 - 2\alpha_A \gamma_5)] \\
&\quad + 8C_T^2 M^2 \mathrm{Tr}(\sigma_j \sigma_k) \mathrm{Tr}[\not{k}\gamma^j \gamma_0 \not{k}' \gamma_0 \gamma^k (1 + \alpha_T^2 - 2\alpha_T \gamma_5)] \\
&\quad + 4C_A C_T M^2 m_e \mathrm{Tr}(\sigma_j \sigma_k) \\
&\quad \times \mathrm{Tr}[\gamma^j \not{k}' \gamma_0 \gamma^k (1 - \alpha_A \alpha_T + (\alpha_A - \alpha_T)\gamma_5)] + \text{c.c.}
\end{aligned} \tag{1.64}$$

The number of gamma matrices under the second trace can be easily reduced to four by using the standard anticommutation relations — one has $\gamma_0 \gamma^j \gamma_0 = -\gamma^j$ and hence $\gamma_0 \not{k} \gamma_0 = \tilde{\not{k}}$ with $\tilde{k} = (k_0, -\vec{k})$. Further, we use the identity $\mathrm{Tr}(\sigma_j \sigma_k) = 2\delta_{jk}$ for the Pauli matrices and some well-known trace properties of Dirac matrices; in particular, one may observe that traces of the type $\mathrm{Tr}(\not{a}\gamma^j \not{b}\gamma^j \gamma_5)$ vanish identically because of antisymmetry of the Levi-Civita tensor $\epsilon_{\alpha\beta\gamma\delta}$. We are thus left with

$$\begin{aligned}
\overline{|\mathcal{M}_{GT}|^2} &= 4C_A^2 (1 + \alpha_A^2) M^2 \mathrm{Tr}(\not{k}\gamma^j \not{k}' \gamma^j) \\
&\quad + 16C_T^2 (1 + \alpha_T^2) M^2 \mathrm{Tr}(\tilde{\not{k}}\gamma^j \not{k}' \gamma^j) \\
&\quad + 8C_A C_T (1 - \alpha_A \alpha_T) M^2 m_e \mathrm{Tr}(\gamma^j \not{k}' \gamma_0 \gamma^j) + \text{c.c.}
\end{aligned} \tag{1.65}$$

The evaluation of the remaining traces is then straightforward and one obtains

$$\begin{aligned}
\overline{|\mathcal{M}_{GT}|^2} &= 16C_A^2 (1 + \alpha_A^2) M^2 (2\vec{k} \cdot \vec{k}' - g^{jj} k \cdot k') \\
&\quad + 64C_T^2 (1 + \alpha_T^2) M^2 (-2\vec{k} \cdot \vec{k}' - g^{jj} \tilde{k} \cdot k') \\
&\quad + 64C_A C_T (1 - \alpha_A \alpha_T) M^2 m_e k_0' g^{jj}
\end{aligned} \tag{1.66}$$

However, $g^{jj} = -3$ and the scalar products of the four-momenta can be written as $k \cdot k' = E_e E_{\bar{\nu}}(1 - \beta_e \cos\vartheta)$ and $\tilde{k} \cdot k' = E_e E_{\bar{\nu}}(1 + \beta_e \cos\vartheta)$.

One thus finally gets, after some simple manipulations,

$$\overline{|\mathcal{M}_{GT}|^2} = 16M^2E_eE_{\bar{\nu}}\left[3C_A^2(1+\alpha_A^2)\left(1-\frac{1}{3}\beta_e\cos\vartheta\right)\right.$$
$$\left.+12C_T^2(1+\alpha_T^2)\left(1+\frac{1}{3}\beta_e\cos\vartheta\right)-12C_AC_T(1-\alpha_A\alpha_T)\frac{m_e}{E_e}\right] \tag{1.67}$$

and (1.46) is thereby proved.

Let us now examine the angular correlation of the electron and antineutrino. As we have noted before, this is described in terms of a differential decay rate of the type (1.55). The angular dependence is contained in the leptonic part of the relevant matrix element squared. The results (1.45) and (1.67) are applicable to the allowed nuclear F and GT transitions, respectively — the nuclear wave functions can only contribute an overall constant factor which of course does not influence the lepton angular distribution in question. We may now also use the conditions (1.57) and (1.58) which express the absence of the Fierz interference. For the F transitions, we thus have

$$\frac{dw_F}{d\Omega_{\bar{\nu}}} = \text{const.} \times (1 + a_F\beta_e\cos\vartheta) \tag{1.68}$$

with the correlation coefficient

$$a_F = \frac{C_V^2(1+\alpha_V^2) - C_S^2(1+\alpha_S^2)}{C_V^2(1+\alpha_V^2) + C_S^2(1+\alpha_S^2)} \tag{1.69}$$

and for GT transitions similarly

$$\frac{dw_{GT}}{d\Omega_{\bar{\nu}}} = \text{const.} \times (1 + a_{GT}\beta_e\cos\vartheta) \tag{1.70}$$

with

$$a_{GT} = -\frac{1}{3}\frac{C_A^2(1+\alpha_A^2) - 4C_T^2(1+\alpha_T^2)}{C_A^2(1+\alpha_A^2) + 4C_T^2(1+\alpha_T^2)} \tag{1.71}$$

It should be stressed that the angular correlations (1.68) or (1.70), respectively, have nothing to do with a possible parity violation: the $\cos\vartheta$ in the considered case is determined by a scalar product of the particle *momenta*, which of course is a $\mathcal{P}$-even quantity (it is also seen that the correlation coefficients do not vanish for $\alpha_j = 0$). The relations (1.69) and (1.71) make it obvious that a pure S coupling in F transitions would lead to the angular correlation coefficient $a_S = -1$, while the V coupling gives $a_V = +1$. Similarly, for the GT transitions, a pure A coupling would produce $a_A = -\frac{1}{3}$, while the T coupling yields $a_T = +\frac{1}{3}$.

The available experimental data show that $a_F \doteq 1$ while $a_{GT} \doteq -\frac{1}{3}$, within some 1–10% accuracy (examples of typical numbers can be found, e.g., in [Ren] or [CoB]). One may interpret this as an indication that the underlying theory of weak interactions yields $a_{\mathrm{F}} = 1$ and $a_{\mathrm{GT}} = -\frac{1}{3}$ *exactly*. Assuming this, the relations (1.69) and (1.71) then immediately imply

$$C_S = 0, \quad C_T = 0 \tag{1.72}$$

Note that (1.72) also automatically satisfies the conditions (1.57), (1.58); the absence of S and T couplings actually provides a simple and natural explanation for the vanishing of the Fierz interference terms.

Of course, at this stage, there are other possible interpretations of the existing data as well: for example, the experimental result $a_F \doteq 1$ (along with the condition (1.57)) can also be reproduced if the C_S and C_V are approximately equal, but $\alpha_S \alpha_V = 1$ with $\alpha_S \ll \alpha_V$ (and similarly for GT transitions). In fact, such a scenario is excluded by further empirical data to be discussed in the next section. For the time being, we adopt — at least tentatively — the straightforward conclusion (1.72), which means that a theoretical description of beta-decay processes can be formulated in terms of the *V and A couplings alone.*[6] Let us emphasize once again that our — rather dramatic — conclusion (1.72) is based on an idealization of the existing empirical data; we simply interpret the real data as a *strong evidence* in favour of an effective theory of weak interactions dominated by V and A couplings without seeking an optimum fit for all possible parameters in (1.27).

Now, it remains to determine the parameters α_V and α_A, which characterize the non-invariance of our effective Lagrangian under space reflection. To this end, one has to examine phenomena which manifest directly the parity violation in beta-decay processes.

1.6 Longitudinal polarization of electrons

There are several observable quantities that may reveal parity violation in weak interactions; a detailed account of the relevant experiments can

[6]Looking back in history, it is amusing to note that in 1950s a prevailing opinion was just opposite: the data available then (mostly before the recognition of parity violation) seemed to favour S and T couplings. In fact, there was a period of confusion and the experimental situation was only clarified in the late 1950s and early 1960s, in a remarkable interplay with some successful theoretical conjectures formulated at that time — this theme we shall discuss later on.

be found, e.g., in [Adv]. Historically, the first example was provided by the celebrated experiment of Wu *et al.* [5] who measured the angular correlation between electron momentum and nuclear spin in the beta decay of the polarized nucleus of Co^{60} (recall that the scalar product of a spin and a momentum is certainly a $\mathcal{P}$-odd quantity). We will discuss this type of angular correlation (for the free neutron) later on, and now, let us examine another parity-violating observable, which can provide the desired information about the parameters α_j in a very straightforward and efficient way. The quantity we have in mind is the *degree of polarization of the electrons* (or positrons) produced in the beta decay of an *unpolarized* nucleon system. In particular, one may consider longitudinal polarization (helicity) and study a relative difference between the rates of emission of a right-handed and a left-handed electron. To put it explicitly, the degree of longitudinal polarization is defined as

$$P_e = \frac{N_R - N_L}{N_R + N_L} \tag{1.73}$$

where N_R and N_L denote the number of emitted electrons with positive and negative helicity, respectively. For a given energy E_e, this can be calculated in terms of the corresponding differential decay rates

$$P_e = \frac{dw_R(E_e) - dw_L(E_e)}{dw_R(E_e) + dw_L(E_e)} \tag{1.74}$$

Intuitively, it should be clear that such a quantity, if non-zero, is a direct manifestation of parity violation in weak interactions. First of all, one should recall that the space inversion transforms a right-handed electron into a left-handed one. An asymmetry between different spin states of the final particles could in fact occur simply as a consequence of the angular momentum conservation if the initial nucleon system is polarized (i.e. if it has a well-defined spin projection). If an asymmetry between N_R and N_L appears in the case of unpolarized initial nucleons, it can only be accounted for by an intrinsic parity violation (the corresponding interaction is able to distinguish between "right" and "left").

Let us now calculate the quantity (1.74), starting from the matrix elements (1.33); in view of our preceding results, we will consider now only the vector (V) and axial-vector (A) couplings. We can perform the calculation for the V and A terms separately, since we know that in the corresponding decay rate, there is no interference between the Fermi and Gamow–Teller parts of the amplitude in case of unpolarized nucleons.

Let us start with the V term. The matrix element corresponding to the emission of a right-handed electron is written as

$$\mathcal{M}_V^R = C_V(U_p^\dagger U_n)[\bar{u}_{eR}(1+\alpha_V\gamma_5)\gamma_0 v_\nu] \tag{1.75}$$

where the right-handed Dirac spinor u_{eR} satisfies

$$u_{eR}(k)\bar{u}_{eR}(k) = (\not{k}+m_e)\frac{1+\gamma_5\not{s}_R}{2} \tag{1.76}$$

with s_R being the longitudinal spin four-vector

$$s_R^\mu(k) = \left(\frac{|\vec{k}|}{m_e}, \frac{E_e}{m_e}\frac{\vec{k}}{|\vec{k}|}\right) \tag{1.77}$$

(see (A.68)). The corresponding matrix element for a left-handed electron is obtained by replacing the s_R with

$$s_L = -s_R \tag{1.78}$$

Looking back at our calculation of the electron energy spectrum carried out in Section 1.4, it is easy to realize that the ratio of differential decay rates shown in (1.74) is in fact equal to

$$P_e^{(V)} = \frac{\int_{-1}^{1} d(\cos\vartheta)\left(\overline{|\mathcal{M}_V^R|^2} - \overline{|\mathcal{M}_V^L|^2}\right)}{\int_{-1}^{1} d(\cos\vartheta)\left(\overline{|\mathcal{M}_V^R|^2} + \overline{|\mathcal{M}_V^L|^2}\right)} \tag{1.79}$$

where the bar over a $|\mathcal{M}|^2$ now indicates summing over the $n, p, \bar{\nu}$ spins and averaging over the initial neutron spin; ϑ denotes, as usual, an angle between the electron and antineutrino directions. Employing the standard trace techniques (including in particular the relations (1.76) and (1.78)), the integrand of the numerator in (1.79) becomes, after some algebra,

$$\begin{aligned} X_V &\equiv \overline{|\mathcal{M}_V^R|^2} - \overline{|\mathcal{M}_V^L|^2} \\ &= 4M^2C_V^2\,\mathrm{Tr}[(\not{k}+m_e)\gamma_5\not{s}_R(1+\alpha_V\gamma_5)\gamma_0\not{k}'(1+\alpha_V\gamma_5)\gamma_0] \end{aligned} \tag{1.80}$$

while the denominator has in fact been calculated before — it is precisely the V part of the result (1.45), namely

$$\overline{|\mathcal{M}_V^R|^2} + \overline{|\mathcal{M}_V^L|^2} = 16M^2E_eE_{\bar{\nu}}C_V^2(1+\alpha_V^2)(1+\beta_e\cos\vartheta) \tag{1.81}$$

Using the well-known properties of Dirac matrices, the expression (1.80) can be further simplified to

$$\begin{aligned} X_V &= 4M^2C_V^2(-2\alpha_V)m_e\,\mathrm{Tr}(\not{s}_R\gamma_0\not{k}'\gamma_0) \\ &= 16M^2C_V^2(-2\alpha_V)m_e(2s_R^0E_{\bar{\nu}} - s_R\cdot k') \end{aligned} \tag{1.82}$$

Using now the explicit expression for the s_R (see (1.77)), one readily gets

$$
\begin{aligned}
X_V &= 16M^2C_V^2(-2\alpha_V)m_e\left(\beta_e\frac{1}{m_e}E_eE_{\bar{\nu}} + \frac{E_e}{m_e}E_{\bar{\nu}}\cos\vartheta\right) \\
&= 16M^2E_eE_{\bar{\nu}}C_V^2(-2\alpha_V)(\beta_e + \cos\vartheta) \qquad (1.83)
\end{aligned}
$$

The terms in (1.81) and (1.83), proportional to $\cos\vartheta$, obviously vanish upon the angular integration indicated in (1.79) and we thus finally obtain

$$
P_e^{(V)} = -\frac{2\alpha_V}{1+\alpha_V^2}\beta_e \qquad (1.84)
$$

For the axial-vector coupling, one can proceed in a similar way. After some simple manipulations, one first gets

$$
\begin{aligned}
X_A &\equiv \overline{|\mathcal{M}_A^R|^2} - \overline{|\mathcal{M}_A^L|^2} \\
&= 2M^2C_A^2\operatorname{Tr}(\sigma_j\sigma_k)\operatorname{Tr}[(\not{k} + m_e)\gamma_5\not{s}_R\gamma^j\not{k}'\gamma^k(1+\alpha_A^2 - 2\alpha_A\gamma_5)] \qquad (1.85)
\end{aligned}
$$

and using the familiar trace identities, this is simplified to

$$
X_A = 4M^2C_A^2(-2\alpha_A)m_e\operatorname{Tr}(\not{s}_R\gamma^j\not{k}'\gamma^j) \qquad (1.86)
$$

Working out the last trace, one obtains

$$
\begin{aligned}
X_A &= 16M^2C_A^2(-2\alpha_A)m_e(2\vec{s}_R\cdot\vec{k}' - g^{jj}s_R\cdot k') \\
&= 16M^2E_eE_{\bar{\nu}}C_A^2(-2\alpha_A)(3\beta_e - \cos\vartheta) \qquad (1.87)
\end{aligned}
$$

On the other hand, a corresponding result for the sum over electron helicities can be retrieved from (1.46); it reads

$$
\overline{|\mathcal{M}_A^R|^2} + \overline{|\mathcal{M}_A^L|^2} = 16M^2E_eE_{\bar{\nu}}C_A^2(1+\alpha_A^2)(3 - \beta_e\cos\vartheta) \qquad (1.88)
$$

Now, the degree of electron polarization can again be evaluated as

$$
P_e^{(A)} = \frac{\int_{-1}^{1} d(\cos\vartheta)\left(\overline{|\mathcal{M}_A^R|^2} - \overline{|\mathcal{M}_A^L|^2}\right)}{\int_{-1}^{1} d(\cos\vartheta)\left(\overline{|\mathcal{M}_A^R|^2} + \overline{|\mathcal{M}_A^L|^2}\right)} \qquad (1.89)
$$

Thus, inserting into (1.89) the expressions (1.87) and (1.88), we immediately get

$$
P_e^{(A)} = -\frac{2\alpha_A}{1+\alpha_A^2}\beta_e \qquad (1.90)
$$

i.e. a result completely analogous to that obtained for the vector coupling.

Formally, we have performed our calculation for a free neutron, but the results (1.84) and (1.90) are in fact also valid for pure F and GT allowed nuclear beta decays, respectively — the energy dependence of the quantity in question is determined solely by the leptonic factor of the matrix element and a constant factor coming from nucleons drops out from the ratio (1.79) or (1.89), respectively. Note also that the full answer in the free-neutron case obviously reads

$$P_e^{(V,A)} = -\beta_e \frac{2\alpha_V C_V^2 + 2\alpha_A\, 3C_A^2}{(1+\alpha_V^2)C_V^2 + (1+\alpha_A^2)\, 3C_A^2} \tag{1.91}$$

since the neutron decay is a mixed transition and, as noted before, there is no F-GT interference for the considered observable.

Now, we are in a position to confront our theoretical results with empirical data. Various measurements of beta-electron helicities (for both the F and GT nuclear transitions) show that — over a wide energy range and with a rather high accuracy — the degree of electron longitudinal polarization is simply related to its velocity:

$$P_e^{(\text{exp})} = -\beta_e \tag{1.92}$$

(for an overview of the data we refer the reader, e.g., to [CoB]). The remarkable result (1.92) means, among other things, that highly relativistic beta electrons are almost completely polarized, being predominantly *left-handed*. Comparing (1.92) with the formulae (1.84) and (1.90), we may then conclude that

$$\alpha_V = 1, \quad \alpha_A = 1 \tag{1.93}$$

We thus see that the parity-violating effects due to weak interactions are substantial; the considered quantity in fact reaches its maximum possible value (obviously, the function $2\alpha/(1+\alpha^2)$ has a maximum for $\alpha = 1$). For this reason, it is usually said that weak interactions exhibit *maximal parity violation*; note that (1.93) also means that the $\mathcal{P}$-even terms and their $\mathcal{P}$-odd counterparts contained in the Lagrangian (1.27) have an equal strength.

When an analogous calculation is carried out for positrons (i.e. starting from the h.c. part of (1.27)), one finds that the overall sign in the relevant results is reversed (a verification of this statement is recommended to the reader as an instructive exercise). The experimental data, though less ample and less accurate than those for electrons, show that indeed

$$P_{\text{positron}} = +\beta_{\text{positron}} \tag{1.94}$$

(cf. [CoB]), i.e. positrons emitted in beta-decay processes are mostly right-handed at relativistic velocities. The Lagrangian (1.27) with

$\alpha_V = \alpha_A = 1$ thus provides a very good description of the data from longitudinal polarization measurements for both electrons and positrons.

The spectacular result (1.93) provides a very important piece of information in our search for a realistic effective theory of beta decay. At present, we are left with two parameters (coupling constants) C_V and C_A that remain to be determined from some further empirical data. We will complete this task later on — an impatient reader may pass immediately to Section 1.8. However, now, we would like to pause for a moment and mention a possible modification of the procedure that has led us to our present position. In particular, we might interchange the last two steps: instead of eliminating plainly the S and T couplings on the basis of the e–$\bar{\nu}$ angular correlation data, we might examine the longitudinal polarization first, keeping for the moment all the parameters C_S, C_V, C_A, C_T in the game. Of course, in any case, we have to account for the observed absence of the Fierz interference; this can be simply achieved by assuming

$$1 - \alpha_S \alpha_V = 0, \quad 1 - \alpha_A \alpha_T = 0 \tag{1.95}$$

(cf. (1.45), (1.46)). Let us consider neutron decay, where all types of couplings may contribute. The evaluation of the degree of electron polarization starts from the full matrix element $\mathcal{M}_S + \mathcal{M}_V + \mathcal{M}_A + \mathcal{M}_T$ (see (1.33)) and proceeds along similar lines as before. As we already know, for the considered quantity, one need not worry about an F–GT interference; moreover, if one makes use of the conditions (1.95), the S–V and A–T interference terms turn out to vanish completely as well. The final result reads

$$P_e = -\beta_e \frac{2\alpha_S C_S^2 + 2\alpha_V C_V^2 + 2\alpha_A \, 3C_A^2 + 2\alpha_T \, 12C_T^2}{(1+\alpha_S^2)C_S^2 + (1+\alpha_V^2)C_V^2 + (1+\alpha_A^2)\, 3C_A^2 + (1+\alpha_T^2)\, 12C_T^2} \tag{1.96}$$

A detailed derivation of the last expression is left to an interested reader as an instructive (though somewhat tedious) exercise. From (1.96), it is also easy to guess the corresponding answers for pure F and GT nuclear transitions. Comparing now our theoretical formula (1.96) with the experimental observation (1.92), one readily gets the condition

$$C_S^2(1-\alpha_S)^2 + C_V^2(1-\alpha_V)^2 + 3C_A^2(1-\alpha_A)^2 + 12C_T^2(1-\alpha_T)^2 = 0 \tag{1.97}$$

Obviously, if one wants to keep momentarily all the C_j non-zero, the last relation can only be satisfied if

$$\alpha_S = \alpha_V = \alpha_A = \alpha_T = 1 \tag{1.98}$$

(note also that the conditions (1.95) are then fulfilled "trivially"). Equipped with this knowledge, we may reconsider the e–$\bar{\nu}$ angular correlations. Remembering the formulae (1.69) and (1.71), it is obvious that the relevant experimental data along with the values of the parameters α_j shown in (1.98) force us to set

$$C_S = 0, \quad C_T = 0 \tag{1.99}$$

in accordance with the option chosen tentatively in Section 1.5. In particular, as a by-product of our analysis, one can see — as we have promised before — that, e.g., a pattern with $C_S \doteq C_V$, $\alpha_S \ll \alpha_V$ is clearly excluded by the available empirical data on the electron longitudinal polarization.

1.7 Neutrino helicity

In view of the preceding arguments, the original Lagrangian (1.27) is now effectively reduced to

$$\begin{aligned}\mathscr{L}_{\text{int}}^{(\beta)} &= C_V(\bar{\psi}_p\gamma_\mu\psi_n)[\bar{\psi}_e(1+\gamma_5)\gamma^\mu\psi_\nu] \\ &\quad + C_A(\bar{\psi}_p\gamma_5\gamma_\mu\psi_n)[\bar{\psi}_e(1+\gamma_5)\gamma^\mu\psi_\nu]\end{aligned} \tag{1.100}$$

(note that one γ_5 factor in the leptonic part of the axial-vector term has been absorbed into the $1+\gamma_5$ because of $\gamma_5^2 = 1$). The hermitian conjugate of (1.100) reads

$$\begin{aligned}\mathscr{L}_{\text{int}}^{(\beta)\dagger} &= C_V(\bar{\psi}_n\gamma_\mu\psi_p)[\bar{\psi}_\nu(1+\gamma_5)\gamma^\mu\psi_e] \\ &\quad + C_A(\bar{\psi}_n\gamma_5\gamma_\mu\psi_p)[\bar{\psi}_\nu(1+\gamma_5)\gamma^\mu\psi_e]\end{aligned} \tag{1.101}$$

It is easy to see that the form (1.100) or (1.101) gives a definite *prediction* for the helicity of the antineutrino or neutrino, respectively. Indeed, making use of the γ_5 anticommutativity, the matrix element for $n \to p + e^- + \bar{\nu}$ corresponding to (1.100) can obviously be written as[7]

$$\begin{aligned}\mathcal{M}_{fi}^{(\bar{\nu})} &= C_V(\bar{u}_p\gamma_\mu u_n)[\bar{u}_e\gamma^\mu(1-\gamma_5)v_\nu] \\ &\quad + C_A(\bar{u}_p\gamma_5\gamma_\mu u_n)[\bar{u}_e\gamma^\mu(1-\gamma_5)v_\nu]\end{aligned} \tag{1.102}$$

[7]Throughout this section, we don't need to use the non-relativistic approximation for nucleons.

and, similarly, for an inverse process $p \to n + e^+ + \nu$, one gets from (1.101)

$$\begin{aligned}\mathcal{M}_{fi}^{(\nu)} &= C_V(\bar{u}_n\gamma_\mu u_p)[\bar{u}_\nu(1+\gamma_5)\gamma^\mu v_e] \\ &\quad + C_A(\bar{u}_n\gamma_5\gamma_\mu u_p)[\bar{u}_\nu(1+\gamma_5)\gamma^\mu v_e] \end{aligned} \tag{1.103}$$

Now, it is obvious that only right-handed antineutrino can be emitted: indeed, v_L satisfies $v_L = \frac{1}{2}(1+\gamma_5)v_L$ in the massless case, so that the factor $1-\gamma_5$ contained in (1.102) makes it vanish. On the other hand, the v_R, satisfying $v_R = \frac{1}{2}(1-\gamma_5)v_R$ clearly survives in (1.102). In other words, the (V, A) structure of the interaction and the presence of the factor $1+\gamma_5$ in (1.100) (enforced by the empirical data on the *electron* helicity) together lead to a definite prediction for the value of antineutrino helicity. In a similar way, from (1.103), it is seen that the neutrino should always be produced as left-handed: indeed, one has

$$\begin{aligned} u_L &= \frac{1-\gamma_5}{2}u_L \quad \Rightarrow \quad \bar{u}_L = \bar{u}_L\frac{1+\gamma_5}{2} \\ u_R &= \frac{1+\gamma_5}{2}u_R \quad \Rightarrow \quad \bar{u}_R = \bar{u}_R\frac{1-\gamma_5}{2} \end{aligned} \tag{1.104}$$

and hence, only u_L can survive in (1.103).

Verification of the above predictions experimentally is an extremely difficult task, since the neutrino has no electromagnetic interactions and hence its helicity cannot be measured directly (as, e.g., that of an electron or photon). Nevertheless, one (indirect) measurement does exist — it has been accomplished in an ingenious experiment by Goldhaber *et al.* [7]. The process investigated in [7] was essentially $e^- + p \to n + \nu$. In particular, Goldhaber *et al.* studied the capture of an electron from an inner atomic orbit in $\mathrm{Eu}^{152}(0^-)$, which produces an excited state $\mathrm{Sm}^{152*}(1^-)$ and a neutrino is emitted (this particular reaction was chosen because of some exceptionally favourable properties of the nuclei involved). The neutrino helicity can then be deduced from the spin and momentum of the daughter nucleus; this is accomplished through a measurement of the circular polarization of the photon emitted (in deexcitation of the samarium nucleus) along the direction of flight of the Sm^{152*}. More details of this unique experiment are described in many places; see, e.g., [CaG], [Gre] and, in particular, [Tel]. Goldhaber *et al.* found that the neutrino was always emitted with negative helicity, i.e. left-handed, which confirms the prediction given above.

Such an independent check of our effective beta-decay theory is gratifying, but we should perhaps add one more remark concerning the importance of the measurement of neutrino helicity. Imagine that we have already exploited the data concerning electron helicity (longitudinal polarization), but all the couplings S, V, A, T are still preserved in the effective Lagrangian — in other words, we set $\alpha_S = \alpha_V = \alpha_A = \alpha_T = 1$, but ignore temporarily the available data on the e–$\bar{\nu}$ angular correlations. The process studied by Goldhaber *et al.* [7] is a pure GT transition (note the spin assignments of the parent and daughter nuclei), so that both A and T couplings can contribute to the relevant matrix element. It is easy to see that a measurement of the neutrino helicity provides, in fact, a clear-cut test of the type of the coupling responsible for the beta transition in question. Indeed, using the hermitian conjugate term in (1.27) with the particular values of the parameters, a general GT matrix element for $e^- + p \to n + \nu$ can formally be written as

$$\begin{aligned} \mathcal{M}_{GT} = {} & C_A(\bar{u}_n\gamma_5\gamma_\mu u_p)[\bar{u}_\nu(1+\gamma_5)\gamma^\mu u_e] \\ & + C_T(\bar{u}_n\sigma_{\mu\nu}u_p)[\bar{u}_\nu(1-\gamma_5)\sigma^{\mu\nu}u_e] \end{aligned} \tag{1.105}$$

Of course, the appearance of $1+\gamma_5$ and $1-\gamma_5$ in the A and T terms, respectively, is due to the different commutation properties of the Dirac matrices involved: γ^μ anticommutes with γ_5 while $\sigma^{\mu\nu}$ commutes. Now, taking into account (1.104), it is clear that neutrino helicity clearly distinguishes between the A and T couplings: neutrinos produced through the A coupling are purely left-handed (the observed case), while the T coupling would yield right-handed ones. These considerations can be easily generalized to the Fermi transitions — the V coupling, as noted before, can only produce left-handed neutrinos while those due to an S coupling would be right-handed. Again, such a "dichotomy" is simply related to the commutation properties of the corresponding matrix structures. However, one should keep in mind that for Fermi transitions, there is no corresponding measurement of the neutrino helicity.

It is important to realize that the above conclusions concerning neutrino helicity and the possible algebraic types of the relevant couplings are intimately related to the empirical data on electron helicity, which tell us that relativistic beta-electrons are left-handed (precisely, this fact has led us to set $\alpha_S = \alpha_V = \alpha_A = \alpha_T = 1$). Clearly, a pattern which thus emerges is the following. The presence of e_L and ν_L reveals a (V, A) structure of the underlying effective theory, while the combination of e_L and ν_R would correspond to an (S, T) model; other equivalent variants are obvious.

Historically, the measurement of neutrino helicity played a very important role in determining the right form of the beta-decay effective

Lagrangian (at least for its GT part). Before the advent of parity violation, there were some controversial results concerning the e–$\bar{\nu}$ angular correlation in GT transitions, which preferred the T rather than A coupling (in this context, see, in particular, the paper by Rustad and Ruby, *Phys. Rev.* 97 (1955) 991, dealing with the decay of He^6). The helicity measurements for electron and neutrino, which followed the discovery of parity violation, provided a powerful argument in favour of the A coupling. In any case — in view of the absence of a measurement of neutrino helicity in Fermi (or mixed) transitions — it is gratifying that the relevant data on the e–$\bar{\nu}$ angular correlation for both F and GT transitions now support the (V, A) effective theory.

1.8 The V and A coupling constants

Let us now show how the remaining free parameters in the Lagrangian (1.100), namely the coupling constants C_V and C_A, can be determined. Our earlier results (1.45) and (1.67) imply that within the effective theory described by (1.100) (and within the usual non-relativistic approximation), the spin-averaged squared matrix element for neutron decay becomes

$$\overline{|\mathcal{M}|^2} = \overline{|\mathcal{M}_V|^2} + \overline{|\mathcal{M}_A|^2} = 32M^2 E_e E_{\bar{\nu}}[C_V^2 + 3C_A^2 + (C_V^2 - C_A^2)\beta_e \cos\vartheta] \tag{1.106}$$

The last expression clearly indicates that a measurement of the e–$\bar{\nu}$ angular correlation in the free neutron decay could fix at least the ratio of the coupling constants squared. Indeed, denoting

$$f = C_A/C_V \tag{1.107}$$

the angular distribution corresponding to (1.106) can obviously be written as

$$\frac{dw}{d(\cos\vartheta)} = \text{const.} \times (1 + a_n \beta_e \cos\vartheta) \tag{1.108}$$

with

$$a_n = \frac{1 - f^2}{1 + 3f^2} \tag{1.109}$$

The experimental value of the correlation coefficient is $a_n = -0.1049 \pm 0.0013$ (the weighted world average [6]). Using this in (1.109), we get roughly

$$|f| \doteq 1.27 \tag{1.110}$$

i.e. the V and A couplings turn out to be of comparable, yet unequal, strength. Note that such a closeness of C_V and C_A is essentially accidental — we will comment on this point in the next chapter.

Of course, the data on the particular angular correlation considered so far can only provide information on the absolute value of the ratio f, since (1.106) does not involve any interference between the V and A couplings — as we know, this is a general feature of the observable quantities calculated for unpolarized nucleons. Thus, in order to find the sign of the f, one obviously has to exploit an observable related to *polarized* nucleons. In particular, a suitable experimentally accessible quantity is the angular correlation between electron momentum and neutron spin in the decay of a polarized neutron. As we have noted at the beginning of Section 1.6, such an angular correlation represents a parity-violating effect, so it would perhaps be also instructive to demonstrate this aspect explicitly in the result of our calculation. For this purpose, let us restore temporarily arbitrary parameters α_V and α_A in our effective Lagrangian; it means that we start the calculation from the matrix element

$$\begin{aligned}\mathcal{M} &= C_V(U_p^\dagger U_n)[\bar{u}_e(1+\alpha_V\gamma_5)\gamma_0 v_\nu] \\ &\quad + C_A(U_p^\dagger \sigma_j U_n)[\bar{u}_e(1+\alpha_A\gamma_5)\gamma_5\gamma^j v_\nu]\end{aligned} \tag{1.111}$$

The coordinate system can be conventionally chosen so that the initial neutron spin is directed along the third axis. For a practical calculation, it then implies that

$$U_n U_n^\dagger = 2M\frac{1+\sigma_3}{2} \tag{1.112}$$

The evaluation of the matrix element squared is somewhat tedious and we have therefore relegated the technical details to Appendix C. Here, let us quote only the result; it reads

$$\begin{aligned}\int \frac{d\Omega_{\bar{\nu}}}{4\pi} \sum_{\text{spin } p,e,\bar{\nu}} |\mathcal{M}|^2 &= 16M^2 E_e E_{\bar{\nu}}\big[C_V^2(1+\alpha_V^2) + 3C_A^2(1+\alpha_A^2) \\ &\quad + \big(2C_V C_A(\alpha_V+\alpha_A) - 4\alpha_A C_A^2\big)\beta_e\cos\theta_e\big]\end{aligned} \tag{1.113}$$

where θ_e denotes the angle between the electron momentum and neutron spin (i.e. the polar angle for the electron direction in our coordinate frame). Now, the parity-violating nature of the considered angular dependence should be obvious — as expected, the term involving the $\mathcal{P}$-odd $\cos\theta_e$

is proportional to the parameters α_V and α_A, and thereby, it is trivial for $\alpha_V = \alpha_A = 0$. Another remarkable feature of the result (1.113) is that the coefficient at $\cos\theta_e$ also vanishes for $C_A = 0$ (for arbitrary values of α_V and α_A); in other words, the effect would be trivial for a pure F transition (recall that Wu *et al.* in their celebrated experiment [5] measured the angular distribution of the considered type for a pure GT transition $\mathrm{Co}^{60} \to \mathrm{Ni}^{60}$).

Let us now proceed to determine the ratio $f = C_A/C_V$, as indicated above. Returning to the known values $\alpha_V = \alpha_A = 1$, the expression (1.113) becomes

$$\int \frac{d\Omega_{\bar{\nu}}}{4\pi} \sum_{\text{spin } p,e,\bar{\nu}} |\mathcal{M}|^2 = 32M^2 E_e E_{\bar{\nu}} C_V^2 [1 + 3f^2 + 2(f - f^2)\beta_e \cos\theta_e] \tag{1.114}$$

The corresponding angular distribution then obviously can be written as

$$\frac{dw}{d(\cos\theta_e)} = \text{const.} \times (1 + A_n \beta_e \cos\theta_e) \tag{1.115}$$

with the coefficient A_n given by

$$A_n = 2\frac{f - f^2}{1 + 3f^2} \tag{1.116}$$

The experimental value of the "β asymmetry parameter" A_n is $= -0.11958 \pm 0.00021$ (the rounded world average [6]). Using this in Eq. (1.116), one obtains two solutions for f, namely $f^{(1)} \doteq 1.27$ and $f^{(2)} \doteq -0.06$. Obviously, the latter possibility is not compatible with our preceding result for the $|f|$ (see (1.110)). Thus, one may conclude that

$$f \doteq 1.27 \tag{1.117}$$

i.e. the coupling constants C_V and C_A have the same sign, within our system of definitions (the reader should be warned, however, that a definition of the axial-vector coupling constant with opposite sign occurs rather frequently in the literature, cf., e.g., [6]). Looking now back at the formula (1.113) (with $\alpha_V = \alpha_A = 1$), it is clear that the F–GT interference acts "destructively" on the magnitude of the correlation coefficient in question — this is one more reason why Wu *et al.* [5] have chosen a pure GT transition for their investigation of parity violation. In any case, the value of the correlation coefficient A_n is negative (similar to the case considered in [5]), which means that the electrons are emitted preferentially in the direction opposite to the neutron spin.

At this place, it is worth noting that the calculation leading to (1.116) (see Appendix C) can be easily modified to yield an analogous result for the coefficient of the correlation of neutron spin and antineutrino momentum. This "$\bar{\nu}$ asymmetry parameter" comes out to be

$$B_n = 2\frac{f + f^2}{1 + 3f^2} \tag{1.118}$$

With the known value of f (fixed by other experiments), the last result represents a *prediction* of our effective beta-decay theory. For $f \doteq 1.27$, one gets from (1.118) $B_n \doteq 0.988$ to be compared with the experimental value $B_n = 0.981 \pm 0.003$ (the weighted world average [6]).

For a complete knowledge of the coupling constants C_V and C_A, it is now sufficient to fix the absolute value of one of them by means of a suitable experiment. Obviously, an appropriate observable quantity would be a fully integrated decay rate (the decay width), which determines the mean lifetime τ of the neutron or of a beta-radioactive nucleus. Such a decay width is obtained by integrating the electron energy distribution function over the whole kinematical range and it obviously comes out to be a linear combination of C_V^2 and C_A^2 with calculable coefficients. (Needless to say, we have in mind the first order of perturbation theory. In the case of a nuclear beta transition, the practical calculability is of course limited by our knowledge of the wave functions of the nuclei involved.) An elementary example of such an integration is given in the next section. Thus, any measured lifetime would do, provided that we are able to carry out a reasonably accurate theoretical calculation indeed. This is possible, e.g., in the case of a free neutron decay, but in fact the most favourite and practical method consists in exploiting the pure F transition $O^{14} \to N^{14*} + e^+ + \nu$, which occurs within an isospin multiplet (isotriplet).[8] The lifetime of O^{14} is known with a very good accuracy (note that the half-life $T_{1/2} = \tau \ln 2$ is about 71 s). One thus gets directly the absolute value of the C_V; by convention, C_V is expressed in terms of a "beta-decay Fermi constant" G_β as

$$C_V = -\frac{G_\beta}{\sqrt{2}} \tag{1.119}$$

[8]Other examples of this kind are $C^{10} \to B^{10}$, $Co^{54} \to Fe^{54}$, etc. (see [CoB] and [Gre]). Such transitions are sometimes called super-allowed. The point is that in such a case the nuclear matrix element is easily calculable — it is determined by the isospin lowering or raising operator since the internal structure of the parent and daughter nuclei is essentially identical up to small electromagnetic corrections. An instructive and rather detailed discussion of the O^{14} decay can be found in [HaM], Section 12.3.

G_β is taken to be positive and the data then yield

$$G_\beta = (1.136 \pm 0.001) \times 10^{-5}\ \text{GeV}^{-2} \tag{1.120}$$

Note that the minus sign in the definition (1.119) is pure convention at the present level, but we shall see that it becomes very natural in the context of weak interaction theory involving an intermediate vector boson. The factor of $\sqrt{2}$ is of historical origin — it serves to reproduce the value of the coupling constant G appearing in the old parity-conserving Fermi theory (cf. (1.14)).

Thus, we have got through the determination of the form of an effective beta-decay Lagrangian. Having fixed the values of all relevant free parameters, let us now return to the original relativistic form (1.27) (with only V and A terms preserved). Making use of anticommutativity of γ_5 and the notation (1.107) and (1.119), it is easy to see that the $\mathscr{L}_{\text{int}}^{(\beta)}$ can now be written as

$$\mathscr{L}_{\text{int}}^{(\beta)} = -\frac{G_\beta}{\sqrt{2}}[\bar{\psi}_p\gamma_\mu(1 - f\gamma_5)\psi_n][\bar{\psi}_e\gamma^\mu(1 - \gamma_5)\psi_\nu] + \text{h.c.} \tag{1.121}$$

Paraphrasing the famous Andersen's work [8], one might say that the original "ugly-duckling form" (1.27) has now matured, through some stringent experimental tests to a "swan-like" appearance (1.121). In fact, the realistic effective Lagrangian now in a way resembles the old Fermi model: the vectorial currents of the Fermi theory are replaced by linear combinations of the V and A currents; in particular, the leptonic part has a pure structure $V - A$. This remarkable feature of the weak interaction Lagrangian will be discussed in detail in the next chapter.

1.9 Mean lifetime of the neutron

With the effective Lagrangian (1.121) at hand, we may now make a *prediction* for another physical observable quantity not exploited within our parameter-fixing procedure. In particular, we can calculate the total decay rate (decay width) for the free neutron, which in turn determines the mean lifetime of such an unstable particle. The decay width is obtained by integrating the original differential rate (1.35) over all kinematical variables of the final-state particles. We have implemented some of the relevant integration steps in Section 1.4 when deriving the form of the electron energy spectrum. To apply our previous results in the case of a free neutron, we may start with the intermediate result (1.55) and employ the expression (1.106) for the matrix element squared. The integration over the angular variables is essentially trivial and one thus arrives at the electron energy spectrum

$$\frac{dw(E_e)}{dE_e} = \frac{1}{\pi^3}(C_V^2 + 3C_A^2)\sqrt{E_e^2 - m_e^2}E_e(\Delta - E_e)^2 \tag{1.122}$$

which agrees, as expected, with the generic form (1.59). The decay width Γ is then obtained by means of an integration over the whole range of electron energies, i.e.

$$\Gamma = \int_{m_e}^{\Delta} \frac{dw(E_e)}{dE_e} dE_e$$

Using (1.122), the last expression becomes

$$\Gamma = \frac{1}{\pi^3}(C_V^2 + 3C_A^2)I_{\rm F} \tag{1.123}$$

where the symbol $I_{\rm F}$ stands for the so-called Fermi integral[9]

$$I_F = \int_{m_e}^{\Delta} (\Delta - E)^2 \sqrt{E^2 - m_e^2} E dE \tag{1.124}$$

The evaluation of the integral (1.124) is straightforward and the result can be written as

$$I_F = \frac{1}{30}\Delta^5\left(\beta_{\rm max}^5 - \frac{5}{2}\frac{m_e^2}{\Delta^2}\beta_{\rm max}^3 - \frac{15}{2}\frac{m_e^4}{\Delta^4}\beta_{\rm max} + \frac{15}{2}\frac{m_e^4}{\Delta^4}\ln\frac{\Delta + \sqrt{\Delta^2 - m_e^2}}{m_e}\right) \tag{1.125}$$

where the $\beta_{\rm max}$ denotes the maximum electron velocity, i.e. $\beta_{\rm max} = (1 - m_e^2/\Delta^2)^{1/2}$ (cf.(1.11)). Numerically, (1.125) means that

$$I_F = \frac{1}{30}\Delta^5 K \tag{1.126}$$

with $K \doteq 0.46$. Thus, within the effective theory (1.121), the decay width of a free neutron is given by a formula

$$\Gamma(n \to p + e^- + \bar{\nu}) = K\frac{G_\beta^2 \Delta^5}{60\pi^3}(1 + 3f^2) \tag{1.127}$$

Putting in numbers, one gets $\Gamma \doteq 6.77 \times 10^{-25}$ MeV. The mean lifetime is the reciprocal value of the Γ, so that $\tau = \Gamma^{-1} \doteq 1.48 \times 10^{24}$ MeV^{-1}.

[9]Note that for nuclear beta transitions, the Fermi integral also includes a coulombic correction factor $F(Z, E_e)$, which may be important, especially for higher atomic numbers Z. For more details, see, e.g., [CoB].

Converting this to ordinary units (using $\hbar = 6.58 \times 10^{-22}$ MeV s), one finally gets

$$\tau_n \doteq 974 \text{ s} \tag{1.128}$$

The experimental value quoted in [6] is (878.4 ± 0.5) s. In order to get from (1.128) closer to the experimental result, one should include some additional minor effects (coulombic and radiative corrections in particular), but this would go beyond the scope of this introductory treatment. Anyway, the agreement between our simple theoretical prediction and the empirical value (within about 10%) is quite satisfactory as it stands.

The formula (1.127) is an example of a rather general rule:

$$\Gamma \propto G^2 \Delta^5 \tag{1.129}$$

which is highly useful for making the order-of-magnitude estimates of the decay rates of allowed beta transitions (and of many other semileptonic decays as well). Let us explain briefly the origin of such a rule. The characteristic form of the electron energy spectrum (1.59) clearly suggests that, at least for $\Delta \gg m_e$, a dominant contribution to the Fermi integral (1.124) amounts to Δ^5 (up to a pure numerical factor). Indeed, neglecting m_e in (1.124), one gets

$$\begin{aligned} I_F &\doteq \int_0^\Delta (\Delta - E)^2 E^2 dE = \Delta^5 \int_0^1 (1-x)^2 x^2 dx \\ &= \frac{1}{30} \Delta^5 \end{aligned} \tag{1.130}$$

and the effects of $m_e \neq 0$ are expected to be of a relative order $O(m_e^2/\Delta^2)$ (cf. (1.125)). On the other hand, the decay rate must include a factor of G^2 (with G being a pertinent Fermi-type coupling constant, $G = G_\beta$ for nuclear beta decays), as the corresponding matrix element is proportional to G when calculated in the first order of perturbation theory. The product $G^2\Delta^5$ already has right dimension of a decay width, so any other factor on the right-hand side of (1.129) can only be a dimensionless number.

Of course, the condition $\Delta \gg m_e$ is not always satisfied sufficiently well (e.g., $m_e/\Delta \doteq 0.4$ for neutron decay) and there may be some particular extra factors present (as, e.g., the $1+3f^2$ in (1.127)), but for a wide variety of beta-decay processes, the "rule $G^2\Delta^5$" does provide quite reasonable order-of-magnitude estimates of the lifetimes — the point is that the usual corrections to the leading behaviour (1.129) do not influence the result dramatically. In fact, one only has to be careful to take into account properly, ubiquitous numerical factors such as $1/(60\pi^3)$ in (1.127), since these typically change a naive guess for a Γ by three orders of magnitude. The safest way of including these numerical effects is to relate the estimated decay rate to some "reference value" (for which one may take, e.g., the neutron lifetime); the large universal factors cancel when a ratio of decay rates is taken and one should thus expect a realistic result within one order of magnitude or so. To put it in explicit terms, let us denote quantities referring to an atomic nucleus and neutron by indices A and n, respectively. For the ratio of the decay rates, we have

$$\Gamma_A/\Gamma_n \doteq (\Delta_A/\Delta_n)^5 \tag{1.131}$$

(the coupling constants squared are cancelled in the ratio as well). A mean lifetime τ is equal to Γ^{-1}, and (1.131) thus implies

$$\tau_A \doteq \tau_n \left(\frac{\Delta_n}{\Delta_A}\right)^5 \tag{1.132}$$

Let us now illustrate by some numerical examples how our rule of thumb (1.132) works in practice. We will consider two processes mentioned before, namely the pure GT transition $\mathrm{He}^6 \to \mathrm{Li}^6 + e^- + \bar{\nu}$ and the pure F transition $\mathrm{O}^{14} \to \mathrm{N}^{14^*} + e^+ + \nu$. In the first case, one has $\Delta_A \doteq 2.3$ MeV (for τ_n, we take approximately 900 s and $\Delta_n \doteq 1.3$ MeV). From (1.132), we then get $\tau_{\mathrm{He}^6} \doteq 3.3$ s; for the corresponding half-life $\tau_{1/2} = \tau \ln 2$, this yields the value of about 2.25 s which is reasonably close to the value 0.81 s found in tables of isotopes (see in particular [9]). For the O^{14} decay, one has $\Delta_A \doteq 4$ MeV and (1.132) then yields an estimate $\tau_{1/2} \doteq 36$ s which agrees, as to the order of magnitude (actually within a factor of 2), with the measured value 71 s. The approximate relation (1.132) is thus seen to be quite reliable and we will appreciate the efficiency of such a rule again in the next chapter in connection with semileptonic decays of baryons (other than nucleons) and mesons.

In concluding this chapter, let us add a remark on the role that weak interactions play in our universe in a somewhat broader context. It is well known that apart from being responsible for the beta radioactivity of atomic nuclei, the weak interaction of nucleons and leptons is also crucial for starting up the thermonuclear reactions occurring in visible stars. In particular, the "proton-burning" process $p + p \to p + n + e^+ + \nu \to D + e^+ + \nu$ (where D denotes the deuteron) constitutes the beginning of a chain of reactions producing most of the energy radiated by the Sun (see, e.g., [CoB]). Thus, one should bear in mind that the weak interaction is in fact of immense *practical* importance — without it, life on the Earth could not exist in its present form. In this connection, one may also say that the character of our environment depends rather dramatically on the weak interaction strength: the magnitude of the weak coupling constant determines the rate of solar energy production and this in turn influences the temperature of the Earth's atmosphere, the intensity of ultraviolet radiation, etc. For more details, see [Cah].

Problems

1.1 Derive the formula (1.96).

1.2 Derive the formula (1.118).

1.3 Using the beta-decay matrix element following directly from (1.121) (without making the quasi-static approximation for proton), one can calculate the proton energy spectrum. Perform such a calculation and show that the distribution function $dw(E_p)/dE_p$ vanishes at both ends of the spectrum, i.e. both for E_p^{min} and for E_p^{max}.
Hint: For the phase-space integration over the e and $\bar{\nu}$ momenta, one can employ the formulae (2.36) and (2.37) quoted in Chapter 2.

1.4 Calculate longitudinal polarization of the proton produced in the decay of a free neutron at rest (employing the same matrix element as in the preceding problem). The degree of longitudinal polarization (P) is defined in analogy with (1.74). In particular, consider the value of $P(E_p)$ at the endpoint of the spectrum, $E_p = E_p^{\mathrm{max}}$. Show that for $m_e = 0$, the result is simplified to

$$P(E_p^{\mathrm{max}})\Big|_{m_e=0} = -\frac{2f}{1+f^2}$$

1.5 Compute the cross-section of the process $\bar{\nu}_e + p \to n + e^+$ for low energies of the incident antineutrino (typically, $1\ \mathrm{MeV} \lesssim E_{\bar{\nu}} \lesssim 10\ \mathrm{MeV}$).

Chapter 2

Universal V—A Theory

2.1 Two-component neutrino

In the preceding chapter, we have arrived at a remarkably simple form of the effective Lagrangian for beta decay. The result (1.121) is written as a product of two "currents" — linear combinations of Lorentz vectors and axial vectors (pseudovectors) and, in particular, the leptonic current has a pure $V-A$ structure. The currents are composed of fermionic fields differing by one unit of electric charge and this is why such objects are usually called "weak charged currents" or simply "charged currents". The $V-A$ form of the leptonic current — deduced from empirical data within our approach — is a rather striking feature of the effective Lagrangian (1.121), and it certainly calls for a theoretical interpretation. Of course, such a problem is intimately related to the remarkable phenomenon of maximal parity violation, revealed, e.g., by the data on the electron longitudinal polarization (see Section 1.6). Historically, the first attempt to formulate a "theory" of parity violation in weak interactions appeared almost simultaneously with its experimental discovery (see [10, 11]). It relied on a revival of the two-component relativistic equation for a massless spin-$\frac{1}{2}$ particle (written first by Weyl in 1929) and it has become known as the "two-component neutrino theory" (more concisely, "the theory of two-component neutrino"). We are now going to summarize briefly this simple idea.

To begin with, let us remember the ordinary Dirac equation for a massive spin-$\frac{1}{2}$ particle. This can be written as

$$i\frac{\partial\psi}{\partial t} = (-i\vec{\alpha}\cdot\vec{\nabla} + \beta m)\psi \tag{2.1}$$

where $\vec{\nabla}$ stands for $\partial/\partial x^j$, $j = 1, 2, 3$. The matrices $\vec{\alpha}$ (i.e. $\alpha^j, j = 1, 2, 3$) and β must satisfy

$$\{\alpha^j, \alpha^k\} = 2\delta^{jk}, \quad \{\beta, \alpha^j\} = 0, \quad \beta^2 = 1 \tag{2.2}$$

in order to reproduce correctly the standard relation between the particle energy and momentum known in special relativity. It is well known that the algebraic conditions (2.2) can only be satisfied by matrices of dimension four (or higher). For $m = 0$, one is left with an equation

$$i\frac{\partial\psi}{\partial t} = -i\vec{\alpha}\cdot\vec{\nabla}\psi \tag{2.3}$$

where the matrices α^j satisfy the anticommutation relations shown in (2.2), i.e.

$$\{\alpha^j, \alpha^k\} = 2\delta^{jk} \tag{2.4}$$

but now, there is no β. The relations (2.4) alone can be satisfied by 2×2 matrices; in fact, there are two inequivalent options, namely

$$\alpha^j = \sigma_j \tag{2.5}$$

and

$$\alpha^j = -\sigma_j \tag{2.6}$$

with σ_j being the standard Pauli matrices. (Of course, it is just the need for a fourth matrix β that forces one to work with 4×4 matrices in the massive case — there is no non-trivial 2×2 matrix anticommuting with all Pauli matrices.) Note that the non-equivalence of the sets (2.5) and (2.6) is obvious for the same technical reason: there is no regular matrix that would implement a similarity transformation between the two sets, since the transformation matrix would have to anticommute with σ_j for any $j = 1, 2, 3$. On the other hand, one has infinitely many equivalent representations of α^j, obtained from (2.5) or (2.6), respectively, by means of arbitrary similarity transformations. The two basic options (2.5) and (2.6) define two possible types of two-component Weyl equations, namely

$$i\frac{\partial\psi}{\partial t} = -i\vec{\sigma}\cdot\vec{\nabla}\psi \tag{2.7}$$

and

$$i\frac{\partial\psi}{\partial t} = +i\vec{\sigma}\cdot\vec{\nabla}\psi \tag{2.8}$$

An experienced reader may note that the last two equations are relativistically invariant and correspond to the spinor representations of Lorentz group denoted usually as $(\frac{1}{2}, 0)$ and $(0, \frac{1}{2})$, respectively, or, in an alternative terminology, to dotted and undotted (Weyl) spinors.

Let us now examine the plane-wave solutions of these equations, corresponding to a positive energy $E = |\vec{p}|$, with $\vec{p}$ being the particle momentum. Such a plane wave can be written as

$$\psi_+ = N(p)u(p)\mathrm{e}^{-ipx} \tag{2.9}$$

where $N(p)$ stands for an appropriate normalization factor and $px = |\vec{p}|t - \vec{p}\cdot\vec{x}$. Inserting now (2.9) into Eq. (2.7), one gets

$$(\vec{\sigma}\cdot\vec{p})u = |\vec{p}|u \tag{2.10}$$

This is a remarkable result, as it obviously means that a solution of the Weyl equation of the type (2.7) with positive energy automatically has positive helicity (for a negative-energy plane wave, we would get negative helicity). In a similar way, for the Weyl equation of the type (2.8), one finds that positive-energy solutions have negative helicity. Of course, such a strict correspondence between energy and helicity is a specific feature of the two-component equations — if we use a four-component Dirac equation, we always have both helicities for a given energy even in the massless case.

Thus, if one assumes that neutrino is strictly massless, it seems to be natural to describe it by means of a two-component Weyl equation (since it is then the most economical choice). To decide which variant is relevant in nature is essentially an experimental problem — one has to determine the neutrino helicity. Here, we may refer to the famous experimental result [7] quoted in the preceding chapter (cf. Section 1.7) which states that the neutrino produced in beta decay is left-handed. This suggests that the relevant Weyl equation is that given by (2.8). It is easy to see that the Weyl equations are not invariant under space inversion — technically, it is again due to the algebraic fact that there is no 2×2 matrix anticommuting with Pauli matrices (remember that for the four-component Dirac equation, the parity transformation is implemented through the matrix β, which is missing in the two-component case).

The parity non-invariance of the Weyl equation was precisely the reason for its rejection in 1929, but it has become a blessing after 1956 when parity violation turned out to be an experimental reality. If a two-component field for negative-helicity neutrino is to be incorporated into an interaction Lagrangian involving four-component Dirac fields of other fermions (electron, proton, etc.), one has to find an equivalent four-dimensional description of Weyl neutrino. This can be achieved by making use of the left-handed part of a four-component neutrino field, which of course is obtained by applying the projector $\frac{1}{2}(1-\gamma_5)$. In other words, a two-component neutrino with negative helicity is taken into account automatically if the corresponding field operator occurs in the form $\psi_{\nu L} = \frac{1}{2}(1-\gamma_5)\psi_\nu$ (note that $\psi_{\nu L}$ then describes left-handed neutrinos and right-handed antineutrinos). When one adopts such a principle, a general parity-violating Lagrangian for beta decay can be written in a straightforward way as

$$\mathscr{L}_{\text{int}}^{(\beta)} = \sum_{j=S,V,A,T,P} C_j(\bar{\psi}_p \Gamma_j \psi_n)[\bar{\psi}_e \Gamma^j (1-\gamma_5)\psi_\nu] + \text{h.c.} \tag{2.11}$$

where C_j are arbitrary Fermi-type constants. Thus, we see that the idea of a two-component massless neutrino automatically yields maximal parity violation in weak interactions (i.e. the parity violation is simply due to left-handed Weyl neutrino), but otherwise, any algebraic type of coupling is possible. Obviously, to restrict further the relevant couplings, one needs data (or an educated guess) concerning the electron helicity.

To conclude this section, one should stress that from today's point of view, the theory of two-component neutrino can hardly be taken seriously as an explanation of parity violation in weak interactions, since it is well known by now that maximal parity violation is observed even for interactions of massive particles (e.g., quarks). Moreover, there are hints from various experiments that neutrinos have non-zero (though tiny) masses. Parity violation thus seems to be simply an inherent property of the interaction itself and, in general, has nothing to do with massless neutrinos. It is perhaps fair to say that its deeper origin still remains rather mysterious — an explanation will hopefully be provided by a future more fundamental theory (note that the present-day standard model of electroweak interactions in fact does not shed much light on this issue). Nevertheless, the idea of a two-component left-handed neutrino played an important heuristic role in the history of weak interactions as it stimulated significantly the development of relevant theory.

2.2 Left-handed chiral leptons: Elimination of the S, P, T couplings

Motivated by the two-component neutrino theory, Feynman and Gell-Mann [12] (and independently Marshak and Sudarshan [13]) set forth the idea that, in general, any elementary fermion (regardless of its mass) can participate in weak interactions only through the left-handed chiral component of the corresponding spinor field, i.e. through $\psi_L = \frac{1}{2}(1-\gamma_5)\psi$.[1] It is not difficult to find that such a simple assumption leads to a radical simplification of the Lagrangian (2.11) — in fact, only the V and A terms then survive. To see this, let us assume that, in addition to the left-handed massless neutrino, the (massive) electron field also appears in the form ψ_{eL}. Instead of (2.11), one can then write a general beta-decay Lagrangian as

$$\mathscr{L}_{\text{int.}}^{(\beta)} = \sum_{j=S,V,A,T,P} \widetilde{C}_j(\bar{\psi}_p\Gamma_j\psi_n)(\bar{\psi}_{eL}\Gamma^j\psi_{\nu L}) + \text{h.c.} \tag{2.12}$$

with $\widetilde{C}_j$ being some Fermi-type coupling constants. Taking into account that $\bar{\psi}_L = \frac{1}{2}\bar{\psi}(1+\gamma_5)$, it becomes clear that the leptonic factors appearing in (2.12) contain the matrix products

$$(1+\gamma_5)\Gamma^j(1-\gamma_5) \tag{2.13}$$

However, the well-known (anti)commutation properties of the Dirac matrices now make it clear that the expression (2.13) vanishes identically for $j = S, P, T$ (remember that Γ_S, Γ_P and Γ_T commute with γ_5, cf. (1.28)). Thus, we are indeed left with only V and A terms in (2.12), as stated above.

The lesson to be learnt from this simple exercise is that the "law of left-handed chiral leptons" obviously represents an extremely efficient organizational principle in weak interaction theory: such a theoretical *tour de force* yields immediately the right structure of the beta-decay effective Lagrangian, which in the preceding chapter was obtained via a rather lengthy systematic investigation of the empirical data. On the other hand, if the (V, A) structure is deduced from the Feynman–Gell-Mann (or Marshak–Sudarshan) conjecture, it must be verified experimentally anyway, so as to ensure that the work we have done in Chapter 1 was certainly not in vain. In any case, one should bear in mind that such a simple theoretical rule is not substantiated (at least at the present level of understanding) by any deeper physical principle and may be perceived as a fortunate educated guess of an effective theory (which may be the manifestation of a more fundamental underlying theory).

Nevertheless, it is quite remarkable that the theoretical construction [12, 13] was proposed at a time, when some respected experimental data preferred the T coupling for Gamow–Teller beta transitions instead of the A coupling predicted by the simple theory. Feynman and Gell-Mann [12] went

[1] Note that the adjective "chiral" used here thus means "with a definite chirality", e.g., $\gamma_5\psi_L = -\psi_L$.

so far as to suggest that these data might be wrong — a guess that turned out to be right somewhat later, when the controversial experiments were repeated independently by other groups. In the meantime, measurements of the electron and neutrino helicities were carried out, with results confirming the $V - A$ theory. One can thus say that the ultimate triumph of the (V, A) scheme for weak interactions in the early 1960s resulted from an interplay between the simple theoretical ideas [12, 13] and a careful analysis of the available experimental data.

Of course, if one adopts the principle of negative chirality for nucleons as well, one gets a pure $V - A$ nucleon current in the beta-decay Lagrangian (i.e. $f = 1$ in (1.121)). As we know now from experiments, f is definitely different from 1 (which was not quite clear in the late 1950s, when the papers [12,13] were published). It seems to suggest that the rule of negative chirality can be reasonably used only for elementary fermions (leptons and quarks). We will discuss the quark interactions and related problems later on, and next, we are going to analyze a "canonical" purely leptonic process — the muon decay, which played a crucial role in establishing the concept of weak interaction as a universal force, not necessarily associated with nuclear beta decay.

2.3 Muon decay

By now, it is well known that muon disintegrates into an electron and two neutrinos according to

$$\mu^- \to e^- + \nu_\mu + \bar{\nu}_e \tag{2.14}$$

In (2.14), we have marked explicitly two different neutrino species; in particular, ν_μ carries a muonic lepton number equal to that of the initial muon. We are not going to review here the historical development of muon physics, but a few remarks concerning (2.14) are in order. The fact that the muon (discovered in 1937) decays into more than two particles was recognized around 1949, simply on the basis of the continuous energy spectrum of the final electron. It was also immediately obvious that the remaining decay products are electrically neutral and rather light — information about masses is contained, e.g., in the maximum electron energy that can be calculated along the same lines as in the case of beta decay. If one assumes that the decay products other than electron are massless, one gets

$$E_e^{\max} = \frac{m_\mu^2 + m_e^2}{2m_\mu} \tag{2.15}$$

in good agreement with observed data (note that the current upper bound [6] is $m_{\nu_\mu} < 0.19\,\mathrm{MeV}$). The idea of the muon-decay scheme of the type (2.14) seems to have been accepted in the late 1940s, but the non-trivial question whether $\nu_\mu \neq \nu_e$ has been answered directly only in the early 1960s (see [14] and also, e.g., [CaG]). In this context, one should also note that muon decays provide impressive evidence in favour of separate conservation of muonic and electronic lepton numbers — let us quote, e.g., the bounds for unseen processes like $\mu^- \to e^-\gamma$ or $\mu^- \to e^-e^+e^-$, with branching ratios less than 1.2×10^{-11} and 1.0×10^{-12}, respectively [6].

Let us now try to describe the decay process (2.14) in quantitative terms. If one takes for granted the theory of left-handed chiral leptons [12, 13] described in the preceding section, one can write immediately the corresponding effective Lagrangian as

$$\mathscr{L}_{\text{int}}^{(\mu)} = -\frac{G_\mu}{\sqrt{2}}[\bar{\psi}_{(\nu_\mu)}\gamma_\rho(1-\gamma_5)\psi_{(\mu)}][\bar{\psi}_{(e)}\gamma^\rho(1-\gamma_5)\psi_{(\nu_e)}] \tag{2.16}$$

where G_μ is an appropriate Fermi-type coupling constant. An important goal of our subsequent analysis will be the determination of the relevant coupling strength — this can be done by comparing the calculated muon lifetime with its measured value. Apart from this task, it would also be interesting to test the (postulated) $V-A$ structure of the currents in (2.16). As a simple example of such a check, we will temporarily modify (2.16) to

$$\mathscr{L}_{\text{int}} = -\frac{G_\mu}{\sqrt{2}}[\bar{\psi}_{(\nu_\mu)}\gamma_\rho(1-\lambda\gamma_5)\psi_{(\mu)}][\bar{\psi}_{(e)}\gamma^\rho(1-\gamma_5)\psi_{(\nu_e)}] \tag{2.17}$$

with λ being an arbitrary real parameter and show that the observed shape of the electron energy spectrum clearly favours the value $\lambda = 1$ corresponding to the $V-A$ theory (note that here we essentially follow the treatment of [BjD], Chapter 10).[2]

Thus, we start our calculation with the lowest-order matrix element corresponding to (2.17), i.e.

$$\mathcal{M} = -\frac{G_\mu}{\sqrt{2}}[\bar{u}(k)\gamma_\rho(1-\lambda\gamma_5)u(P)][\bar{u}(p)\gamma^\rho(1-\gamma_5)v(k')] \tag{2.18}$$

[2]Note that we are not trying to perform here a general analysis of the muon-decay effective Lagrangian that would be analogous to the procedure applied to neutron decay in the preceding chapter. Such an analysis is in a sense more difficult for muon decay, as there is only one charged particle in the final state, e.g., the simple electron–antineutrino angular correlations cannot be studied experimentally. For a detailed discussion of muon physics from this point of view, see [15].

where the four-momenta of $\mu, e, \nu_\mu, \bar{\nu}_e$ are denoted by P, p, k, k', respectively. The spin-averaged matrix element squared then becomes, after some simple algebraic manipulations,

$$\overline{|\mathcal{M}|^2} = \frac{1}{2}\sum_{\text{spins}} |\mathcal{M}|^2$$
$$= \frac{1}{2} G_\mu^2 \text{Tr}[\not{k}\gamma_\rho(\not{P} + m_\mu)\gamma_\sigma(1 + \lambda^2 - 2\lambda\gamma_5)]\text{Tr}[(\not{p} + m_e)\gamma^\rho \not{k}'\gamma^\sigma(1 - \gamma_5)] \tag{2.19}$$

Obviously, the mass terms appearing in (2.19) in fact do not contribute and the product of traces can then be easily evaluated with the help of the formulae

$$\begin{aligned}
\text{Tr}(\not{a}\gamma_\rho \not{b}\gamma_\sigma)\text{Tr}(\not{c}\gamma^\rho \not{d}\gamma^\sigma) &= 32[(a \cdot c)(b \cdot d) + (a \cdot d)(b \cdot c)] \\
\text{Tr}(\not{a}\gamma_\rho \not{b}\gamma_\sigma\gamma_5)\text{Tr}(\not{c}\gamma^\rho \not{d}\gamma^\sigma\gamma_5) &= 32[(a \cdot c)(b \cdot d) - (a \cdot d)(b \cdot c)] \\
\text{Tr}(\not{a}\gamma_\rho \not{b}\gamma_\sigma)\text{Tr}(\not{c}\gamma^\rho \not{d}\gamma^\sigma\gamma_5) &= 0
\end{aligned} \tag{2.20}$$

(see (A.51)). We thus get finally

$$\overline{|\mathcal{M}|^2} = 16G_\mu^2 \left[(1 + \lambda)^2(k \cdot p)(k' \cdot P) + (1 - \lambda)^2(k \cdot k')(p \cdot P)\right] \tag{2.21}$$

The differential decay rate is given by the standard formula

$$dw = \frac{1}{2m_\mu}\overline{|\mathcal{M}|^2}\frac{d^3p}{2E(p)(2\pi)^3}\frac{d^3k}{2E(k)(2\pi)^3}\frac{d^3k'}{2E(k')(2\pi)^3}(2\pi)^4\delta^4(P - p - k - k') \tag{2.22}$$

where of course $E(p) = \sqrt{\vec{p}^{\,2} + m_e^2}$, $E(k) = |\vec{k}\,|$ and $E(k') = |\vec{k'}\,|$. To obtain the electron energy spectrum, the expression (2.22) could be integrated in a similar fashion as in the case of neutron beta decay, but now we are not allowed to make the simplifying kinematical approximations used before — in muon decay, all final-state particles may be relativistic, so that no momentum can be neglected. A straightforward integration of (2.22) is left to the reader as a useful (though somewhat tedious) exercise; here, we offer an alternative method, which may be of a more general interest. From the structure of the expression (2.21), it is clear that one needs the integral

$$I_{\alpha\beta}(Q) = \int \frac{d^3k}{2E(k)}\frac{d^3k'}{2E(k')}k_\alpha k'_\beta \delta^4(Q - k - k') \tag{2.23}$$

where we have denoted $Q = P - p$. Now, the crucial observation is that the $I_{\alpha\beta}$ is a second rank tensor under Lorentz transformations. Indeed, using some simple tricks for the integration involving the delta functions, the expression (2.23) can be recast as

$$I_{\alpha\beta}(Q) = \int d^4k \, \theta(k_0)\delta(k^2)\delta\left((Q-k)^2\right) k_\alpha (Q-k)_\beta \tag{2.24}$$

which makes the tensor character of the $I_{\alpha\beta}$ rather obvious. The most general second rank tensor $I_{\alpha\beta}(Q)$ has the form

$$I_{\alpha\beta}(Q) = A g_{\alpha\beta} + B Q_\alpha Q_\beta \tag{2.25}$$

with A and B being arbitrary functions of Q^2. To determine these coefficients, it suffices to evaluate two independent components of the tensor (2.23) in an arbitrary reference frame. The most convenient choice is the c.m. system of the two neutrinos, where Q has components $Q = (Q_0, \vec{0}\,)$, with $Q_0 = 2k_0 = 2|\vec{k}\,|$. Let us consider, e.g., the tensor components I_{00} and I_{33}. According to (2.25), these are related to A and B by

$$\begin{aligned} I_{00} &= A + B Q_0^2 \\ I_{33} &= -A \end{aligned} \tag{2.26}$$

and a direct integration of the original form (2.23) in the c.m. system gives

$$I_{00} = \frac{\pi}{8} Q_0^2, \quad I_{33} = -\frac{\pi}{24} Q_0^2 \tag{2.27}$$

This result, together with (2.26), then yields

$$I_{\alpha\beta}(Q) = \frac{\pi}{24}(Q^2 g_{\alpha\beta} + 2 Q_\alpha Q_\beta) \tag{2.28}$$

With (2.21) and (2.28) at hand, the evaluation of the electron energy spectrum is reduced to purely algebraic manipulations. These are elementary but somewhat lengthy, so we give only the final answer

$$\begin{aligned} \frac{dw(E_e)}{dE_e} &= \frac{1}{3\pi^3} G_\mu^2 \frac{1+\lambda^2}{2} m_\mu |\vec{p}\,| E_e \\ &\quad \times \left[3(W - E_e) + \frac{1}{4} \frac{(1+\lambda)^2}{1+\lambda^2} \left(4E_e - 3W - \frac{m_e^2}{E_e} \right) \right] \end{aligned} \tag{2.29}$$

where we have used the symbol W for the maximum electron energy (see (2.15)). Note that for some historical reasons, it has become customary to denote

$$\frac{1}{4}\frac{(1+\lambda)^2}{1+\lambda^2} = \frac{2}{3}\rho \tag{2.30}$$

where ρ is the so-called Michel parameter.[3] For energies $E_e \gg m_e$, one can neglect the term m_e^2/E_e in (2.29). In terms of the dimensionless variable $x = E_e/W$, the high-energy part of the spectrum (2.29) can then be approximately written as

$$\frac{dw}{dx} = \frac{1}{48\pi^3} G_\mu^2 m_\mu^5 \frac{1+\lambda^2}{2} x^2 \left[3(1-x) + \frac{2}{3}\rho(4x-3)\right] \tag{2.31}$$

(note that in the last expression, we have also set $W \doteq \frac{1}{2}m_\mu$). Obviously, the value of the Michel parameter

$$\rho = \frac{3}{8}\frac{(1+\lambda)^2}{1+\lambda^2} \tag{2.32}$$

determines the shape of the energy spectrum near its endpoint. From (2.32), it is clearly seen that for $\rho = 0$, the distribution function dw/dx would vanish at $x = 1$, but in general, it is non-zero at the endpoint (in contrast to the case of neutron beta decay). Some illustrative examples are shown in Fig. 2.1 (note that $0 \le \rho \le 3/4$ for any λ in (2.32)). The different character of muon-decay spectrum in comparison with the beta decay is due to the different kinematical conditions in both processes; we will comment on this point later on. The current experimental value is $\rho = 0.74979 \pm 0.00026$ (the world average according to [6]), i.e. $\rho \doteq 3/4$ with high accuracy. In view of (2.32), this immediately implies $\lambda = 1$, which confirms the anticipated validity of the $V - A$ theory for muon decay.

To determine the coupling constant G_μ, let us calculate the full decay width. For simplicity, we will employ the approximate expression (2.31) over the whole electron energy range, as one may thus presumably lose

[3] ρ is in fact one of the four or five parameters used for the description of muon decay in a general case when one also takes into account particle polarizations. In the unpolarized case, two parameters are usually introduced — apart from ρ, there is another one denoted by η, which characterizes the shape of the low-energy end of the electron spectrum. We will be mostly interested in the upper endpoint of the spectrum, so that only ρ is relevant for our further considerations. The parameters are named after Michel, who in the 1950s performed a comprehensive analysis of the muon decay within the framework of a general Fermi-type model involving all the S, V, A, T, P couplings [15]. For details, see also [CoB] and [Gre].

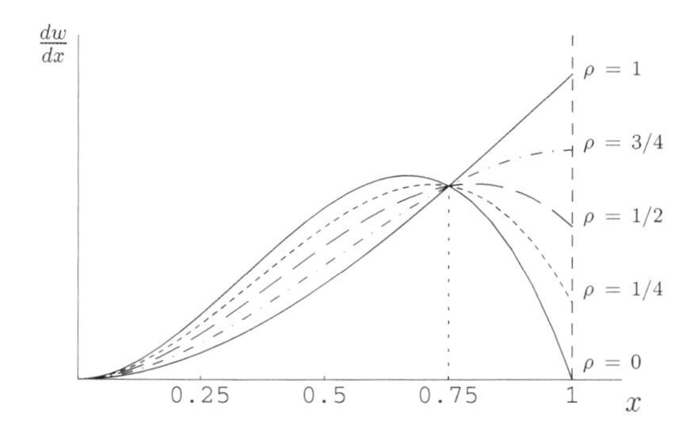

Figure 2.1. Variation of the shape of muon-decay spectrum with respect to the value of the Michel parameter ρ, as described by the approximate formula (2.31).

only small correction terms of the relative order $O(m_e^2/m_\mu^2)$. Setting $\lambda = 1$ in (2.31), one has

$$\frac{dw}{dx} = \frac{1}{48\pi^3} G_\mu^2 m_\mu^5 x^2 \left(\frac{3}{2} - x\right) \tag{2.33}$$

and the decay width is then obtained by integrating (2.33) over x from 0 to 1. One thus readily gets the result

$$\Gamma = \frac{G_\mu^2 m_\mu^5}{192\pi^3} \tag{2.34}$$

Let us remark that the characteristic dependence on m_μ^5 can be easily understood on dimensional grounds: the decay width must have dimension of a mass, the coupling constant squared supplies a mass to minus four and for $m_e = 0$, m_μ is the only mass scale left in the game. Since there are essentially no other relevant decay modes for muon, the inverse of the width (2.34) gives the muon lifetime. This is measured with a rather high accuracy, $\tau_\mu = (2.1969811 \pm 0.0000022) \times 10^{-6}$ s. The experimental number is to be compared with the result of our calculation, corrected for the electron mass effects and for QED effects (radiative corrections). Such a detailed calculation goes beyond the scope of our treatment, so let us only quote the result of such an analysis. The value of the coupling constant G_μ corresponding to the measured muon lifetime is

$$G_\mu = (1.1663787 \pm 0.0000006) \times 10^{-5}\ \text{GeV}^{-2} \tag{2.35}$$

and is identified with the "canonical" Fermi constant G_F recorded in the *Review of Particle Physics* [6].

The value (2.35) almost coincides with the beta-decay constant G_β (cf. (1.120)) and this clearly indicates that muon decay is a manifestation of essentially the same basic force that is responsible for the nuclear beta decay. In other words, the results of the analysis of muon decay strongly support the idea that the original "weak nuclear force" discovered in connection with beta decay represents in fact only one aspect of a universal weak interaction acting on widely different subatomic particles. On the other hand, although G_μ is very close to G_β, these two coupling parameters clearly differ by many standard deviations, so one obviously needs an additional small parameter to characterize the universality of weak interactions properly. As we shall see in subsequent sections, it makes sense to describe the difference between G_μ and G_β by means of the so-called Cabibbo angle — this observation lies at the basis of the "Cabibbo universality" formulated in the early 1960s. In any case, it should be stressed that through our analysis, we have arrived at a simple explanation of the widely different muon and neutron lifetimes (about 10^{-6} seconds for muon and 15 minutes for neutron): such a difference of many orders of magnitude is entirely due to the phase-space factors being proportional to the fifth power of the relevant energy scales — the coupling strengths (i.e. the basic dynamics) are essentially the same.

Before closing this section, let us return briefly to the problem of the shape of electron energy spectrum. As we have seen in (2.29), the distribution function $dw(E_e)/dE_e$ in general does not vanish at the endpoint $E_e = W$, unless $\rho = 0$. It is not difficult to realize that such a peculiar dissimilarity to the case of neutron beta decay is due to the assumption that both neutrinos produced in muon decay are massless. Indeed, when the electron energy reaches its maximum value W, the neutrinos carry off the remaining part $m_\mu - W$. In the massless case, there are infinitely many ways how to divide it between ν_μ and $\bar{\nu}_e$: any kinematical configuration such that both neutrinos are emitted in the direction opposite to the electron momentum, with otherwise arbitrary energies satisfying $E(k) + E(k') = m_\mu - W$, fulfills the required simultaneous conservation of energy and momentum. Thus, the endpoint of the electron energy spectrum corresponds to infinitely many degenerate states and, consequently, the volume of the phase space at $E_e = W$ can be non-zero. On the other hand, if at least one of the neutrinos, say ν_μ, is massive, there is only one possible kinematical configuration corresponding to $E_e = W$ and the phase-space volume then vanishes at the endpoint of the spectrum. This can be nicely illustrated if one calculates explicitly the integral (2.23) for a massive ν_μ

(leaving the ν_e massless for simplicity). Denoting the ν_μ mass by m, one gets

$$I_{\alpha\beta}(Q;m) = A(Q^2;m)g_{\alpha\beta} + B(Q^2;m)Q_\alpha Q_\beta \tag{2.36}$$

with

$$\begin{aligned} A(Q^2;m) &= \frac{\pi}{24}\frac{1}{(Q^2)^2}(Q^2-m^2)^3 \\ B(Q^2;m) &= \frac{\pi}{12}\frac{1}{(Q^2)^3}(Q^2-m^2)^2(Q^2+2m^2) \end{aligned} \tag{2.37}$$

(a derivation of (2.37) is left to the reader as an exercise). Note that the expressions (2.37) reduce to (2.28) for $m = 0$. Now, the maximum electron energy corresponds to $Q^2 = m^2$, where the form factors A and B are seen to vanish and this confirms our previous considerations concerning the phase space at the endpoint of the electron spectrum. Of course, if the neutrino mass is very small, one cannot practically distinguish the shape of a spectrum falling steeply to zero near the endpoint from the case where the energy distribution function is truly non-vanishing for $E_e = W$.

2.4 Universal interaction of $V - A$ currents

In the preceding discussion, we have seen two examples of physical processes — the nuclear beta decay and the decay of muon — that involve quite different particles and also have widely different lifetimes, yet they turn out to be governed by essentially the same force. We have found that neutron beta decay (and associated processes) can be successfully described by an effective Lagrangian of the form

$$\mathscr{L}_{\text{int}}^{(\beta)} = -\frac{G_\beta}{\sqrt{2}}[\bar{\psi}_p\gamma^\rho(1-f\gamma_5)\psi_n][\bar{\psi}_e\gamma_\rho(1-\gamma_5)\psi_\nu] + \text{h.c.} \tag{2.38}$$

while the muon decay corresponds to

$$\mathscr{L}_{\text{int}}^{(\mu)} = -\frac{G_\mu}{\sqrt{2}}[\bar{\psi}_{\nu_\mu}\gamma^\rho(1-\gamma_5)\psi_\mu][\bar{\psi}_e\gamma_\rho(1-\gamma_5)\psi_{\nu_e}] + \text{h.c.} \tag{2.39}$$

and the coupling constants G_β and G_μ nearly coincide.

Another important reaction, studied experimentally since the 1940s, is the so-called "weak muon capture":

$$\mu^- + p \to n + \nu_\mu \tag{2.40}$$

Processes of the type (2.40) occur (at the level of atomic nuclei) with both μ^- and μ^+; as for an effective interaction, the experimental data show that the $V - A$ leptonic current $\bar{\psi}_{\nu_\mu}\gamma^\rho(1 - \gamma_5)\psi_\mu$ must be involved and an overall strength of the coupling of nucleons to muon-type leptons practically coincides with G_β (for details, see, e.g., [CoB]).[4]

Observations of such diverse processes that occur among different particles, yet with an essentially equal coupling strength, led soon to the idea of a universal weak interaction connecting leptons with nucleons and leptons with themselves. In those early days, such a universal coupling scheme was symbolized by the so-called "Tiomno–Wheeler triangle" (see [17] and [Jac]) — an equilateral triangle with pairs (n, p), (e, ν), and (μ, ν) at vertices. In the 1950s, it gradually became clear that the decays of charged pions and of the newly discovered strange mesons (to say nothing of new baryons) could also be accounted for by a force of the "Fermi strength". Thus, it appeared desirable to incorporate mesons (i.e. spin-zero particles) into the weak interaction scheme as well, although this was originally conceived as a model of direct four-fermion coupling. Anyway, with the number of observed decay processes proliferating rapidly and with the accumulating evidence for a universal magnitude of the corresponding couplings, there was obviously a growing need for a coherent unified picture of weak interaction phenomenology that would involve all known leptons and hadrons. In this context, one might perhaps use the following pictorial description of the status of the provisional weak interaction models discussed so far: the simple effective Lagrangians, directly applicable in particular cases mentioned before, look rather like individual pieces of a "jigsaw puzzle", presumably with some missing parts to be added in order to get a complete pattern.

Feynman and Gell-Mann [12] made an important step forward by postulating a universal current–current form of weak interaction, namely

$$\mathscr{L}_{\text{int}}^{(w)} = -\frac{G_F}{\sqrt{2}} J^\rho J_\rho^\dagger \tag{2.41}$$

where the "universal Fermi constant" G_F is to be identified with the muon decay constant G_μ (cf., (2.35)) and the weak current J^ρ consists of leptonic

[4]Of course, it took some time to establish the V–A nature of the relevant interaction, but the observation that an overall coupling strength is of the order of the beta-decay Fermi constant has already been made in the late 1940s, cf., e.g., [17] and [Jac]. Note, however, that a simple effective Lagrangian analogous to (2.38) would provide only a rough description of the nucleonic part of a corresponding scattering matrix element, since the momentum transfer in the muon capture processes can be relatively high in comparison with the neutron beta decay. We will discuss the general structure of nucleonic matrix elements later on.

and hadronic parts

$$J^\rho = \bar{\psi}_{\nu_e}\gamma^\rho(1-\gamma_5)\psi_e + \bar{\psi}_{\nu_\mu}\gamma^\rho(1-\gamma_5)\psi_\mu + J^\rho_{(\text{hadron})} \tag{2.42}$$

The hadronic part is assumed to have the structure $V - A$ with respect to Lorentz transformations, but it need not be expressed explicitly in terms of the field operators of physical hadrons. One should only require that $J^\rho_{(\text{hadron})}$ is an operator having non-trivial matrix elements between physical hadronic states and, eventually, between a meson state and vacuum (as we shall see later, this last property is an important prerequisite for describing, e.g., the leptonic decays of charged pions). Such a phenomenological device, which bypasses the field-theoretic treatment of the hadronic sector, seems to be quite reasonable in a situation when one faces the rich spectrum of hadrons with different spins. Needless to say, a simple organizing principle for hadronic world emerged somewhat later with the advent of the quark model — more about this later. By combining various pieces of the two currents in (2.42), one is obviously able to reproduce the weak processes discussed before (i.e. beta decay, muon decay, weak muon capture, etc.). Moreover, the form (2.42) also predicts some new ("diagonal") processes, in particular the elastic (anti)neutrino–electron scattering.[5] It is perhaps in order to remark here that by considering in (2.41) the interaction of the weak current (2.42) *with itself*, Feynman and Gell-Mann clearly envisaged a possible alternative description of weak interactions in terms of an exchange of a massive charged "intermediate vector boson" — a scheme that leads to essentially identical predictions as the current–current form at sufficiently low energies (this is a point to be discussed in detail in Chapter 3).

In any case, one important aspect of the weak hadronic current was missing in the pioneering treatment [12]. Owing to the lack of sufficiently accurate measurements at that time, Feynman and Gell-Mann were not aware of the subtle difference between the muon-decay constant $G_\mu = G_F$ and G_β that we know now (cf. (2.35) and (1.120)) and it was also not clear what is precisely the relative strength of the strangeness-changing weak transitions in comparison with the ordinary beta decay. In this respect, the Feynman–Gell-Mann model [12] was significantly improved by Cabibbo [20].

[5]Note that the elastic scattering processes $\bar{\nu}_e e \to \bar{\nu}_e e$ and $\nu_e e \to \nu_e e$ were in fact first observed experimentally only many years after their prediction by Feynman–Gell-Mann theory — see [18, 19].

2.5 Cabibbo angle and selection rules for strangeness

In order to explain the development that led to the notion of "Cabibbo universality", let us focus on the difference between G_F and G_β. As we noted before, $G_F (= G_\mu) \doteq 1.166 \times 10^{-5}$ GeV^{-2} with high accuracy, whereas $G_\beta \doteq 1.136 \times 10^{-5}$ GeV^{-2} with about one per mille accuracy. Thus, since $G_\beta < G_F$, one may introduce a parametrization

$$G_\beta / G_F = \cos\theta_C \tag{2.43}$$

with θ_C being the so-called "Cabibbo angle". Putting in numbers, one has $\cos\theta_C \doteq 0.974$, so that

$$\theta_C \doteq 13^\circ \tag{2.44}$$

At this stage, such a parametrization may seem somewhat artificial, as it is not clear why a particular *angle* should be appropriate for describing the simple fact that $G_\beta < G_F$. The true relevance of the parameter θ_C can be appreciated only when one considers more fancy weak processes, namely the hadron decays in which strangeness is not conserved. As an instructive example, let us consider the semileptonic decay[6]

$$\Sigma^- \to n + e^- + \bar{\nu}_e \tag{2.45}$$

The baryon Σ^- carries the strangeness $S = -1$, while for neutron, one has, of course, $S = 0$. The process (2.45) may be viewed as a "strangeness-changing beta decay" of Σ^-. Measurements analogous to those performed for ordinary neutron beta decay lead to the conclusion that a corresponding decay matrix element can approximately be written as

$$\mathcal{M}_{fi} \doteq -\frac{G_F}{\sqrt{2}} \sin\theta_C [\bar{u}_n \gamma_\rho (1 - \tilde{f}\gamma_5) u_{\Sigma^-}][\bar{u}_e \gamma^\rho (1 - \gamma_5) v_\nu] \tag{2.46}$$

with $\tilde{f} \doteq -0.34$ (cf. [6]); the meaning of the other symbols is obvious. Let us stress that in writing (2.46), we have neglected the effects associated with the corresponding momentum transfer (i.e. effects of the order of

[6] It should be noted that the considered process is one of the relatively rare decay modes of the Σ^-, as its branching ratio is about 10^{-3} (see [6]). The Σ^- decays predominantly via the non-leptonic mode $\Sigma^- \to n + \pi^-$, with the corresponding branching ratio being 99.85%. Nevertheless, the relevant characteristics of decay (2.45) are measured with rather good accuracy and the data constitute a valuable source of information on the weak interactions of strange particles. For completeness, let us recall that the Σ^- mass is 1197 MeV and the mean lifetime $\tau_{\Sigma^-} \doteq 1.48 \times 10^{-10}$s.

$(m_{\Sigma^-} - m_n)/m_{\Sigma^-})$; of course, such an approximation is of poorer quality than in the case of neutron beta decay. From (2.46), one may infer readily a corresponding effective Lagrangian of the form analogous to (1.121). The crucial (experimental) result embodied in (2.46) is that, instead of the $G_\beta = G_\text{F} \cos\theta_\text{C}$ appearing in (1.121), the relevant Fermi coupling constant is now $G_\text{F} \sin\theta_\text{C}$! (In view of (2.44), it practically means that the coupling strength now constitutes only about 23% of the G_β.) This is the essence of the crucial observation made by Cabibbo [20] (who analyzed primarily the strangeness-changing kaon decays) — now, it is at least clear that it makes sense to parametrize the subtle difference between G_β and G_F in terms of an angle. Nevertheless, the origin of such an empirical angle remained entirely obscure within the framework of the weak interaction theory in the 1960s. Looking ahead, let us remark at this place that the situation is slightly better now, since within the present-day electroweak Standard Model, the appearance of an angle like θ_C is quite natural; it turns out to be intimately related to the mechanism of fermion mass generation. While the reader may be pleased by this encouraging news, it is fair to admit, on the other hand, that the numerical value (2.44) remains mysterious even within SM (in fact, a prediction of the θ_C value constitutes one of the major challenges for theories attempting to go beyond SM).

Up to now, we have only compared the strangeness-changing process (2.45) with the ordinary beta decay, where the strangeness does not play any role. In fact, it turns out that while all strangeness-changing decays proceed with strength $G_\text{F} \sin\theta_\text{C}$, the relevant Fermi constant for any strangeness-conserving decay is $G_\text{F} \cos\theta_\text{C}$ (irrespective of whether the hadrons involved are strange or not). For example, if the decay $\Sigma^- \to \Lambda e^- \bar{\nu}_e$ is analyzed, one finds that the corresponding coupling strength is $G_\text{F} \cos\theta_\text{C}$ as for the ordinary beta decay, although both the Σ^- and Λ have strangeness $S = -1$. Since θ_C is numerically small, one may thus conclude that, at a phenomenological level, the role of the Cabibbo angle consists in suppressing the strangeness-changing weak processes relative to the strangeness-conserving ones.

The previous considerations may be summarized by writing the hadronic weak current schematically as

$$J_\rho^{(\text{hadron})} = \cos\theta_C J_\rho^{(\Delta S=0)} + \sin\theta_C J_\rho^{(\Delta S\neq 0)} \tag{2.47}$$

where the operators $J_\rho^{(\Delta S=0)}$ and $J_\rho^{(\Delta S\neq 0)}$ have the form $V - A$ and it is tacitly assumed that they do not incorporate any other suppression factors related to strangeness. To get a deeper insight into the structure of the operators appearing in (2.47), we have to consider some further empirical selection rules that hold for weak transitions. There are essentially two such rules that should be taken into account.

First, it is a well-established fact that processes in which the strangeness is changed by more than one unit are very strongly suppressed. As an example, one may consider the process $\Xi^- \to n + \pi^-$, where $S(\Xi^-) - S(n) = -2$. The experimental upper bound for the corresponding branching ratio is about 1.9×10^{-5}, though the available phase space for the decay products is certainly more favourable than in the case of the dominant mode $\Xi^- \to \Lambda + \pi^-$ (which has the branching ratio of 99.88%). There are other similar examples for the decays of Ξ^0 and Ω^- ($S = -3$) as well.

The second non-trivial phenomenological constraint is provided by the well-known "rule $\Delta S = \Delta Q$". This holds for strangeness-changing semileptonic decays of hadrons (both mesons and baryons) and can be formulated concisely as follows. Let us denote by h_i and h_f the initial and final hadrons, respectively. For the process

$$h_i \to h_f + \text{lepton pair} \tag{2.48}$$

define $\Delta S = S(h_f) - S(h_i)$ and $\Delta Q = Q(h_f) - Q(h_i)$ (with Q being, as usual, a charge expressed in units of e). Then, if $\Delta S \neq 0$, an allowed transition satisfies

$$\Delta S = \Delta Q \tag{2.49}$$

whereas the processes with $\Delta S \neq \Delta Q$ are strongly suppressed.

As an illustrative example of validity of this second rule one may consider, e.g., the decay $\Sigma^- \to n e^- \bar{\nu}_e$ discussed earlier in this section. For this rare yet well-established process, one has $\Delta S = \Delta Q = 1$. On the other hand, its counterpart $\Sigma^+ \to n e^+ \nu_e$ (which naively would be conceivable) has not been observed; there is an upper bound for its decay width that can be expressed as

$$\Gamma(\Sigma^+ \to n e^+ \nu_e)/\Gamma(\Sigma^- \to n e^- \bar{\nu}_e) < 0.04 \tag{2.50}$$

Another example that should be quoted here is the spectacular suppression of weak decays with $\Delta S \neq 0$ and $\Delta Q = 0$ ("weak neutral-current processes"). In particular, the decay $K^+ \to \pi^+ e^+ e^-$ (for which $\Delta S = -1$ and $\Delta Q = 0$) has been measured to have the branching ratio of about 3×10^{-7}, while the branching ratio of its natural counterpart $K^+ \to \pi^0 e^+ \nu_e$ (called K_{e3}) is roughly 5%. There are many other examples of unseen processes with $\Delta S \neq 0$, $\Delta Q = 0$, for which rather stringent upper bounds are available.[7]

Of course, in a model involving only charged currents, there is no place for $\Delta Q = 0$ weak transitions in the lowest order (though they are conceivable as higher-order effects, e.g., at the level of one-loop Feynman graphs). The original assumption of Feynman and Gell-Mann [12] (adopted by Cabibbo [20] as well) actually was that in the weak interaction Lagrangian, there were *no* neutral currents at all, since at that time there had been no phenomenological need for them. Such an assumption is obviously correct as far as the strangeness-changing neutral currents are concerned; however, as we know now, *strangeness-conserving* weak neutral currents do play an important role in the standard model of electroweak unification — this issue will be discussed in detail in Chapter 7.

Now, we are going to focus on the structure of hadronic weak charged currents complying with the above-mentioned phenomenological constraints. In fact, it is quite easy to construct currents satisfying the empirical rules $|\Delta S| \leq 1$ and $\Delta S = \Delta Q$ if one adopts the quark model [22]. Within such a framework, it is natural to view a hadronic weak transition as a process involving a pair of quarks (or antiquarks, or a quark–antiquark pair), possibly with some other ones playing the role of "spectators". One is then led to write the operators in (2.47) in terms of the Dirac spinor fields of quarks u (up), d (down) and s (strange) simply as

$$
\begin{aligned}
J_\rho^{(\Delta S=0)} &= \bar{\psi}_u \gamma_\rho (1-\gamma_5)\psi_d \\
J_\rho^{(\Delta S\neq 0)} &= \bar{\psi}_u \gamma_\rho (1-\gamma_5)\psi_s
\end{aligned}
\tag{2.51}
$$

Taking into account the charge assignments $Q_u = 2/3$ and $Q_d = Q_s = -1/3$, as well as the strangeness of the s-quark being -1 (the u and d quarks of course have zero strangeness), it is easy to see that weak hadronic transitions mediated by the currents (2.51) indeed obey automatically the necessary empirical rules: at the quark level, the basic transitions are $d \to u$ and $s \to u$ (or conjugated processes) so that one has clearly $\Delta S = 0$ or $\Delta S = 1$ and for $s \to u$, the relation $\Delta S = \Delta Q$ holds obviously.

An important aspect of the representation of weak hadronic currents in terms of quark fields is that the original concept of a universal *four-fermion* interaction is thereby restored — as we shall see later, this plays

[7]Note that two other weak neutral strangeness-changing processes have been observed recently, namely $K^+ \to \pi^+ \nu\bar{\nu}$ and $K^+ \to \pi^+\mu^+\mu^-$ (see [21]), with branching ratios of about 10^{-10} and 10^{-8}, respectively.

a crucial role in the formulation of the electroweak SM. Within the context of the provisional phenomenological weak interaction theory, it is certainly gratifying that the expressions (2.51) reproduce automatically the empirically established selection rules, but they in fact represent more than mere mnemonics. The point is that the quark currents (2.51) can be conveniently recast in a form exhibiting their transformation properties under the approximate "flavour $SU(3)$" symmetry (actually, this was originally done by Cabibbo [20] before the emergence of the quark model). In particular, one has

$$\bar{\psi}_u \gamma_\rho (1-\gamma_5)\psi_d = \bar{\psi}_q \gamma_\rho (1-\gamma_5)\frac{\lambda^1 + i\lambda^2}{2}\psi_q \tag{2.52}$$

and

$$\bar{\psi}_u \gamma_\rho (1-\gamma_5)\psi_s = \bar{\psi}_q \gamma_\rho (1-\gamma_5)\frac{\lambda^4 + i\lambda^5}{2}\psi_q \tag{2.53}$$

where ψ_q denotes the triplet

$$\psi_q = \begin{pmatrix} \psi_u \\ \psi_d \\ \psi_s \end{pmatrix} \tag{2.54}$$

(belonging to the fundamental representation of $SU(3)_{\text{flavour}}$) and λ^a, $a = 1, 2, 4, 5$ are Gell-Mann matrices

$$\lambda^1 = \begin{pmatrix} 0 & 1 & 0 \\ 1 & 0 & 0 \\ 0 & 0 & 0 \end{pmatrix} \quad \lambda^2 = \begin{pmatrix} 0 & -i & 0 \\ i & 0 & 0 \\ 0 & 0 & 0 \end{pmatrix}$$
$$\lambda^4 = \begin{pmatrix} 0 & 0 & 1 \\ 0 & 0 & 0 \\ 1 & 0 & 0 \end{pmatrix} \quad \lambda^5 = \begin{pmatrix} 0 & 0 & -i \\ 0 & 0 & 0 \\ i & 0 & 0 \end{pmatrix} \tag{2.55}$$

The verification of the identities (2.52) and (2.53) is a trivial algebraic exercise. Thus, the weak hadronic current operator can be written as

$$J_\rho^{(\text{hadron})} = \cos\theta_C J_\rho^{(\Delta S=0)} + \sin\theta_C J_\rho^{(\Delta S=1)} \tag{2.56}$$

where

$$J_\rho^{(\Delta S=0)} = V_\rho^{1+i2} - A_\rho^{1+i2}, \quad J_\rho^{(\Delta S=1)} = V_\rho^{4+i5} - A_\rho^{4+i5} \tag{2.57}$$

(the superscripts in (2.57) represent an obvious shorthand notation for the combinations of Gell-Mann matrices appearing in (2.52) and (2.53)).

The utility of such an algebraic form becomes clear when one takes into account the familiar classification of known hadrons within (approximate) $SU(3)$ flavour multiplets (the famous "eightfold way" [GeN]): in the limit of exact $SU(3)$ flavour symmetry, one can calculate matrix elements of weak currents between hadronic states by means of the general Wigner–Eckart theorem; it turns out that, e.g., all relevant matrix elements between baryon octet states can be expressed in terms of only two independent "reduced matrix elements", which must eventually be measured (these phenomenological parameters are directly related to the coefficients f and $\tilde{f}$ describing neutron beta decay and the process $\Sigma^- \to ne^-\bar{\nu}_e$). In this way, one is able to get some non-trivial predictions for semileptonic baryon decays. Of course, the $SU(3)$ flavour symmetry is in fact broken, so that such a calculational scheme provides only an approximate description of the weak decays of real hadrons. Nevertheless, the results agree reasonably well with empirical data, so one can say that the weak interaction theory formulated in terms of the quark fields does have some predictive power at the level of physical hadrons. More details concerning the Cabibbo theory of semileptonic baryon decays can be found, e.g., in [CoB] or [Geo].

In any case, it should be stressed that for weak processes of the beta-decay type (i.e. for semileptonic decays of mesons and baryons), one may use again the "rule $G^2\Delta^5$" to estimate approximately the corresponding partial decay rates; one only has to include correctly the relevant coupling strength, namely $G_\text{F}\cos\theta_\text{C}$ or $G_\text{F}\sin\theta_\text{C}$ for strangeness-conserving or strangeness-changing decays, respectively, and relate the estimated quantity to that of an appropriate "reference" process (e.g. neutron beta decay) — cf. the discussion around the formula (1.131).

Let us now summarize the model of universal weak interaction established in the 1960s and generally accepted before the advent of the modern gauge theories. The interaction Lagrangian (due to Feynman, Gell-Mann, Cabibbo, etc.) can be written as

$$\mathscr{L}_\text{int}^{(w)} = -\frac{G_F}{\sqrt{2}} J^\rho J_\rho^\dagger \tag{2.58}$$

with the charged current

$$J_\rho = \bar{\nu}_e\gamma_\rho(1-\gamma_5)e + \bar{\nu}_\mu\gamma_\rho(1-\gamma_5)\mu + \bar{u}\gamma_\rho(1-\gamma_5)(d\cos\theta_\text{C} + s\sin\theta_\text{C}) \tag{2.59}$$

(for the sake of brevity, we have denoted here all fermion fields by means of the corresponding particle labels — we will stick to this shorthand notation henceforth). The last expression indicates that we actually have to do with a rotation in the space of quark fields. One can also say that the Cabibbo angle describes a mixing between the quarks d and s that carry the same electric charge but differ in flavour. As we shall see in Chapter 7, these hints become transparent within the framework of the electroweak SM.

The reader should bear in mind that the simple and elegant form (2.58) and (2.59) resulted primarily from intricate confrontation of the earlier provisional Fermi-type models with experimental data; nevertheless, at some stages of the development, brilliant insight and intuition of theorists played an important role as well. The current–current four-fermion Lagrangian (2.58) will serve later on as an appropriate starting point of our path towards the electroweak unification, but now we are going to examine further phenomenological applications of this low-energy effective theory of weak interactions.

2.6 Pion decays into leptons

A familiar weak process that can be calculated rather easily is the decay of a charged pion into a pair of leptons, i.e.

$$\pi^- \to \ell^- + \bar{\nu}_\ell \tag{2.60}$$

(or $\pi^+ \to \ell^+ + \nu_\ell$), where $\ell = e$ or μ. For the sake of brevity, we will denote such a decay process as $\pi_{\ell 2}$. The evaluation of the lowest-order matrix element for (2.60) may proceed as follows. The final state in (2.60) can be obtained by applying the appropriate creation operators to vacuum, namely $|f\rangle = b^+(p)d^+(k)|0\rangle$, where k and p denote the antineutrino and charged lepton four-momenta, respectively. As for the initial state, we will write this rather symbolically as $|i\rangle = |\pi^-(q)\rangle$ with $q = k + p$; without introducing an effective pion field, one can hardly do more. The first-order S-matrix element is expressed through $\langle f|\mathscr{L}_{\text{int}}|i\rangle$, which involves the conjugated (bra) vector $\langle f| = \langle 0|d(k)b(p)$. Thus, it becomes clear that the part of the weak interaction Lagrangian responsible for (2.60) should certainly contain a piece $\bar{\ell}\gamma_\rho(1-\gamma_5)\nu_\ell$ (descending from the hermitian conjugate current $J_\rho^\dagger$ in (2.58)). On the other hand, (2.60) is clearly a strangeness-conserving process — π^- can be considered as a $\bar{u}d$ state within the quark model. Hence, the relevant interaction Lagrangian describing the decay (2.60) is

$$\mathscr{L}_{\text{int}}^{(\pi_{\ell 2})} = -\frac{G_\text{F}}{\sqrt{2}} \cos\theta_C [\bar{\ell}\gamma_\rho(1-\gamma_5)\nu_\ell][\bar{u}\gamma^\rho(1-\gamma_5)d] \tag{2.61}$$

Of course, we are not able to take straightforwardly a matrix element of the operator (2.61) between the initial pion state and the final leptonic state — the hadronic current is expressed in terms of quark fields while the pion is a composite state involving strong-interaction dynamics that cannot be treated by means of perturbative methods. Nevertheless, we may at least try to write the hadronic part of the matrix element on general grounds; as we shall see, even so one is able to make some interesting predictions for the corresponding decay rates. To this end, we will naturally assume that the quark current connects only hadronic states (including vacuum) and, similarly, that the lepton current has no non-trivial matrix elements between leptonic and hadronic states (with the only possible exception of hadronic vacuum). Now, when evaluating the matrix element of the interaction Lagrangian (2.61) in question, let us imagine inserting a complete set of states between the quark and lepton currents. Taking into account the above-mentioned assumptions, one is then obviously led to the conclusion that out of the whole infinite sum of such intermediate states, the only non-trivial contribution is provided by the vacuum insertion, i.e. one may write

$$\begin{aligned}\langle \ell^-(p)\bar{\nu}_\ell(k)|\mathscr{L}_{\text{int}}^{(\pi_{\ell 2})}(x)|\pi^-(q)\rangle &= -\frac{G_F}{\sqrt{2}}\cos\theta_C \\ &\quad\times \langle \ell^-(p)\bar{\nu}_\ell(k)|\bar{\ell}(x)\gamma_\rho(1-\gamma_5)\nu_\ell(x)|0\rangle \\ &\quad\times \langle 0|\bar{u}(x)\gamma^\rho(1-\gamma_5)d(x)|\pi^-(q)\rangle \qquad (2.62)\end{aligned}$$

Proceeding from (2.62) to the lowest-order S-matrix element S_{fi} and subsequently to the usual relativistically invariant matrix element $\mathcal{M}_{fi}$, it is easy to see that the leptonic part will contribute a factor of the form $\bar{u}(p)\gamma_\rho(1-\gamma_5)v(k)$ to $\mathcal{M}$, with u and v being the corresponding Dirac spinors for the ℓ^- and $\bar{\nu}_\ell$, respectively. The hadronic part of (2.62) must then supply the necessary further factors making the $\mathcal{M}$ Lorentz-invariant — in other words, it must be a four-vector. Of course, this can only depend on the pion four-momentum q, so that the most general covariant hadronic contribution entering the decay matrix element $\mathcal{M}$ can be written as $F(q^2)q^\rho$ with the "form factor" F being an essentially arbitrary function. However, we consider the decay of a physical (on-shell) pion, i.e. $q^2 = m_\pi^2$, and therefore, $F(q^2 = m_\pi^2)$ is simply a constant, which we denote as F_π. As a result of these considerations, the matrix element for the decay $\pi^-(q) \to \ell^-(p) + \bar{\nu}_\ell(k)$ can be written as

$$\mathcal{M}_{\pi_{\ell 2}} = -\frac{G_F}{\sqrt{2}}\cos\theta_C F_\pi q^\rho \bar{u}(p)\gamma_\rho(1-\gamma_5)v(k) \qquad (2.63)$$

The last expression suggests a convenient change of notation, namely

$$F_\pi = f_\pi\sqrt{2} \tag{2.64}$$

that we will use in the sequel. f_π, called "pion decay constant", is the only free parameter entering our description of the $\pi_{\ell 2}$ decay processes and its value must be determined experimentally (from the measured lifetime of the charged pion).

The main message of the preceding simple considerations should be that the relevant matrix element for a $\pi_{\ell 2}$ decay can in fact be written almost by heart. Nevertheless, before proceeding further, a brief commentary on the definition of f_π may be useful. In formal terms, this is actually defined as follows. First, using translational covariance of the field operators, the x-dependence of the matrix element of the $V-A$ hadronic current in (2.62) is easily factorized as

$$\langle 0|V_\rho^{1+i2}(x) - A_\rho^{1+i2}(x)|\pi^-(q)\rangle = \mathrm{e}^{-iqx}\langle 0|V_\rho^{1+i2}(0) - A_\rho^{1+i2}(0)|\pi^-(q)\rangle$$

Now, since the pion is a pseudoscalar meson, only the axial-vector current in the last expression can give a non-zero contribution and this will be a true Lorentz vector (apart from the conventional normalization factor for the one-pion state) depending on the pion four-momentum q only. Note that a corresponding matrix element of the vector current would have to be a pseudovector, but obviously, there is no such thing that could be written in terms of a single four-momentum q. For the on-shell pion, one may thus write finally

$$\langle 0|A_\rho^{1+i2}(0)|\pi^-(q)\rangle = -N(q)\sqrt{2}f_\pi q_\rho$$

where the normalization factor $N(q)$ is taken to be $(2\pi)^{-3/2}(2E(q))^{-1/2}$ in accordance with our conventions (cf. Appendix B). This relation can eventually be used in the calculation of the lowest-order S-matrix element and the result (2.63) for $\mathcal{M}$ is thus reproduced. Let us remark that our definition of the constant f_π differs from the convention used in the literature by an inessential phase factor of $-i$. Finally, one should also note that f_π is in fact a fundamental parameter describing the spontaneous breakdown of chiral symmetry in the theory of strong interactions (see, e.g., [Geo]), but we will not elaborate here on this profound aspect of the pion decay constant.

Let us now proceed to calculate the decay width corresponding to (2.63). To this end, it is useful first to simplify the matrix element by means of the equations of motion. In particular, setting in (2.63) $q = k + p$, one may utilize the Dirac equations $\bar{u}(p)\not{p} = m_\ell \bar{u}(p)$ and $\not{k}v(k) = 0$ (as usual, neutrino is taken to be massless for simplicity) and the $\pi_{\ell 2}$ matrix element thus becomes

$$\mathcal{M}_{\pi_{\ell 2}} = -G_\mathrm{F} \cos\theta_C f_\pi m_\ell \bar{u}(p)(1-\gamma_5)v(k) \tag{2.65}$$

Squaring the last expression, summing over the lepton spins and employing the usual trace techniques, one gets

$$\begin{aligned}\overline{|\mathcal{M}|^2} &= G_F^2 \cos^2\theta_\mathrm{C} f_\pi^2 m_\ell^2 \mathrm{Tr}[(\not{p}+m_\ell)(1-\gamma_5)\not{k}(1+\gamma_5)] \\ &= 2G_\mathrm{F}^2 \cos^2\theta_\mathrm{C} f_\pi^2 m_\ell^2 \mathrm{Tr}(\not{k}\not{p}) \\ &= 8G_\mathrm{F}^2 \cos^2\theta_\mathrm{C} f_\pi^2 m_\ell^2 (k\cdot p)\end{aligned} \tag{2.66}$$

and using $2k\cdot p = (k+p)^2 - k^2 - p^2 = m_\pi^2 - m_\ell^2$, this is recast as

$$\overline{|\mathcal{M}_{\pi_{\ell 2}}|^2} = 4G_\mathrm{F}^2 \cos^2\theta_\mathrm{C} f_\pi^2 m_\ell^2 (m_\pi^2 - m_\ell^2) \tag{2.67}$$

To get the decay rate, we also need the two-body phase space for the final-state leptons. According to the general formula shown in Appendix B, this is $\mathrm{LIPS}_2 = (4\pi m_\pi)^{-1}|\vec{p}|$, where $\vec{p}$ is a lepton momentum in the rest system of the decaying pion. The energy conservation $\sqrt{|\vec{p}|^2 + m_\ell^2} + |\vec{p}| = m_\pi$ yields the solution $|\vec{p}| = (m_\pi^2 - m_\ell^2)/(2m_\pi)$, so that

$$\mathrm{LIPS}_2 = \frac{1}{8\pi}\left(1 - \frac{m_\ell^2}{m_\pi^2}\right) \tag{2.68}$$

Putting all the necessary factors together, the result for the decay width can be written as

$$\Gamma(\pi^- \to \ell^- + \bar{\nu}_\ell) = \frac{G_F^2}{4\pi}\cos^2\theta_C f_\pi^2 m_\ell^2 m_\pi \left(1 - \frac{m_\ell^2}{m_\pi^2}\right)^2 \tag{2.69}$$

It is easy to see that the same result holds for a process $\pi^+ \to \ell^+\nu_\ell$ as well. Of course, the formula (2.69) does not represent a pure prediction, as it involves the hitherto arbitrary pion decay constant f_π. Rather, it can be used for the determination of f_π by comparing the calculated pion decay rate with its measured lifetime. Using the result (2.69) for $\ell = e, \mu$ and the experimental value $\tau_{\pi^\pm} \doteq 2.6 \times 10^{-8}$ s, one gets

$$f_\pi \doteq 93\,\mathrm{MeV} \tag{2.70}$$

(needless to say, for an accurate measurement of f_π, one should also take into account the electromagnetic radiative corrections, etc., but in fact, the approximate value (2.70) already represents the right number to be remembered for further applications). On the other hand, the result (2.69) does entail a clear-cut prediction for the ratio of the decay rates corresponding to the electronic and muonic modes. Indeed, taking

$$R_{e/\mu} = \frac{\Gamma(\pi^- \to e^- + \bar{\nu}_e)}{\Gamma(\pi^- \to \mu^- + \bar{\nu}_\mu)} \tag{2.71}$$

f_π drops out and one readily gets

$$R_{e/\mu} = \frac{m_e^2}{m_\mu^2} \frac{(m_\pi^2 - m_e^2)^2}{(m_\pi^2 - m_\mu^2)^2} \tag{2.72}$$

Putting in numbers, namely $m_\pi \doteq 139.6\,\text{MeV}$, $m_\mu \doteq 105.6\,\text{MeV}$ and $m_e \doteq 0.5\,\text{MeV}$, the relation (2.72) yields $R_{e/\mu} \doteq 1.28 \times 10^{-4}$, which agrees very well with the experimental result $R_{e/\mu}^{\text{exp.}} = (1.230 \pm 0.004) \times 10^{-4}$ (note that the branching ratio for the muonic decay mode thus constitutes about 99.99%).

Such a dramatic suppression of the electronic decay mode relatively to the muonic mode may be somewhat surprising at first sight, since the phase-space volume obviously prefers the π_{e2} mode:

$$\frac{\text{LIPS}_2(\pi_{e2})}{\text{LIPS}_2(\pi_{\mu 2})} = \frac{m_\pi^2 - m_e^2}{m_\pi^2 - m_\mu^2} \doteq 2.34 \tag{2.73}$$

Of course, the suppression of the π_{e2} decay mode is due to the factorization of the squared lepton mass in the formula (2.69) and this in turn can be easily traced back to the factor of q^ρ in the basic matrix element (2.63). This factor clearly reflects the (pseudo)vector character of the weak current; one may thus say that the observed suppression of the π_{e2} decay mode provides a stringent test of the nature of weak interactions.[8] Indeed, if the weak interaction had, e.g., "scalar character" (i.e. the weak currents were a combination of Lorentz scalars and pseudoscalars), the pion-to-vacuum matrix element of the (pseudoscalar) hadronic current would essentially

[8]The beautiful prediction (2.72) was made first by Feynman and Gell-Mann in their fundamental paper [12]. It is amusing to note that they considered the predicted numerical value of $R_{e/\mu}$ a serious problem for their theory, since no π_{e2} decay was observed experimentally at that time and the estimated upper bound for the π_{e2} branching ratio was consequently much too low to fit in (2.72). Feynman and Gell-Mann also remarked that they "had no idea on how such a discrepancy could be resolved". Needless to say, the problem was solved by the experimentalists later on and the result (2.72) became a triumph for the theory of weak interactions based on vector and axial-vector currents.

reduce to a constant f_π alone and thus one would be led to a prediction

$$R_{e/\mu}^{(\text{scalar})} = \frac{(m_\pi^2 - m_e^2)^2}{(m_\pi^2 - m_\mu^2)^2} \doteq 5.5 \tag{2.74}$$

On the other hand, the reader should realize that in fact one does not need precisely the $V-A$ currents to achieve (2.72); from our derivation, it should be obvious that any combination of V and A currents would give the same result, provided that the A component is non-trivial (let us recall once again that A is needed because the intrinsic parity of pion is -1).

Two remarks are in order here. First, the proportionality of the $\pi_{\ell 2}$ matrix element to m_ℓ and the ensuing suppression of the electronic decay mode, shown to be characteristic feature of a weak interaction model of the (V, A) type, can also be easily understood on the basis of chirality and angular momentum conservation. Indeed, let us consider the limit $m_\ell = 0$. Both V and A interactions preserve chirality (formally, this is due to γ_α and $\gamma_\alpha\gamma_5$ anticommuting with γ_5). For massless leptons, this entails a simple helicity selection rule: ℓ^- and $\bar{\nu}_\ell$ can only be produced with opposite helicities (i.e. a left-handed ℓ must be accompanied by right-handed $\bar{\nu}_\ell$ and vice versa). On the other hand, since the pion spin is zero, angular momentum conservation obviously requires that ℓ^- and $\bar{\nu}_\ell$ helicities be equal in their c.m. system. Thus, the $\pi_{\ell 2}$ decay is forbidden for $m_\ell = 0$ within a (V, A) weak interaction theory. For $m_\ell \neq 0$, a helicity flip is possible (though the chirality selection rules remain valid) and the matrix element corresponding to ℓ^- and $\bar{\nu}_\ell$ with like helicities is then naturally proportional to m_ℓ. In particular, within the $V - A$ theory, the (massless) $\bar{\nu}_\ell$ is of course produced as right-handed in $\pi^- \to \ell^- \bar{\nu}_\ell$ and hence the (massive) ℓ^- must also be emitted with positive helicity, owing to the angular momentum conservation (the reader is recommended to verify, by means of an explicit calculation, that the probability of an emission of left-handed electron or negative muon indeed vanishes).

Second, looking towards the present-day Standard Model, it is important to emphasize that experimental confirmation of the remarkable result (2.72) implies a definite message for weak interaction models involving an intermediate boson (which were "on the market" since the early days of weak interaction theory): *vector* W boson (i.e. that of spin 1) coupled to V and A currents is thereby clearly favoured over the other possibilities (e.g., W of spin 0 or 2 corresponding to scalar or tensor couplings).

The simple calculational techniques discussed above can also be successfully applied to other processes closely related to the $\pi_{\ell 2}$ decays. There are at least two "classic" applications that should be mentioned before closing this section, namely the leptonic decays of a charged kaon (the $K_{\ell 2}$ decays with $\ell = e$ or μ) and the decay process $\tau^- \to \pi^- + \nu_\tau$ (or $\tau^- \to K^- + \nu_\tau$, respectively). Let us start with the $K_{\ell 2}$ decays. In such a case, one may repeat essentially all the steps that led us previously to the formula (2.63), except that now one has to replace $\cos\theta_C$ by $\sin\theta_C$ (we are dealing with a strangeness-changing process) and the corresponding "kaon decay constant" f_K may in general be different from f_π. It is useful to remember that under an assumption of exact $SU(3)$ flavour symmetry, f_K and f_π would be equal — although such a statement may not be immediately obvious, it is not difficult to realize that the kaon carries the same quantum numbers as the $\Delta S = 1$ weak current (cf. (2.53)). Thus, knowing the f_π value from $\pi_{\ell 2}$ decays, one can make quite reasonable predictions for the $K_{\ell 2}$ decay rates at least in the $SU(3)$ flavour symmetry limit. Of course, in the real world, $SU(3)_{\text{flavour}}$ is broken and such a symmetry prediction may be reliable only with an accuracy of about 20%. One may best assess this accuracy by comparing a measured value of f_K with f_π found before. f_K can be determined by matching the formula for the $K_{\ell 2}$ decay rate

$$\Gamma(K^- \to \ell^- + \bar{\nu}_\ell) = \frac{G_F^2}{4\pi} \sin^2\theta_C f_K^2 m_\ell^2 m_K (1 - m_\ell^2/m_K^2)^2 \tag{2.75}$$

(cf. (2.69)) with relevant experimental data. In particular, one may use the branching ratio $\text{BR}(K^- \to \mu^- + \bar{\nu}_\mu) \doteq 63.5\%$ together with the kaon mean lifetime $\tau_{K^\pm} \doteq 1.24 \times 10^{-8}\text{s}$ (for completeness, let us also recall that $m_{K^\pm} \doteq 494\,\text{MeV}$). One thus gets $f_K \doteq 111\,\text{MeV}$, i.e. $f_K/f_\pi \doteq 1.2$.

To describe the process $\tau^- \to \pi^- + \nu_\tau$, one adds a term $\bar{\nu}_\tau \gamma_\rho (1 - \gamma_5)\tau$ to the leptonic current in (2.59). The corresponding matrix element can then be written on similar grounds as that obtained earlier for the $\pi_{\ell 2}$ decays. Armed with our previous experience, we may guess the relevant result rather easily; this obviously reads

$$\mathcal{M}(\tau^- \to \pi^- + \nu_\tau) = -G_F \cos\theta_C f_\pi \bar{u}(k) \not{p} (1 - \gamma_5) u(q) \tag{2.76}$$

where we have denoted τ, π and ν_τ four-momenta by q, p and k, respectively (the reader is recommended to recover the formal steps leading to the last expression). For the corresponding decay rate, we then have

$$\Gamma(\tau^- \to \pi^- + \nu_\tau) = \frac{G_F^2}{8\pi} \cos^2\theta_C f_\pi^2 m_\tau^3 \left(1 - \frac{m_\pi^2}{m_\tau^2}\right)^2 \tag{2.77}$$

Having fixed the f_π value through the $\pi_{\ell 2}$ decays, the last result now represents a definite prediction for the partial decay width in question.

Putting in numbers (in particular, $m_\tau = 1.777$ GeV) and taking into account that the τ lifetime is about 2.9×10^{-13}s, the formula (2.77) yields the branching ratio $\mathrm{BR}(\tau^- \to \pi^- + \nu_\tau) \doteq 10.8\%$, in good agreement with the experimental value, which is $10.82 \pm 0.05\%$ according to [6]. In a similar way, we can calculate the branching ratio for the mode $\tau^- \to K^- + \nu_\tau$, or, alternatively, the ratio

$$\frac{\Gamma(\tau^- \to K^- + \nu_\tau)}{\Gamma(\tau^- \to \pi^- + \nu_\tau)} = \tan^2 \theta_C \frac{f_K^2}{f_\pi^2} \frac{(1 - m_K^2/m_\tau^2)^2}{(1 - m_\pi^2/m_\tau^2)^2} \tag{2.78}$$

Numerically, the last expression gives approximately 0.06, which agrees well with experimental data (note that $\mathrm{BR}(\tau^- \to K^- \nu_\tau) \doteq 7 \times 10^{-3}$).

Thus, we have seen that the meson decay constants f_π and f_K measured in $\pi_{\ell 2}$ and $K_{\ell 2}$ decays can be used in other places as well, enabling one to make useful physical predictions, e.g., for τ decays. The argument could be reversed — one may, e.g., imagine determining f_π from $\tau^- \to \pi^- \nu_\tau$ and employing it to predict the $\pi^\pm$ lifetime. In other words, although the pion lifetime cannot be simply calculated from the first principles (i.e. by using the Lagrangian (2.58) only), one additional measurement (of another process) is sufficient for accomplishing such a prediction.

In this context, one last remark should be added. The phenomenological parameter f_π also plays an important role in the decay of the neutral pion. It is well known that the π^0 decays predominantly into a pair of photons, i.e. through an electromagnetic interaction (note that the branching ratio for $\pi^0 \to \gamma\gamma$ constitutes about 98.8% and the mean lifetime $\tau_{\pi^0} \doteq 8.4 \times 10^{-17}$s). The matrix element for the $\pi^0 \to \gamma\gamma$ decay can be written with good accuracy as

$$\mathcal{M}(\pi^0 \to \gamma\gamma) = -\frac{1}{f_\pi} \frac{\alpha}{\pi} \epsilon_{\mu\nu\rho\tau} k^\rho p^\tau \varepsilon^{*\mu}(k) \varepsilon^{*\nu}(p) \tag{2.79}$$

where $\varepsilon^\mu(k)$, and $\varepsilon^\nu(p)$ denote the polarization vectors of the final-state photons with four-momenta k, p and α is the electromagnetic fine-structure constant, $\alpha \doteq 1/137$. The uninitiated reader should be warned that the remarkable formula (2.79) is by no means obvious — a comprehensive treatment of this subject can be found, e.g., in [ChL], [Geo], [Don]. Here, let us only add that (2.79) holds in the limit of zero pion mass (the "soft pion limit") and eventual corrections due to finite m_π may be of the order of 1%. The decay rate corresponding to (2.79) is then

$$\Gamma(\pi^0 \to \gamma\gamma) = \frac{1}{64\pi} m_\pi^3 \left(\frac{\alpha}{\pi}\right)^2 \frac{1}{f_\pi^2} \tag{2.80}$$

Numerically, the formula (2.80) yields the prediction 7.65 eV for the decay width in question, in good agreement with the experimental value. In view of the preceding considerations, one may now say that, alternatively, the experimental value of the π^0 lifetime could provide the necessary input for making a prediction for the charged pion lifetime. In principle, it is true, yet one should bear in mind that the $\pi^\pm$ lifetime is measured with much better accuracy than the π^0 lifetime ($\tau_{\pi^\pm} = (2.6033 \pm 0.0005) \times 10^{-8}$s, to be compared with $\tau_{\pi^0} = (8.43 \pm 0.13) \times 10^{-17}$s), i.e. the charged pion decays provide the most accurate data for determination of the f_π value.

2.7 Beta decay of charged pion

There is another possible decay channel for the charged pion, namely

$$\pi^+ \to \pi^0 + e^+ + \nu_e \tag{2.81}$$

(or, equivalently, $\pi^- \to \pi^0 e^- \bar{\nu}_e$). This may be naturally called "pion beta decay" and is sometimes denoted as π_{e3}. Its muonic analogue is obviously precluded by energy conservation. The decay mode (2.81) is in fact very rare — its branching ratio constitutes only about 10^{-8}, as one can easily guess on the basis of the "rule $G^2\Delta^5$" (cf. Section 2.5); the essential point is that masses of the charged and neutral pion are rather close ($m_{\pi^+} - m_{\pi^0} \doteq 4.6\,\text{MeV}$) so that the available phase space for the decay products is small. Nevertheless, despite being so rare, the pion beta decay is extremely interesting, since it serves as a "test bench" for some basic ideas of the theory of hadronic weak interactions. In particular, it provides an important check on the properties of the vector part of the weak current. As we shall see, the π_{e3} decay width is calculable in a theoretically clean way, without introducing further phenomenological parameters (in contrast to the $\pi_{\ell 2}$ decays discussed in the preceding section), i.e. one essentially gets a pure prediction based on the Lagrangian (2.58). Such a theoretical result can then be compared with the corresponding experimental value that has been measured with good accuracy; according to [6], one has

$$\text{BR}_{\text{exp}}(\pi^+ \to \pi^0 e^+ \nu_e) = (1.036 \pm 0.006) \times 10^{-8} \tag{2.82}$$

In the remainder of this section, we will explain how the calculation can be done. From the lepton content of the final state in (2.81), it is easy to guess that the relevant part of the weak interaction Lagrangian (2.58) is

$$\mathscr{L}_{\text{int}}^{(\pi_{e3})} = -\frac{G_F}{\sqrt{2}} \cos\theta_C [\bar{d}\gamma_\mu (1-\gamma_5) u][\bar{\nu}_e \gamma^\mu (1-\gamma_5) e] \tag{2.83}$$

In analogy with the discussion of preceding section, it should be clear that the S-matrix element in the first order of perturbation theory will be factorized into the hadronic and leptonic parts. Thus, one needs to know matrix elements of the currents appearing in (2.83), namely

$$\langle \pi^0(p)|\bar{d}(x)\gamma_\mu(1-\gamma_5)u(x)|\pi^+(P)\rangle \tag{2.84}$$

and

$$\langle e^+(k)\nu_e(l)|\bar{\nu}_e(x)\gamma^\mu(1-\gamma_5)e(x)|0\rangle \tag{2.85}$$

Note that in (2.84) and (2.85), we have marked explicitly the corresponding particle momenta. Of course, the leptonic term can be evaluated in a straightforward way, so let us focus first on the hadronic matrix element. Before we proceed to work it out, the reader should be warned that we are going to employ some tricks characteristic for the so-called "current algebra" — a basic and powerful technique in hadron physics developed mostly in the 1960s, which, however, is not commonly used in the bulk of this text. A brief review of the current-algebra ideas can be found, e.g., in [ChL].

For our purpose, the basic observations are as follows. First of all, one should realize that both π^+ and π^0 are pseudoscalar mesons and therefore only the *vector* part of the weak quark current in (2.84) can contribute to the matrix element in question.[9] Next, one may recall that the algebraic structure of the quark current corresponds to

$$\bar{d}\gamma_\mu u = V_\mu^1 - iV_\mu^2 \tag{2.86}$$

where V_μ^1 and V_μ^2 are defined in accordance with Section 2.5 (cf. the discussion around the relation (2.57)). In the present context, one actually does not have to use the Gell-Mann matrices — the Pauli matrices will do. The last relation thus explicitly reads

$$\begin{aligned}\bar{d}\gamma_\mu u &= (\bar{u},\bar{d})\,\gamma_\mu\left[\frac{1}{2}\begin{pmatrix}0 & 1\\ 1 & 0\end{pmatrix} - i\frac{1}{2}\begin{pmatrix}0 & -i\\ i & 0\end{pmatrix}\right]\begin{pmatrix}u\\ d\end{pmatrix}\\ &= \bar{\psi}_q\gamma_\mu\frac{\tau^1}{2}\psi_q - i\bar{\psi}_q\gamma_\mu\frac{\tau^2}{2}\psi_q \end{aligned}\tag{2.87}$$

(the notation should be self-explanatory). The meaning of the symbolic identity (2.86) consists in exhibiting the isospin properties of the

[9] The matrix element of the axial-vector part would be a pseudovector in the considered case, but a pseudovector obviously cannot be constructed from two independent four-vectors (the four-momenta p and P). Thus, in a sense, we now have a situation opposite to that encountered previously in the case of the $\pi_{\ell 2}$ decays.

strangeness-conserving weak current: it is seen that we are working with components of an isospin triplet $V_\mu^a = \bar{\psi}_q \gamma_\mu \frac{\tau^a}{2} \psi_q$, $a = 1, 2, 3$, in particular with the "isospin-lowering" combination $V_\mu^1 - iV_\mu^2$. Now, we come to a point that is crucial for our calculation. It turns out that the weak current in (2.86) can be expressed as a commutator of the electromagnetic current and a pertinent combination of the isospin charges, namely

$$V_\mu^1 - iV_\mu^2 = [Q^1 - iQ^2, J_\mu^{(\mathrm{em})}] \tag{2.88}$$

where $J_\mu^{(\mathrm{em})} = \frac{2}{3}\bar{u}\gamma_\mu u - \frac{1}{3}\bar{d}\gamma_\mu d$ and

$$Q^a = \int V_0^a(\vec{x}, x_0) d^3x \tag{2.89}$$

Some technical details of the derivation of the important relation (2.88) can be found in [Bai]. At this place, let us only remark that such a relation is in fact quite natural: it is easy to realize that the electromagnetic current of the quarks u and d can be expressed in terms of the third component of the isotopic triplet and an isosinglet as

$$\begin{aligned} J_\mu^{(\mathrm{em})} &= \frac{2}{3}\bar{u}\gamma_\mu u - \frac{1}{3}\bar{d}\gamma_\mu d \\ &= \bar{\psi}_q \gamma_\mu \frac{\tau^3}{2} \psi_q + \frac{1}{6}\bar{\psi}_q \gamma_\mu \mathbb{1} \psi_q \\ &= V_\mu^3 + \frac{1}{6} V_\mu^0 \end{aligned} \tag{2.90}$$

The commutators are trivial for the singlet term (since this involves the unit matrix) and the identity (2.88) thus essentially corresponds to the familiar algebraic relations among the isospin $SU(2)$ generators. Last but not least, let us emphasize that the isospin currents may be considered, with a rather good accuracy, as *conserved* quantities (isospin is a good approximate symmetry of strong interactions); consequently, the charges (2.89) can be taken (approximately) as time-independent generators of a corresponding $SU(2)$ algebra.

The above considerations form a basis of what has been called, historically, the "CVC hypothesis" — the acronym stands for "conserved vector current". Within our quark picture, this emerges quite naturally and almost automatically, but in the early days of the weak interaction theory, it was a rather non-trivial assumption (an essential "leap of faith" was precisely placing the vector part of a weak transition operator into the same multiplet with the relevant part of the electromagnetic current — this undoubtedly also represents a major step towards a conceptual unification of both forces).[10] The concept of CVC is originally due to Gershtein and Zeldovich [23] and Feynman and Gell-Mann [12]; for a rather detailed discussion, see, e.g., [BjD], [MRR].

Let us now show how the matrix element (2.84) can be evaluated. In view of the preceding discussion, this is equal to

$$\langle \pi^0(p)|V^1_\mu - iV^2_\mu|\pi^+(P)\rangle = \langle \pi^0(p)|[Q^1 - iQ^2, J^{(\mathrm{em})}_\mu]|\pi^+(P)\rangle \quad (2.91)$$

Note that here and in what follows, we may take the current operators at the point $x = 0$; as ever, the coordinate dependence of the matrix element in question is essentially trivial — it is carried by a usual exponential factor obtained through an appropriate space–time translation. To work out the last expression, one has to realize that the pion states are (with good accuracy) isospin eigenstates, while the combination $Q^1 - iQ^2$ is an isospin-lowering operator. One may then utilize the familiar relations known, e.g., from the quantum-mechanical theory of angular momentum (remember that algebraic properties of the isospin and ordinary spin are formally the same); we thus get, in particular[11]

$$\begin{aligned}(Q^1 - iQ^2)|\pi^+(P)\rangle &= \sqrt{2}|\pi^0(P)\rangle \\ \langle \pi^0(p)|(Q^1 - iQ^2) &= \sqrt{2}\langle \pi^+(p)| \end{aligned} \quad (2.92)$$

Using this, the last expression in (2.91) becomes

$$\sqrt{2}\left(\langle \pi^+(p)|J^{(\mathrm{em})}_\mu|\pi^+(P)\rangle - \langle \pi^0(p)|J^{(\mathrm{em})}_\mu|\pi^0(P)\rangle\right) \quad (2.93)$$

but the last term vanishes identically (to see this formally, one should realize that the electromagnetic current changes its sign under charge conjugation

[10] There is another important conceptual aspect of the above discussion that should perhaps be mentioned here. The currents entering the commutators of the current algebra originate in strong-interaction symmetries (such as the isospin) and these "symmetry currents" are subsequently identified with physical currents participating in weak interaction dynamics. Before the advent of quark-model Lagrangians (that gradually led to the present-day Standard Model), such an identification was by no means obvious.

[11] For reader's convenience, let us remark that in order to arrive at (2.92), one employs the relation

$$Q_\pm|T, T_3\rangle = \sqrt{T(T+1) - T_3(T_3 \pm 1)}|T, \quad T_3 \pm 1\rangle$$

with $Q_\pm = Q^1 \pm iQ^2$, $T = 1$ for pion isotriplet, $T_3 = 0$ and 1 for π^0 and π^+, respectively.

$\mathcal{C}$ while the neutral pion is a $\mathcal{C}$ eigenstate). We thus arrive at the result

$$\langle\pi^0(p)|V_\mu^1(0) - iV_\mu^2(0)|\pi^+(P)\rangle = \sqrt{2}\langle\pi^+(p)|J_\mu^{(\mathrm{em})}(0)|\pi^+(P)\rangle \quad (2.94)$$

which embodies, technically, the essence of the "CVC relation" relevant for the considered process. Up to conventional normalization factors associated with the one-particle states, the quantity (2.94) is a Lorentz vector that must be made of two independent four-momenta. Clearly, the most general form of such a vector can be described as

$$\langle\pi^+(p)|J_\mu^{(\mathrm{em})}(0)|\pi^+(P)\rangle = F_+(q^2)(P+p)_\mu + F_-(q^2)(P-p)_\mu \quad (2.95)$$

where $q = P - p$ and the coefficients $F_\pm(q^2)$ are essentially arbitrary functions (form factors). Note that we consider the on-mass-shell pions, i.e. $p^2 = m_{\pi^0}^2$ and $P^2 = m_{\pi^+}^2$. However, one should not forget that the assumption of exact isospin symmetry actually means that $m_{\pi^0} = m_{\pi^+}$, so one has to set $p^2 = P^2$ whenever the current conservation is used explicitly. In terms of the parametrization (2.95), the current conservation is tantamount to

$$0 = q^\mu[F_+(q^2)(P+p)_\mu + F_-(q^2)(P-p)_\mu] \quad (2.96)$$

Setting there $p^2 = P^2$, the first term in (2.96) drops out automatically and one is left with the condition $q^2F_-(q^2) = 0$, i.e. the form factor F_- has to vanish (in the considered symmetry limit). Thus, the matrix element of the electromagnetic current is given by

$$\langle\pi^+(p)|J_\mu^{(\mathrm{em})}(0)|\pi^+(P)\rangle = F_\pi(q^2)(P+p)_\mu \quad (2.97)$$

where we have denoted $F_+ = F_\pi$ so as to introduce a standard symbol for the pion electromagnetic form factor. For small q^2 (which is our case), $F_\pi(q^2)$ is a slowly varying function and the value $F_\pi(0)$ is determined by the pion charge, i.e.

$$F_\pi(0) = 1 \quad (2.98)$$

Thus, we may now state our main result as follows. With a rather good accuracy, the matrix element (2.84) is given by

$$\langle\pi^0(p)|\bar{d}\gamma_\mu(1-\gamma_5)u|\pi^+(P)\rangle = \langle\pi^0(p)|V_\mu^1 - iV_\mu^2|\pi^+(P)\rangle \doteq \sqrt{2}(p+P)_\mu \quad (2.99)$$

This result is indeed remarkable, since — as we indicated earlier in this section — one needs no extra phenomenological parameter to describe the hadronic matrix element in question. Obviously, an essential point was that we were able to recast the whole problem in terms of the electromagnetic current (see (2.94)) whose properties are well known. It is also useful to realize that, since only the vector part of the weak current contributes in the considered case, the pion beta decay is in fact an example of a pure Fermi transition within the domain of particle physics. With the result (2.99) at hand, we are in a position to write down the complete matrix element for the pion beta decay, corresponding to the Lagrangian (2.83) in the lowest order. The evaluation of the leptonic factor descending from (2.85) is essentially trivial and one thus readily gets

$$\mathcal{M}_{\pi_{e3}} = -G_F \cos\theta_C (p+P)^\mu [\bar{u}(l)\gamma_\mu(1-\gamma_5)v(k)] \tag{2.100}$$

The calculation of the corresponding decay rate is then routine. For the squared matrix element summed over the lepton spins one gets, after some algebra,

$$\begin{aligned}\overline{|\mathcal{M}_{\pi_{e3}}|^2} &= 2G_F^2 \cos^2\theta_C \mathrm{Tr}[(\not{k} - m_e)\not{Q}\not{l}\not{Q}(1-\gamma_5)] \\ &= 2G_F^2 \cos^2\theta_C \mathrm{Tr}(\not{k}\not{Q}\not{l}\not{Q})\end{aligned} \tag{2.101}$$

where we have denoted $Q = P + p$ for brevity. In what follows, we shall work in the rest frame of the decaying π^+. Kinematically, the considered process is similar to the neutron beta decay, as one of the decay products (π^0) has a mass that is very close to m_{π^+}. The maximum positron energy is $\Delta = (m_{\pi^+}^2 - m_{\pi^0}^2 + m_e^2)/(2m_{\pi^+}) = 4.52\,\mathrm{MeV}$ ($\Delta \doteq m_{\pi^+} - m_{\pi^0}$), which means that the positron can be highly relativistic near the endpoint of its spectrum ($\beta_e^{\max} \doteq 0.99$). On the other hand, the recoil π^0 is safely non-relativistic over the whole kinematical range — it is easy to check that the maximum π^0 momentum is of the order of the mass difference $m_{\pi^+} - m_{\pi^0}$ and constitutes thus only about 3% of its rest mass. Using the static approximation for π^0 (i.e. setting $Q \doteq (m_{\pi^+} + m_{\pi^0}, \vec{0})$), one gets from (2.101), after some simple manipulations,

$$\overline{|\mathcal{M}_{\pi_{e3}}|^2} \doteq 32G_F^2 \cos^2\theta_C m_\pi^2 E_e E_\nu (1 + \beta_e \cos\vartheta) \tag{2.102}$$

where we have also introduced (in the spirit of our kinematical approximation) an average pion mass $m_\pi = \frac{1}{2}(m_{\pi^+} + m_{\pi^0})$. The expression (2.102) illustrates explicitly the pure Fermi character of the considered transition — the coefficient of the $e^+-\nu$ angular correlation is seen to be equal $+1$ (for a comparison with nuclear beta decay, see, e.g., (1.45)).

To obtain the π_{e3} decay rate, one can now proceed in full analogy with the case of neutron decay described in detail in Section 1.4. Within our kinematical approximation, the positron energy spectrum has the familiar "statistical" form

$$\frac{dw_{\pi_{e3}}(E)}{dE} \doteq \frac{1}{\pi^3} G_F^2 \cos^2\theta_C |\vec{k}| E(\Delta - E)^2 \tag{2.103}$$

and this can be integrated over E from m_e to Δ. m_e can be neglected with reasonable accuracy (note that such an approximation is better here than for neutron decay, since in the present case, $m_e/\Delta \doteq 0.1$). One thus has

$$\begin{aligned} \Gamma(\pi^+ \to \pi^0 e^+ \nu_e) &= \int_{m_e}^{\Delta} \frac{dw_{\pi_{e3}}(E)}{dE} dE \\ &\doteq \frac{1}{\pi^3} G_F^2 \cos^2\theta_C \int_0^{\Delta} E^2(\Delta - E)^2 dE \end{aligned} \tag{2.104}$$

so that our final answer reads

$$\Gamma(\pi^+ \to \pi^0 e^+ \nu_e) \doteq \frac{G_F^2 \cos^2\theta_C}{30\pi^3} \Delta^5 \tag{2.105}$$

Putting in numbers (in particular, $\Delta = 4.52\,\text{MeV}$), the last expression yields

$$\Gamma_{\text{theor}}(\pi^+ \to \pi^0 e^+ \nu_e) \doteq 2.62 \times 10^{-22}\,\text{MeV} \tag{2.106}$$

Taking into account the measured π^+ lifetime, which corresponds to the total width of about 2.53×10^{-14} MeV, one gets from (2.106) a prediction for the branching ratio

$$\text{BR}_{\text{theor}}(\pi^+ \to \pi^0 e^+ \nu_e) \doteq 1.03 \times 10^{-8} \tag{2.107}$$

in agreement with the experimental value (2.82). It should be stressed that, in view of the approximations made in the course of our calculation, one should in general only expect an accuracy at the level of several per cent (say, up to 10%) in our theoretical prediction. Indeed, for possible corrections to the basic approximation (2.105), one would have to take into account the isospin violation effects, the recoil π^0 motion and also the $m_e \neq 0$ effects neglected in the evaluation of the Fermi integral in (2.104); each of these corrections may typically represent several per cent for the calculated decay width. The possible corrections to (2.105) seem to be well under control and the agreement between theory and experiment is very good; for more details, see, e.g., [Bai], [MRR]. In any case, the successful theoretical prediction (2.105) certainly represents a remarkable test of our ideas about the structure of weak currents and the rare process π_{e3} therefore occupies a very important niche in theoretical particle physics.

2.8 Nucleon matrix elements of the weak current

Let us now return to the process that has been the starting point of our discussion in Chapter 1 — the neutron beta decay. When this is to be calculated within the theory described by the Lagrangian (2.58), one has to know the relevant matrix elements of the weak current (2.59). The evaluation of the leptonic part is straightforward as ever, but a matrix element of the hadronic current between nucleon states is not directly calculable. The reason is clear: in contrast to the phenomenological approach adopted in Chapter 1, the currents appearing now in our deeper theory are not expressed explicitly in terms of nucleon fields and, at the same time, we do not know a precise quantitative connection between the quark fields and the nucleon states.[12] One might say that this is the price we have to pay for a more elegant formulation of the weak interaction theory in terms of fundamental degrees of freedom. Nevertheless, it should be clear that from the point of view of practical phenomenology, we in fact do not lose anything. Indeed, for the purpose of a practical beta-decay calculation, we may resort to the method used in preceding sections in connection with pion decays. In particular, we can write down the most general form of the nucleon matrix element in question, compatible with some obvious requirements, such as Lorentz covariance etc., as usual, the corresponding expression then involves a few phenomenological coefficients that have to be measured anyway. As we shall see, the resulting picture represents a natural generalization of our old treatment (that was based on an effective Lagrangian involving local nucleon fields) and also incorporates some new subtle phenomena, absent within the old provisional framework; the description developed in Chapter 1 is only recovered in the limit of vanishing nucleon momentum transfer.

Thus, how can one write the desired matrix element on general grounds? For definiteness, let us start with the corresponding vector part.[13] Up to the normalization factors associated with one-particle nucleon states, the matrix element must transform as a Lorentz four-vector and this should be constructed in terms of the relevant Dirac spinors for nucleons and the corresponding four-momenta (we will restrict ourselves to the space–time

[12]The nucleon is a composite state made of confined quarks and its description would involve complicated strong-interaction bound-state dynamics. A corresponding *ab initio* calculation is thus beyond the reach of the present-day techniques of quantum field theory.

[13]The discussion given here essentially follows the lecture notes by Jarlskog [24].

point $x = 0$, since an appropriate shift can be performed trivially). Denoting the neutron and proton four-momenta as p_1 and p_2, respectively, it is not difficult to realize that the matrix element $\langle p|V_\mu|n\rangle$ can then be written in a most general way as

$$\begin{aligned}
&\langle p(p_2)|V_\mu(0)|n(p_1)\rangle \\
&\quad = N_p N_n \bar{u}_p(p_2)(A\gamma_\mu + \gamma_\mu B + C\gamma_\mu D + Ep_{1\mu} + Fp_{2\mu})u_n(p_1) \quad (2.108)
\end{aligned}$$

where the coefficients E and F are Lorentz scalars (they may depend on p_1^2, p_2^2 and $p_1 \cdot p_2$ only) while A, B, C and D are to be understood as matrices made of products of $\not{p}_1$ and $\not{p}_2$ (consequently, they do not necessarily commute with γ_μ). Having in mind the identity $\not{a}\not{b} + \not{b}\not{a} = 2a \cdot b$, it is easy to see that without loss of generality, one may take

$$A = a_0 + a_1\not{p}_1 + a_2\not{p}_2 + a_3\not{p}_1\not{p}_2 \tag{2.109}$$

with $a_j, j = 0, \ldots, 3$ being some scalar coefficients (and similarly for B, C and D). By using (2.109) and the Dirac equations for $\bar{u}_p$ and u_n, one can then get rid of the matrix coefficients in (2.108). Indeed, one has, e.g.,

$$\begin{aligned}
\bar{u}_p\not{p}_1\gamma_\mu u_n &= 2p_{1\mu}\bar{u}_p u_n - \bar{u}_p\gamma_\mu\not{p}_1 u_n \\
&= 2p_{1\mu}\bar{u}_p u_n - m_n\bar{u}_p\gamma_\mu u_n
\end{aligned}$$

and other relevant relations of such a type can be obtained easily. Needless to say, the anticommutation relation $\{\gamma_\mu, \gamma_\nu\} = 2g_{\mu\nu}$ is amply used throughout all these calculations. Thus, one arrives at the form

$$\langle p|V_\mu|n\rangle = \text{N.f.} \times \bar{u}_p[K_0(t)\gamma_\mu + K_1(t)p_{1\mu} + K_2(t)p_{2\mu}]u_n \tag{2.110}$$

where the symbol N.f. stands for the normalization factors $N_p N_n$ and the coefficients $K_j, j = 0, 1, 2$ are scalars. Since the nucleons are taken on the mass shell, K_j in fact depend only on the kinematical variable $t = (p_2 - p_1)^2$ and we may call them form factors. In (2.110), one can replace the variables p_1 and p_2 by

$$P = p_1 + p_2, \quad q = p_2 - p_1$$

and employ the well-known Gordon identity, which in the present case reads

$$\bar{u}(p_2)P_\mu u(p_1) = (m_p + m_n)\bar{u}(p_2)\gamma_\mu u(p_1) - i\bar{u}(p_2)\sigma_{\mu\nu}q^\nu u(p_1) \tag{2.111}$$

(cf. (A.80)). One thus immediately gets the standard representation

$$\langle p|V_\mu|n\rangle = \text{N.f.} \times \bar{u}_p \left[\gamma_\mu f_1^{pn}(t) - i\frac{1}{2M}\sigma_{\mu\nu}q^\nu f_2^{pn}(t) + \frac{1}{2M}q_\mu f_3^{pn}(t)\right] u_n \tag{2.112}$$

where M denotes the average nucleon mass, introduced on dimensional grounds (the correspondence between the old form factors K_j and the new ones is straightforward). In a similar way, we would find that a general form of the axial-vector matrix element can be written as

$$\langle p|A_\mu|n\rangle = \text{N.f.} \times \bar{u}_p \left[\gamma_\mu g_1^{pn}(t) - i\frac{1}{2M}\sigma_{\mu\nu}q^\nu g_2^{pn}(t) + \frac{1}{2M}q_\mu g_3^{pn}(t)\right]\gamma_5 u_n \tag{2.113}$$

The terminology used in the literature for the form factors appearing in (2.112) and (2.113) is as follows:

$$\begin{aligned} f_1^{pn}(t) &= \text{vector form factor}, \\ f_2^{pn}(t) &= \text{weak magnetism}, \\ f_3^{pn}(t) &= \text{induced scalar}, \\ g_1^{pn}(t) &= \text{axial vector form factor}, \\ g_2^{pn}(t) &= \text{pseudotensor form factor}, \\ g_3^{pn}(t) &= \text{induced pseudoscalar}. \end{aligned}$$

Thus, the nucleon matrix element of the weak current has, in general, a more complicated structure than that appearing within the effective Lagrangian approach Chapter 1 (cf. (1.111)). However, for nucleon beta decay, one has the kinematical limits

$$(0.51\,\text{MeV})^2 = m_e^2 \leq t \leq (m_n - m_p)^2 = (1.29\ \text{MeV})^2 \tag{2.114}$$

and thus obviously $|q_\mu|/2M \ll 1$. Then it is natural to expect that the form factors f_2, f_3, g_2 and g_3 do not give sizable contributions to observable quantities (unless they are anomalously large near zero — but this does not seem to be the case). Of course, their contributions would be of the same order of magnitude as the proton recoil effects ignored throughout our previous discussion in Chapter 1. Such effects can also be neglected for most of nuclear beta decays, with only a few exceptions.[14] For the free neutron

[14]In some cases, the relevant momentum transfer (the energy release) may be large enough so that, e.g., the weak magnetism does give a measurable effect. In particular, such effects were studied for the processes $B^{12} \to C^{12} + e^- + \bar{\nu}_e$ and $N^{12} \to C^{12} + e^+ + \nu_e$, where the energy release can be as large as 13 and 16 MeV, respectively. Gell-Mann [25] was the first who calculated the effect of the weak magnetism in the electron (or positron) energy spectra for these beta decays and subsequent experiments [26] confirmed the theoretical results. For further details, the interested reader is referred, e.g., to [CoB].

decay, one may thus safely assume that the only essential contributions are due to the vector and axial vector form factors f_1 and g_1. Further, according to (2.114), the variable t is restricted to be near zero and one can thus presumably neglect the t-dependence of these form factors altogether (barring some unexpectedly wild behaviour), setting simply $f_1(t) \doteq f_1(0)$ and $g_1(t) \doteq g_1(0)$. Thus, in the limit of zero momentum transfer, we arrive at

$$\langle p|V_\mu - A_\mu|n\rangle = \text{N.f.} \times \bar{u}_p\gamma_\mu[f_1(0) - g_1(0)\gamma_5]u_n \tag{2.115}$$

which corresponds formally to the approximate description employed in Chapter 1.

Returning to the interaction Lagrangian (2.58) with the current (2.59) and using (2.115), it is clear that the beta-decay matrix element can now be approximately written as

$$\mathcal{M}^{(\beta)} = -\frac{G_F}{\sqrt{2}}\cos\vartheta_C[\bar{u}_p\gamma_\mu(f_1(0) - g_1(0)\gamma_5)u_n][\bar{u}_e\gamma^\mu(1-\gamma_5)v_\nu] \tag{2.116}$$

The measurements of the effective beta-decay constants that we have discussed earlier (see (1.117), (1.120) and Section 2.5) now clearly show that

$$f_1(0) \doteq 1 \tag{2.117}$$

with high accuracy, while $g_1(0) \doteq 1.27$. It should be emphasized that the result (2.117) does not represent merely a convenient normalization of the weak hadronic matrix element. In fact, it is a rather non-trivial statement: the coupling constant for the vector part of the weak current appearing in the Lagrangian at the quark level is essentially left unchanged when one passes to physical nucleons.[15] In other words, the strong interactions binding quarks in nucleons do not renormalize the vectorial weak coupling. On the other hand, $g_1(0)$ is seen to differ significantly from the axial-vector coupling appearing in the Lagrangian. The remarkable relation (2.117) is due to the conserving nature of the vectorial weak current (which lies in a common isospin multiplet with the electromagnetic current) and to the fact that neutron and proton are classified as members of an isospin doublet. Thus, we have another example of a "CVC relation" — it is basically of the same origin as those discussed in the preceding section in

[15]Note that the weak coupling strength appearing in the quark-level interaction Lagrangian manifests itself directly, e.g., in high-energy processes of deep-inelastic neutrino scattering, where we can "look inside" the target nucleons.

connection with pion beta decay. It should also be stressed that the result (2.117) is analogous to the equality of electric charges of, e.g., electron and proton: strong interactions forming the proton do not renormalize the "bare charge", i.e. the parameter appearing in the electromagnetic Lagrangian and associated with conserved current written in terms of elementary fields (charged leptons and quarks).

Although this topic goes slightly beyond the basic framework of our treatment, let us summarize here, for completeness and for reader's convenience, a set of CVC relations valid for the vector weak form factors appearing in (2.112). For this purpose, let us define first the electromagnetic formfactors of a nucleon (e.g., proton):

$$\langle p(p')|J_\mu^{em}(0)|p(p)\rangle = \text{N.f.} \times \bar{u}(p')\left[\gamma_\mu F_1^p(t) - i\frac{1}{2M}\sigma_{\mu\nu}q^\nu F_2^p(t)\right]u(p) \tag{2.118}$$

Note that the general form (2.118) follows from considerations analogous to those employed for weak current; we have discarded the third form factor in order to satisfy the current conservation. The relevant CVC relations can be written as

$$\begin{aligned} f_1^{pn}(t) &= F_1^p(t) - F_1^n(t) \\ f_2^{pn}(t) &= F_2^p(t) - F_2^n(t) \\ f_3^{pn}(t) &= 0 \end{aligned} \tag{2.119}$$

where the electromagnetic form factors of neutron are defined in complete analogy with the proton case. A derivation of these relations is outlined, e.g., in [FaR]. The normalization of nucleon electromagnetic form factors is

$$F_1^p(0) = 1, \quad F_2^p(0) = \mu_p \tag{2.120}$$

for the proton and

$$F_1^n(0) = 0, \quad F_2^n(0) = \mu_n \tag{2.121}$$

for the neutron, where $\mu_{p,n}$ are the corresponding magnetic moments (given in units of nuclear magneton, i.e. $\mu_p = 2.79$, $\mu_n = -1.91$). From (2.119), one thus first gets $f_1^{pn}(0) = 1$ (cf. (2.117)) and

$$f_2^{pn}(0) = \mu_p - \mu_n \tag{2.122}$$

The remarkable prediction (2.122) has been confirmed experimentally in the beta decays of B^{12} and N^{12} that we have mentioned earlier.

In closing this section, we should perhaps recapitulate briefly the circular path we have gone through in our theoretical description of the free neutron beta decay. In Chapter 1, we have started with an effective Lagrangian written directly in terms of nucleon fields and involving some unknown constants that have to be determined experimentally. The universal weak interaction theory of the 1960s, which crystallized from the wealth of empirical data as a masterpiece of theoretical insight, is certainly more elegant and includes only the Fermi constant (inferred from muon lifetime) and the Cabibbo angle. However, the other phenomenological parameters enter through the back door when physical matrix elements are considered: instead of constant parameters of the effective Lagrangian, we first get a set of form factors and the expression for the decay amplitude acquires a more general structure that can, in principle at least, be tested experimentally. At low energies, the form factors can be replaced by constants, the novel effects (as the weak magnetism, etc.) may be safely neglected and the result stemming originally from the nucleonic effective Lagrangian is thus recovered. Of course, some phenomenological parameters must be fixed experimentally anyway, so that for most practical purposes, both approaches are essentially equivalent.

2.9 $\mathcal{C}, \mathcal{P}$ and $\mathcal{CP}$

The last topic to be discussed in this chapter concerns discrete symmetries — the space reflection (parity) $\mathcal{P}$, charge conjugation $\mathcal{C}$ and their combination $\mathcal{CP}$. We will show that the provisional weak interaction Lagrangian (2.58) is invariant under the "combined parity" transformation $\mathcal{CP}$, though both $\mathcal{P}$ and $\mathcal{C}$ are violated maximally. On the other hand, there is a long-standing and well-established experimental evidence for tiny $\mathcal{CP}$-violating effects in the neutral kaon system [27]. The standard model (SM) of electroweak interactions that we will discuss in detail later on can incorporate $\mathcal{CP}$ violation in a rather natural way — it is quite remarkable that the occurrence of $\mathcal{CP}$-violating terms in the SM interaction Lagrangian is due to the existence of the third generation of quarks (b and t). Relevant experimental data are accumulating from measurements on B-mesons (i.e. those containing b-quarks) and the underlying theoretical picture should consequently be further clarified. All this makes the $\mathcal{CP}$ violation one of the most prominent open problems in modern particle physics. This present section serves as a prelude to our later discussion of this issue within the standard electroweak model.

The non-invariance of weak interactions under the space reflection $\mathcal{P}$ (parity violation) has already been described earlier in this chapter. As we know, the chiral $V - A$ structure of the current (2.59) corresponds to maximum parity violation — the $\mathcal{P}$-odd terms of the types VA descending from the product of currents enter the interaction Lagrangian with the same strength as the $\mathcal{P}$-even contributions of the type VV and AA. It is quite remarkable that the chiral structure of weak currents also leads to non-invariance of the Lagrangian under a discrete internal symmetry — the charge conjugation $\mathcal{C}$. To see this, let us work out the corresponding transformations of vector and axial-vector fermionic currents explicitly. Throughout our calculation, we will use the standard representation of gamma matrices. A Dirac spinor transforms under the charge conjugation as

$$\psi_c = C\bar{\psi}^T \tag{2.123}$$

where the superscript T denotes transposed matrix and C is defined by

$$C^{-1}\gamma_\mu C = -\gamma_\mu^T \tag{2.124}$$

(let us recall that the free-field Dirac equation is then invariant under (2.123)). It is well known that within the standard representation, the matrix C can be written as

$$C = i\gamma^2\gamma^0 \tag{2.125}$$

From the last expression, some useful relations follow immediately, in particular

$$\begin{aligned} C^{-1} = C^\dagger = C^T = -C, \\ \{C, \gamma_0\} = 0, \quad [C, \gamma_5] = 0 \end{aligned} \tag{2.126}$$

Now, it is easy to evaluate the Dirac conjugate of ψ_c. One obtains first

$$\bar{\psi}_c = (C\bar{\psi}^T)^\dagger\gamma_0 = (C\gamma_0\psi^*)^\dagger\gamma_0 = \psi^T\gamma_0 C^\dagger\gamma_0$$

and with the help of (2.126), the last result is easily recast as

$$\bar{\psi}_c = \psi^T C \tag{2.127}$$

From (2.123) and (2.127), one obtains readily the transformation of a fermionic current. For a vector (made generally of two different Dirac fields), one has

$$\bar{\psi}_{1c}\gamma_\mu\psi_{2c} = \psi_1^T C\gamma_\mu C\bar{\psi}_2^T = -\psi_1^T C^{-1}\gamma_\mu C\bar{\psi}_2^T = \psi_1^T\gamma_\mu^T\bar{\psi}_2^T \tag{2.128}$$

For classical fields, one may write

$$\psi_1^T\gamma_\mu^T\bar{\psi}_2^T = (\bar{\psi}_2\gamma_\mu\psi_1)^T \tag{2.129}$$

and the last expression is, of course, equal to $\bar{\psi}_2\gamma_\mu\psi_1$. We thus arrive at the following result:

$$\text{for classical fields,} \quad \bar{\psi}_1\gamma_\mu\psi_2 \xrightarrow{\mathcal{C}} \bar{\psi}_2\gamma_\mu\psi_1 \tag{2.130}$$

For quantized Dirac fields, one has to take into account their anticommuting nature (if $\psi_1 = \psi_2$, we assume that the current is normal-ordered) and the relation (2.129) then obviously acquires a negative sign. Thus, we have

$$\text{for quantum fields:} \quad \bar{\psi}_1\gamma_\mu\psi_2 \xrightarrow{\mathcal{C}} -\bar{\psi}_2\gamma_\mu\psi_1 \tag{2.131}$$

(in particular, for $\psi_1 = \psi_2$, this leads to the desirable result that electromagnetic current changes its sign upon charge conjugation). Similarly, for the axial-vector current, one first gets

$$\begin{aligned}\bar{\psi}_{1c}\gamma_\mu\gamma_5\psi_{2c} &= \psi_1^T C\gamma_\mu\gamma_5 C\bar{\psi}_2^T = -\psi_1^T C^{-1}\gamma_\mu C C^{-1}\gamma_5 C\bar{\psi}_2^T \\ &= \psi_1^T\gamma_\mu^T C^{-1}\gamma_5 C\bar{\psi}_2^T\end{aligned} \tag{2.132}$$

Using (2.124), it is easy to find that $C^{-1}\gamma_5 C = \gamma_5^T$ (note also that $\gamma_5^T = \gamma_5$ in the standard representation) and we thus have the following result:

$$\text{for classical fields:} \quad \bar{\psi}_1\gamma_\mu\gamma_5\psi_2 \xrightarrow{\mathcal{C}} -\bar{\psi}_2\gamma_\mu\gamma_5\psi_1 \tag{2.133}$$

For quantum fields, there is an extra minus sign due to anticommutators, so that one has

$$\text{for quantum fields:} \quad \bar{\psi}_1\gamma_\mu\gamma_5\psi_2 \xrightarrow{\mathcal{C}} \bar{\psi}_2\gamma_\mu\gamma_5\psi_1 \tag{2.134}$$

Thus, we see that the vector and axial-vector current have an opposite $\mathcal{C}$-parity and this in turn means that the VA term in the interaction Lagrangian is $\mathcal{C}$-odd while the AA or VV terms are $\mathcal{C}$-even. Since the overall strength of all these terms is the same, one can say that the Lagrangian (2.58) exhibits maximum $\mathcal{C}$-violation (similarly to $\mathcal{P}$); moreover, one may also observe that the $\mathcal{C}$ and $\mathcal{P}$ violation have the same algebraic origin in the chiral structure of the weak current.

Our next goal is finding the $\mathcal{CP}$ transformation law for currents. For this purpose and for reader's convenience, let us first summarize here the relevant formulae for the space inversion $\mathcal{P}$. Starting with the well-known transformation law for Dirac spinors,

$$\psi_P(x) = \gamma_0 \psi(\tilde{x}) \tag{2.135}$$

where $\tilde{x} = (x^0, -\vec{x}\,)$, it is straightforward to obtain

$$\bar{\psi}_1(x)\gamma_\mu \psi_2(x) \xrightarrow{\mathcal{P}} \bar{\psi}_1(\tilde{x})\gamma^\mu \psi_2(\tilde{x}) \tag{2.136}$$

for the vector current and

$$\bar{\psi}_1(x)\gamma_\mu \gamma_5 \psi_2(x) \xrightarrow{\mathcal{P}} -\bar{\psi}_1(\tilde{x})\gamma^\mu \gamma_5 \psi_2(\tilde{x}) \tag{2.137}$$

for the axial-vector current. Note that these results follow from simple properties of the Dirac gamma matrices and hold for both classical and quantum fields.

Putting now together (2.130), (2.133), (2.136) and (2.137), it is clear that for the combined transformation $\mathcal{CP}$, one has

$$\text{for classical fields:} \quad \begin{cases} \bar{\psi}_1(x)\gamma_\mu \psi_2(x) \xrightarrow{\mathcal{CP}} \bar{\psi}_2(\tilde{x})\gamma^\mu \psi_1(\tilde{x}) \\ \bar{\psi}_1(x)\gamma_\mu \gamma_5 \psi_2(x) \xrightarrow{\mathcal{CP}} \bar{\psi}_2(\tilde{x})\gamma^\mu \gamma_5 \psi_1(\tilde{x}) \end{cases} \tag{2.138}$$

For quantum fields, there is an extra overall minus sign descending from the $\mathcal{C}$ transformation, so that

$$\text{for quantum fields:} \quad \begin{cases} \bar{\psi}_1(x)\gamma_\mu \psi_2(x) \xrightarrow{\mathcal{CP}} -\bar{\psi}_2(\tilde{x})\gamma^\mu \psi_1(\tilde{x}) \\ \bar{\psi}_1(x)\gamma_\mu \gamma_5 \psi_2(x) \xrightarrow{\mathcal{CP}} -\bar{\psi}_2(\tilde{x})\gamma^\mu \gamma_5 \psi_1(\tilde{x}) \end{cases} \tag{2.139}$$

The $\mathcal{CP}$-invariance of the weak interaction Lagrangian (2.58) should now be clear. Indeed, using, e.g., (2.138), the transformation of the current (2.59) can be written as

$$J^\mu(x) \xrightarrow{\mathcal{CP}} J^\dagger_\mu(\tilde{x}) \tag{2.140}$$

since $(\bar{\psi}_1 \Gamma_\mu \psi_2)^\dagger = \bar{\psi}_2 \Gamma_\mu \psi_1$ for Γ_μ equal to γ_μ or $\gamma_\mu \gamma_5$. However, the Lagrangian density is given by the product $J^\mu J^\dagger_\mu$, where, of course, raising of the Lorentz index in (2.140) becomes irrelevant; $\mathscr{L}_{\text{int}}$ is thus scalar under $\mathcal{CP}$, i.e.

$$\mathscr{L}_{\text{int}}(x) \xrightarrow{\mathcal{CP}} \mathscr{L}_{\text{int}}(\tilde{x}) \tag{2.141}$$

Note that the restriction to classical fields in the present context has not been important — for quantum fields, there is a twofold sign change in the transformed currents and (2.141) is recovered anyway. Another remark is in order here. It is useful to realize that for the relation (2.140) to be valid, it is essential that the current J_μ involves, apart from Dirac matrices, only real coefficients like $\cos\theta_C$ and $\sin\theta_C$. This, however, need not be the case in the world built upon three generations of quarks. It turns out that the simple Cabibbo mixing ("rotation") is then naturally replaced by elements of a 3×3 unitary matrix that may be imaginary and give rise to $\mathcal{CP}$-violating terms in the interaction Lagrangian. As we noted earlier in this section, a natural framework for such a discussion is provided by the standard electroweak model and we will have more to say about this in Chapter 7.

Problems

2.1 Derive the result (2.31) by means of a straightforward integration of the differential decay rate (2.22) (i.e. without using the "tensor trick" employed in Section 2.3).

2.2 Within the $V - A$ theory of weak interactions, calculate the degree of polarization of electrons in the decay of an unpolarized muon at rest.

2.3 Calculate the angular distribution of electrons in the decay of a polarized muon at rest. For simplicity, neglect the electron mass throughout the calculation (for an instructive discussion of this problem, see [BjD], Chapter 10 therein).

2.4 Calculate the probability of production of left-handed electron in the pion decay $\pi^- \to e^-\bar{\nu}_e$ within $V - A$ theory, assuming that the neutrino is massless. How would the result change if the lepton weak current had the form $V - \lambda A$, with λ being an arbitrary real parameter? How is the result obtained within the $V - A$ theory changed, when the neutrino has a non-zero mass?

2.5 Consider scattering processes $\bar{\nu}_e + e^- \to \bar{\nu}_\mu + \mu^-$ and $\nu_\mu + e^- \to \mu^- + \nu_e$ in a high-energy domain, i.e. for $E_{\text{c.m.}} \gg m_\mu$ (thus, lepton masses can be neglected). Suppose that the weak charged current has the structure $V - aA$ for electron-type leptons and $V - bA$ for muon-type leptons, where a and b are essentially arbitrary real parameters. Show that the ratio

$$R = \frac{\sigma(\nu_\mu e^- \to \mu^- \nu_e)}{\sigma(\bar{\nu}_e e^- \to \bar{\nu}_\mu \mu^-)}$$

satisfies inequality $1 \leq R \leq 3$.

2.6 Within the $V - A$ theory of weak interactions, calculate the cross-section of the process $e^+e^- \to \nu_e\bar{\nu}_e$. Calculate also (at the tree level) the QED cross-section $\sigma(e^+e^- \to \mu^+\mu^-)$. In both cases, assume that the collision energy is sufficiently large ($E_{\text{c.m.}} \gg m_\mu$) and neglect lepton masses. Evaluate the ratio $\sigma(e^+e^- \to \nu_e\bar{\nu}_e)/\sigma(e^+e^- \to \mu^+\mu^-)$ as a function of energy in the considered domain. For which energy the two cross-sections become comparable?

2.7 Employing the "$G^2\Delta^5$ rule", estimate the branching ratios for

$$\Sigma^- \to n + e^- + \bar{\nu}_e$$

$$K^- \to \pi^0 + e^- + \bar{\nu}_e$$

$$\Sigma^- \to \Lambda + e^- + \bar{\nu}_e$$

$$\Xi^- \to \Sigma^0 + e^- + \bar{\nu}_e$$

$$\Omega^- \to \Xi^0 + e^- + \bar{\nu}_e.$$

2.8 The decay amplitude for $\Sigma^- \to \Lambda e^- \bar{\nu}_e$ can be written approximately as

$$\mathcal{M}_{fi} = \frac{G_F}{\sqrt{2}} \cos\theta_C \sqrt{\frac{2}{3}}\, a \left[\bar{u}(p)\gamma_\mu\gamma_5 u(P)\right] \left[\bar{u}(k)\gamma^\mu(1-\gamma_5)v(k')\right]$$

where P, p, k, k' denote consecutively the four-momenta of $\Sigma^-, \Lambda, e^-, \bar{\nu}_e$ and the constant a reflects non-perturbative nature of the hadronic matrix element (numerically, $a \doteq 0.81$); the remaining symbols have a standard meaning. Calculate the decay width for the considered process as a function of the maximum electron energy Δ and of the other parameters. Compute also the corresponding branching ratio and compare the result with the estimate obtained in solving Problem 2.7. Throughout the calculation, neglect the Λ momentum wherever it is possible and set also $m_e = 0$ (why is such an approximation good?).

Remark: It is amusing to note that the above amplitude corresponds to a pure Gamow–Teller transition for *hyperons* (only the axial-vector term contributes to the hadronic matrix element). Further details concerning the Cabibbo theory of semileptonic decays of baryons can be found in [CoB].

2.9 Consider the beta decay of charged kaon, $K^- \to \pi^0 e^- \bar{\nu}_e$. Compute the electron energy spectrum and partial decay width.

Hint: For the necessary current-algebra background, see [FaR].

2.10 Consider the decay $\pi^- \to e^- + \bar{\nu}$ involving the neutrino with a non-zero mass. Let us denote as $|\mathcal{M}_L|^2$ the squared matrix element for the production of left-handed (i.e. negative-helicity) electron, with the neutrino spin states summed over. The analogous quantity for the production of right-handed electron is denoted as $|\mathcal{M}_R|^2$ and for the full decay matrix element squared, the usual symbol $\overline{|\mathcal{M}|^2}$ is employed. Show that

$$\frac{|\mathcal{M}_L|^2}{\overline{|\mathcal{M}|^2}} = \frac{(a-b)^2}{2(a^2+b^2)}, \qquad \frac{|\mathcal{M}_R|^2}{\overline{|\mathcal{M}|^2}} = \frac{(a+b)^2}{2(a^2+b^2)}$$

where

$$a = (m_e - m_\nu)\sqrt{m_\pi^2 - (m_e + m_\nu)^2}$$
$$b = (m_e + m_\nu)\sqrt{m_\pi^2 - (m_e - m_\nu)^2}$$

For a consistency check, one may note immediately that $a = b$ if $m_\nu = 0$; the result $|\mathcal{M}_L|^2 = 0$, anticipated *a priori* in the case of massless neutrino, is thus recovered. Further, as a simple algebraic exercise, show that expanding the above result for $|\mathcal{M}_L|^2$ in powers of m_ν, one obtains

$$\frac{|\mathcal{M}_L|^2}{\overline{|\mathcal{M}|^2}} = \frac{m_\nu^2}{m_e^2}\left(\frac{m_\pi^2}{m_\pi^2 - m_e^2}\right)^2 (1 + \mathcal{O}(m_\nu))$$

Such a result demonstrates clearly that the possibility of producing a left-handed electron in the considered π_{e2} decay process is due to the distinction between helicity and chirality for massive neutrino.

Chapter 3

Intermediate Vector Boson *W*

3.1 Difficulties of Fermi-type theory

The weak interaction theory built according to Fermi's paradigm, discussed at length in the preceding two chapters, was certainly one of the highlights of the particle physics in 1960s. The simple and elegant Feynman–Gell-Mann interaction Lagrangian (2.58) was capable of describing a lot of experimental data available then and — as we have demonstrated in several examples — it also had a considerable predictive power. Since its early days, the theory was successfully tested for a variety of decay processes and, with less accuracy, also for some particular scattering processes at low energies (the famous reaction $\bar{\nu} + p \to n + e^+$ used for the first direct neutrino detection [28] can serve as one such example). In any case, the relevant theoretical predictions were verified at that time within a rather limited kinematical region, corresponding to low energy and low momentum transfer — certainly less than 1 GeV or so. Having established an effective weak interaction theory, phenomenologically successful at low energies, we should scrutinize its behaviour at higher energies as well. To this end, one must naturally consider scattering processes, as these can be studied (at least in principle) at an arbitrarily high collision energy. We shall see below that the usual Feynman-diagram methods become rather problematic for sufficiently high energies and it will also be immediately clear that such difficulties are common to all Fermi-type models, i.e. to those involving a direct interaction of four fermionic fields.

The problem we have in mind can be demonstrated on an example of any binary reaction proceeding in lowest order through the interaction Lagrangian (2.58). For definiteness, we may consider, e.g., the neutrino–electron elastic scattering. In the lowest order of perturbation expansion, this process is described by the simple Feynman diagram shown in Fig. 3.1.

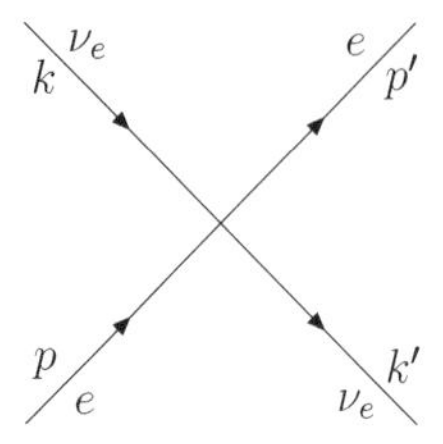

Figure 3.1. Tree-level diagram for ν_e–e elastic scattering within a Fermi-type weak interaction theory.

Before carrying out a technical calculation of the corresponding cross-section, it will be instructive to estimate its high-energy behaviour on dimensional grounds. According to the arguments given in Chapter 1, the Fermi constant G_F has dimension of inverse squared mass (see (1.15)). In the first order of perturbation expansion, the relevant matrix element is proportional to G_F; consequently, the corresponding cross-section must contain the factor G_F^2, which is of dimension M^{-4}. On the other hand, the dimension of a cross-section is (length)^2, i.e. M^{-2}. Thus, one needs an additional factor of energy (mass) squared to balance the dimensionality of G_F^2 and get the quantity with right dimension of a cross-section. At high energies ($E \gg m_e$), i.e. in the ultrarelativistic limit, the effects of electron mass can be neglected (such a guess is indeed confirmed by an explicit calculation — see below) and the only quantity with dimension of mass that remains in the game is the collision energy. Since the cross-section is Lorentz invariant, it can only depend on the Mandelstam invariant $s = E_{\text{c.m.}}^2$, where $E_{\text{c.m.}}$ is the full centre-of-mass energy. Thus, we arrive at the following "rule of thumb" for the high-energy behaviour of the neutrino-electron cross-section:

$$\sigma \simeq G_F^2 E_{\text{c.m.}}^2 \tag{3.1}$$

Similarly, one can estimate the behaviour of the scattering amplitude. Within our normalization conventions, the matrix element $\mathcal{M}_{fi}$ is dimensionless for any binary process (see Appendix B), and in the first order of perturbation theory, it is proportional to G_F. Thus, barring the irrelevant mass dependence, one can expect that

$$\mathcal{M}_{fi} \simeq G_F E_{\text{c.m.}}^2 f(\Omega) \tag{3.2}$$

in the high-energy limit, with $f(\Omega)$ being a (dimensionless) function of scattering angles. Obviously, it is the dimensionality of the Fermi constant that plays a crucial role in preceding considerations. Consequently, the "scaling laws" (3.1) and (3.2) should be valid for any binary reaction involving four elementary fermions and within any particular model of the Fermi type. We will explain shortly what is wrong with such a high-energy behaviour, but now let us verify — just to be sure — the results of our simple dimensional analysis by means of an explicit calculation.

The scattering amplitude corresponding to Fig. 3.1 can be written as

$$\mathcal{M}_{fi} = -\frac{G_F}{\sqrt{2}}[\bar{u}(p')\gamma_\mu(1-\gamma_5)u(k)][\bar{u}(k')\gamma^\mu(1-\gamma_5)u(p)] \tag{3.3}$$

where we have suppressed, for the sake of brevity, the spin labels of the Dirac spinors. For the spin-averaged matrix element squared, one then gets, by means of the usual trace techniques,

$$\begin{aligned}\overline{|\mathcal{M}_{fi}|^2} &= \frac{1}{2}\sum_{spins}|\mathcal{M}_{fi}|^2 \\ &= G_F^2\mathrm{Tr}[\not{p}'\gamma^\rho\not{k}\gamma^\sigma(1-\gamma_5)]\cdot\mathrm{Tr}[\not{k}'\gamma_\rho\not{p}\gamma_\sigma(1-\gamma_5)]\end{aligned} \tag{3.4}$$

where we have used $(1-\gamma_5)^2 = 2(1-\gamma_5)$ and other familiar properties of the gamma matrices (needless to say, we have set $m_\nu = 0$ from the very beginning). The spinor traces in (3.4) can be evaluated most economically with the help of the formulae (A.51). One thus gets immediately

$$\overline{|\mathcal{M}_{fi}|^2} = 64G_F^2(k\cdot p)(k'\cdot p') \tag{3.5}$$

and this can be further recast in terms of the Mandelstam variable $s = (k+p)^2$ as

$$\overline{|\mathcal{M}_{fi}|^2} = 16G_F^2(s-m_e^2)^2 \tag{3.6}$$

For the differential cross-section (angular distribution in the c.m. system), one then has

$$\frac{d\sigma^{(\nu e)}}{d\Omega_{\text{c.m.}}} = \frac{G_F^2}{4\pi^2}\frac{(s-m_e^2)^2}{s} \tag{3.7}$$

The angular integration of the last expression is trivial and yields the result

$$\sigma^{(\nu e)} = \frac{G_F^2}{\pi}\frac{(s-m_e^2)^2}{s} \tag{3.8}$$

which makes it clear that the effect of electron mass can indeed be neglected in the high-energy limit. For $s \gg m_e^2$, thus (3.8) becomes simply

$$\sigma^{(\nu e)}|_{s\gg m_e^2} \approx \frac{G_F^2}{\pi}s \tag{3.9}$$

which confirms our previous estimate (3.1) made on simple dimensional grounds.[1] Other examples of this kind (such as the $\bar{\nu} - e$ scattering and the annihilation process $e^+e^- \to \nu\bar{\nu}$) can be worked out easily, but they will not be immediately necessary here and the corresponding calculation can be left as an instructive exercise to the interested reader (some technical details can also be found in [Hor]).

Now, let us explain what is wrong with the high-energy behaviour shown in (3.1) and (3.2), respectively. To put it briefly, such a power-like growth of a scattering amplitude leads to rapid violation of the S-matrix unitarity. This statement may be understood rather easily at an intuitive level: the absolute value of an element of a unitary matrix is bounded from above (it must be less than unity) and one thus naturally expects that the scattering amplitude $\mathcal{M}_{fi}$ should not rise indefinitely with energy. For an explicit discussion of such a "unitarity bound", one has to invoke the technique of partial-wave expansion (some basic formulae can be found in Appendix B). Using (3.6), we note that in the considered case the relevant scattering amplitude $\mathcal{M}_{fi}$ does not depend on the scattering angle[2] and this in turn means that the whole partial-wave expansion is reduced to the lowest term carrying the angular momentum $j = 0$. The amplitude of the partial wave with $j = 0$ can then be easily inferred from (3.6); for its absolute value, one gets

$$|\mathcal{M}^{(0)}(s)| = \frac{1}{2\pi\sqrt{2}} G_F s \tag{3.10}$$

In the high-energy limit, an $\mathcal{M}^{(j)}$ can be written as $\mathcal{M}^{(j)} = (S^{(j)} - 1)/2i$, where the $S^{(j)}$ is an element of a finite-dimensional unitary matrix (the S-matrix restricted to the subspace characterized by a given value of j)

[1] The reader may wonder why we emphasize the verification of an intuitively plausible claim that the electron mass effects in the considered cross-section can be neglected at high energy. The point is that in some other cases (within other field-theory models), one can get results that appear, in this sense, rather counter-intuitive. In particular, as we shall see later in this chapter, for processes involving a physical massive charged particle with spin 1 (the vector boson W), the corresponding mass effects persist even at high energies — the limit of taking the vector boson mass to zero becomes singular.

[2] Obviously, only the negative-helicity states of electron and neutrino contribute in (3.6) in the high-energy limit and the considered scattering amplitudes should consequently be labelled, e.g., as $\mathcal{M}_{----}$. In what follows, we will usually suppress the helicity indices for the sake of brevity.

and this obviously yields the bound

$$|\mathcal{M}^{(j)}(s)| \leq 1 \tag{3.11}$$

Applying now the constraint (3.11) to our result (3.10), it becomes clear that only for energies within the range $s \leq 2\pi\sqrt{2}G_F^{-1}$, i.e. for

$$E_{\text{c.m.}} \leq \left(\frac{2\pi\sqrt{2}}{G_F}\right)^{\frac{1}{2}} \doteq 870 \text{ GeV} \tag{3.12}$$

one avoids a manifest violation of unitarity — in other words, outside the domain (3.12), our calculation cannot be reliable. A restriction of the type (3.12) is usually called "unitarity bound". The unitarity violation at high energy within weak interaction theory of Fermi type has been first emphasized in the early 1960s (see [29]). The power-law rise of the cross-section (3.9) has often been referred to as the weak interaction "becoming strong" at high energies.

Several remarks are in order here. The numerical value of the "critical energy" shown in (3.12) is in fact rather high; one can hardly expect that the $\nu - e$ collisions would be studied experimentally at such energies in foreseeable future (note that $E_{\text{c.m.}} = 870$ GeV corresponds to the incident neutrino energy of about 7.5×10^5 TeV in the electron rest system!). However, it is not the particular value of the unitarity bound that really matters. The important point is that the considered scattering amplitude *grows as a positive power of energy* — this is precisely what we have in mind when saying that there is a "rapid violation" of unitarity. This feature distinguishes the Fermi-type theory of weak interactions from, e.g., quantum electrodynamics of electrons and photons (spinor QED), where the high-energy behaviour of lowest-order Feynman diagrams is much softer (so that a possible conflict with unitarity is deferred to the realm of astronomically high energies). Thus, at least from a technical point of view, the weak interaction theory of Fermi type seems to be inferior to some other field theory models that work successfully in different areas of particle physics. On the other hand, one may object that the violation of unitarity discussed here is not of fundamental nature: since the interaction Hamiltonian is hermitian, the *exact* S-matrix must be unitary and the offending behaviour (3.10) is just an artefact of the lowest-order perturbation theory. This argument is perfectly true, but rather academic. Indeed, nobody can solve exactly a quantum field theory model of considered type to see that the full scattering amplitude behaves decently. One may, e.g., try to calculate higher orders of perturbation theory, but then one obviously runs into more severe difficulties than in the basic approximation — as there are higher powers of the Fermi constant G_F, there must also be higher powers of the energy in order to get a dimensionless scattering amplitude. Thus, although there is no fundamental inconsistency in the Fermi-type models, their practical applicability is limited, as it is notoriously difficult to go beyond the framework of perturbation theory.

There is another important aspect of perturbation expansion that should be mentioned separately. In our discussion of the neutrino–electron elastic scattering, we have used the lowest-order approximation, which corresponds to the simple Feynman graph shown in Fig. 3.1. In the standard terminology of perturbative quantum field theory, the diagrams of such a type are called **tree diagrams** as they do not contain closed loops of internal lines. In higher orders of perturbation expansion, closed-loop diagrams necessarily appear; some examples are depicted in Fig. 3.2. Contributions of the closed loops are expressed in terms of integrals over the four-momenta of "virtual particles" associated with the internal lines. Unfortunately, such integrals usually diverge in the ultraviolet region (i.e. in the neighbourhood of infinity) — this phenomenon is in fact typical for most of the quantum field theory models. These ultraviolet (UV) divergences can be tamed successfully within some QFT models by means of the renormalization procedure, which essentially consists in a redefinition of a certain (finite) number of parameters of the model. Such a procedure has been first formulated in the late 1940s for quantum electrodynamics, where one is then able to calculate explicitly some finite higher-order "radiative" corrections to observable quantities (which are tiny but measurable). A discussion of the renormalization techniques can be found in any textbook on quantum field theory (for a concise summary, see, e.g., [ChL]). As for the QED, this became one of the most precise physical

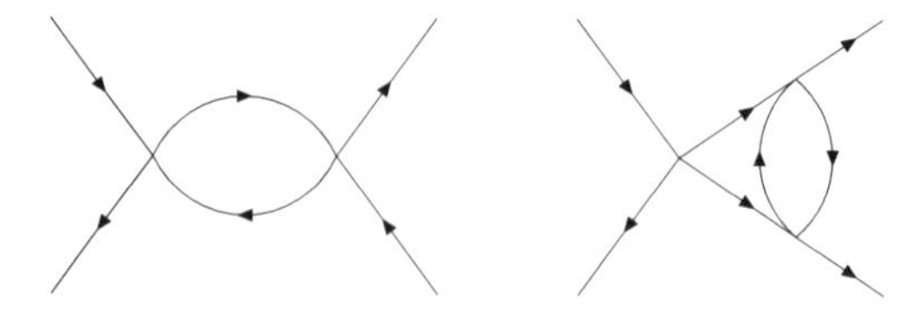

Figure 3.2. Examples of closed-loop Feynman graphs representing higher-order contributions to the ν–e scattering.

theories ever conceived — for a relatively recent overview of the successes of QED, see, e.g., [30]. However, the Fermi-type weak interaction theory is *not* renormalizable in such a manner. A detailed analysis shows that there are infinitely many types of UV divergences that would require introducing an infinite number of parameters in the interaction Lagrangian — needless to say, the theory thus loses considerably its predictive power. The point is that according to a standard "power counting" for Feynman graphs, a QFT model can only be renormalizable if its interaction Lagrangian incorporates terms with dimension less than or equal to 4.[3] The dimension of the four-fermion interaction is obviously equal to *six* and this becomes fatal for renormalizability of any Fermi-type theory of weak interactions. We will not go into further technical details here and rather refer the interested reader to standard textbooks on quantum field theory (see, e.g., [ItZ]).

Finally, let us emphasize what is perhaps the most interesting moment of the considered situation. It turns out that — for a general QFT model — the non-renormalizability of UV divergences in higher orders of perturbation expansion is closely connected with the character of high-energy behaviour of scattering amplitudes at lowest order (i.e. at the tree level): **the power-like growth of a tree-level scattering amplitude implies non-renormalizability in higher orders.** This statement is perhaps more useful in the reverse direction: **absence of a power-like growth of tree-level scattering amplitudes with energy is a necessary condition for perturbative renormalizability at higher orders of perturbation expansion.** In view of its relation to unitarity, the absence of a power-like high-energy rise of scattering amplitudes is usually termed technically as "tree unitarity" (cf., e.g., [31]). Let us note that this remarkable connection of two different aspects of perturbative QFT — the tree unitarity and UV renormalizability — has never been proved quite rigorously, but still it seems to be valid beyond any reasonable doubt. The point is that there is no known exception from this rule and, beside that, there is a rather plausible hand-waving argument in its favour based on the technique of dispersion relations (for a more detailed discussion, the interested reader is referred, e.g., to [Hor] and to the relevant

[3]The dimension we have in mind here does not include the corresponding coupling constant and is to be understood as the pertinent power of a mass; thus, fermionic and bosonic fields have dimensions 3/2 and 1, respectively, and a derivative carries dimension 1 (inverse length has a dimension of mass in the natural system of units). The dimension of a given term in (polynomial) interaction Lagrangian is then the sum of dimensions of all fermion and boson fields and derivatives occurring therein.

literature quoted therein, in particular [31]). Throughout this chapter, we adopt the criterion of "tree-level unitarity" as a simple and practical necessary condition for perturbative renormalizability and it will often serve as a subsidiary guiding principle in our road toward the unified theory of weak and electromagnetic interactions.

3.2 The case for intermediate vector boson

Having described the "splendeurs et misères" of the Fermi-type weak interaction theory, one should now seek a viable alternative, which would lead to a more satisfactory high-energy behaviour of scattering amplitudes already in lowest approximation. To this end, it is important to realize that the source of all difficulties arising within a Fermi-type model is the *dimensionality* of the relevant coupling constant G_F. Indeed, as we have seen, this leads to the quadratic growth of tree-level scattering amplitudes with c.m. energy and is also obviously related to the fact that the dimension of any four-fermion interaction is equal to six. Formally, one can get rid of the dimensionful coupling constant if the original "current × current" interaction is replaced by a coupling of the weak current (2.59) to a vector field

$$\mathscr{L}_{\text{int}}^{(w)} = \frac{g}{2\sqrt{2}}(J^{\mu}W_{\mu}^{+} + J^{\mu\dagger}W_{\mu}^{-}) \tag{3.13}$$

Obviously, the new coupling constant g is dimensionless in analogy with spinor electrodynamics (note that dim $J^{\mu} = 3$ and dim $W_{\mu} = 1$). The numerical factor $1/2\sqrt{2}$ in (3.13) is purely conventional; its origin will become clear in the context of the gauge theory of weak interactions. The field W_{μ}^{+} must be complex (non-hermitian) as it is coupled to the charged current; of course, we use a natural notation $W_{\mu}^{-} = (W_{\mu}^{+})^{\dagger}$. The corresponding quanta (vector bosons $W^{\pm}$) are spin-1 particles carrying electric charge ± 1. Taking into account the structure of the current J^{μ}, it is not difficult to realize that the field W_{μ}^{+} must contain annihilation operator for the W^{+} boson and creation operator for the W^{-} if the charge conservation is to be maintained in (3.13). In the theory described by (3.13), the vector field W_{μ} mediates weak interactions of fermions and the particle W^{+} or W^{-} is therefore usually called **intermediate vector boson** (IVB). Note that in the first order of perturbation expansion, the Lagrangian (3.13) gives rise to the two-fermion decays of the W boson (for example, the first term produces $W^{+} \to e^{+} + \nu_e$, while its hermitian conjugate leads to $W^{-} \to e^{-} + \bar{\nu}_e$).

Some additional remarks are perhaps in order here. It should be clear that the intermediary of an interaction between fermion pairs must be a boson — this is an obvious general consequence of angular momentum (spin) conservation. Historically, theorists contemplated the idea of an intermediate weak boson (in analogy with the description of strong and electromagnetic interactions) since the late 1930s, i.e. long before the generic technical flaws of four-fermion models have been appreciated. As we know now, it took more than 20 years to clarify that such a hypothetical particle must carry *spin one* (this, of course, was tantamount to establishing the dominance of *vector and axial-vector currents* in weak interactions). Since the early 1960s, the IVB concept has been taken quite seriously, and over the years, it was discussed in numerous theoretical papers. At the same time, direct experimental searches have shown soon that if a W boson exists, its mass must be larger than, e.g., 1 GeV, a typical hadronic mass. Still further 20 years were then necessary to prove its real existence (for a rather detailed survey of the IVB history and discovery see, e.g., [Wat]).

Coming back to the IVB interaction Lagrangian (3.13), we now have to find out whether such a model can indeed reproduce the successes of the Fermi-type theory at low energies and whether it is able to remedy its maladies in the high-energy limit. In order to examine the correspondence between the two versions of weak interaction theory in the low-energy limit, let us consider a particular decay process involving four fermions, e.g., muon decay. Within the IVB model (3.13), one needs at least one W exchange to connect the lepton pairs of muon and electron type — in other words, the lowest approximation in which such a process can appear is the second order of perturbation expansion. The corresponding tree diagram is shown in Fig. 3.3 along with its counterpart arising within the Fermi-type theory.

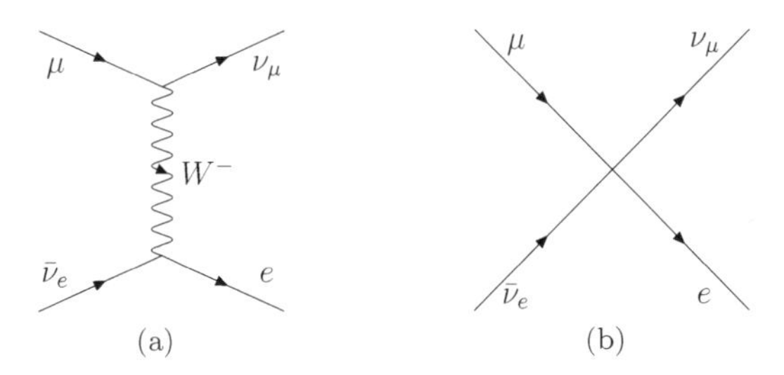

Figure 3.3. Tree-level diagrams for the muon decay: (a) within the W boson model (b) within a Fermi-type model.

Note that one must assume *a priori* that the W boson is massive since the weak interaction is known to be of a (very) short range. Thus, the W exchange in Fig. 3.3(a) is described by the vector boson propagator with $m_W \neq 0$ and the decay matrix element can then be written as

$$i\mathcal{M}_a = i^3 \left(\frac{g}{2\sqrt{2}}\right)^2 [\bar{u}(k)\gamma_\rho(1-\gamma_5)u(P)][\bar{u}(p)\gamma_\sigma(1-\gamma_5)v(k')]$$
$$\times \frac{-g^{\rho\sigma} + m_W^{-2} q^\rho q^\sigma}{q^2 - m_W^2} \tag{3.14}$$

(for a concise summary of basic properties of the massive vector field see Appendix D). On the other hand, for Fig. 3.3(b), we have

$$\mathcal{M}_b = -\frac{G_F}{\sqrt{2}}[\bar{u}(k)\gamma_\rho(1-\gamma_5)u(P)][\bar{u}(p)\gamma^\rho(1-\gamma_5)v(k')] \tag{3.15}$$

Now, how can one reduce — at least approximately — the form (3.14) to (3.15)? In fact, this can be done quite easily. First of all, one has to realize that the kinematical limits for the four-momentum of the virtual W boson are given by

$$m_e^2 \leq q^2 \leq m_\mu^2 \tag{3.16}$$

(proving (3.16) is left to the reader as a simple exercise). Then, taking $m_W^2 \gg m_\mu^2$ (as we have noted earlier, such a bound for the W mass has been established long before its discovery), one can safely neglect the q-dependence in the denominator of the propagator in (3.14). Further, it is easy to see that the second term in the numerator becomes in fact proportional to $m_e m_\mu / m_W^2$; indeed, using the four-momentum conservation and equations of motion for the u and v spinors, one readily gets

$$q^\rho q^\sigma [\bar{u}(k)\gamma_\rho(1-\gamma_5)u(P)][\bar{u}(p)\gamma_\sigma(1-\gamma_5)v(k')]$$
$$= [\bar{u}(k)(\not{P} - \not{k})(1-\gamma_5)u(P)][\bar{u}(p)(\not{p} + \not{k}')(1-\gamma_5)v(k')]$$
$$= m_e m_\mu [\bar{u}(k)(1+\gamma_5)u(P)][\bar{u}(p)(1-\gamma_5)v(k')] \tag{3.17}$$

Thus, the effect of the $q^\rho q^\sigma$ term in the W propagator can be reliably neglected as well. Putting all this together, we see that the matrix element

(3.14) is approximately equal to

$$\mathcal{M}_a \approx -\frac{g^2}{8m_W^2}[\bar{u}(k)\gamma_\rho(1-\gamma_5)u(P)][\bar{u}(p)\gamma^\rho(1-\gamma_5)v(k')] \tag{3.18}$$

which is indeed of the form (3.15). Matching the two expressions, one gets a condition for the parameters of the IVB model, namely

$$\frac{G_F}{\sqrt{2}} = \frac{g^2}{8m_W^2} \tag{3.19}$$

One should also note that the origin of the minus sign in the Fermi-type Lagrangian (2.58) becomes transparent through our calculation: such a convention is necessary for the correspondence relation (3.19) to be valid with a positive value of the G_F.

Thus, we have shown that for the considered process, the IVB model (3.13) leads to the same result as the original Fermi-type theory (up to corrections of the relative order $O(m_\mu^2/m_W^2)$ or less), provided that parameters of the IVB Lagrangian satisfy the relation (3.19). In fact, our reasoning makes it clear that such an equivalence should hold for any process involving four light fermions, whenever the relevant momentum transfer (energy) is small in comparison with the W boson mass. In any particular example of that kind, the steps described above can be repeated and one may consequently ignore all momentum dependence in the IVB propagator, which is thereby effectively reduced to a constant with the dimension of $(\text{mass})^{-2}$; a Fermi-type matrix element thus emerges as a low-energy approximation to the original expression. A generic structure of the relation (3.19) is also transparent: the IVB model at second order (g^2) and at low energy (m_W^{-2} replacing the propagator) is equivalent to a corresponding Fermi-type model in the first order (G_F). Our preceding considerations can now be concisely summarized as follows. If (3.19) is valid, then predictions of the IVB model (3.13) and those of the current–current model (2.58) are practically indistinguishable for energies and momentum transfers much smaller than the W boson mass; the four-fermion Lagrangian (2.58) thus represents a low-energy effective theory corresponding to the underlying IVB model (3.13).

Now that we have made sure of the right low-energy properties of the IVB model, let us investigate its behaviour in the high-energy limit. To this end, we will consider again the neutrino–electron scattering, now described by the tree-level (i.e. second-order) Feynman diagram shown in Fig. 3.4.

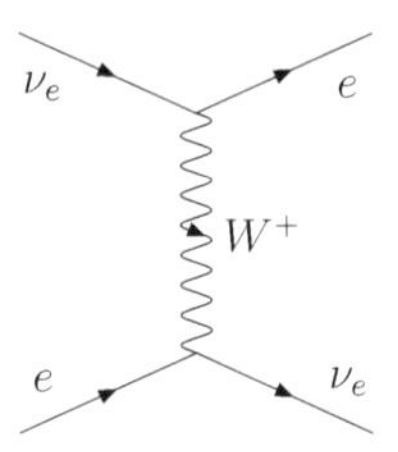

Figure 3.4. Tree-level W-exchange diagram for the ν_e–e elastic scattering.

The corresponding matrix element can be written as

$$i\mathcal{M}^{(\nu e)}_{\text{IVB}} = i^3 \left(\frac{g}{2\sqrt{2}}\right)^2 [\bar{u}(p')\gamma_\rho(1-\gamma_5)u(k)][\bar{u}(k')\gamma_\sigma(1-\gamma_5)u(p)] \\ \times \frac{-g^{\rho\sigma} + m_W^{-2} q^\rho q^\sigma}{q^2 - m_W^2} \tag{3.20}$$

At first glance, one might worry that we have actually won nothing in comparison with Fermi theory: the W boson propagator contains a piece proportional to m_W^{-2} and one could thus expect, on dimensional grounds, a quadratic growth of the (dimensionless) matrix element (3.20) for $E \to \infty$. However, a closer look reveals that it is not so. As in the previous example, one may employ the equations of motion to factorize m_e^2 from the potentially dangerous $q^\rho q^\sigma$ term. Thus, it becomes in fact strongly suppressed (by the factor of m_e^2/m_W^2) in comparison with the $g^{\rho\sigma}$ term and we will drop it in subsequent manipulations. Using the standard trace techniques, the spin-averaged square of the matrix element (3.20) then comes out to be

$$\overline{|\mathcal{M}^{(\nu e)}_{\text{IVB}}|^2} = \frac{1}{2} g^4 \frac{(s - m_e^2)^2}{(u - m_W^2)^2} \tag{3.21}$$

where we have denoted, as usual, $s = (k+p)^2$ and $u = (k-p')^2$ (an astute reader may note that (3.21) can in fact be obtained essentially without any calculation by utilizing our previous result (3.6)). With the high-energy limit in mind, we will of course ignore the effects of electron mass. Then, when recast in terms of the c.m. scattering angle, the expression (3.21) becomes

$$\overline{|\mathcal{M}^{(\nu e)}_{\text{IVB}}|^2} = 2g^4 \frac{1}{(1 + \cos\vartheta_{\text{c.m.}} + 2m_W^2/s)^2} \tag{3.22}$$

The corresponding differential cross-section is then integrated easily; one obtains

$$\begin{aligned}\sigma_{\text{IVB}}^{(\nu e)} &= \frac{g^4}{16\pi}\frac{1}{s}\int_{-1}^{1}\frac{1}{(1+\cos\vartheta_{\text{c.m.}}+2m_W^2/s)^2}d(\cos\vartheta_{\text{c.m.}}) \\ &= \frac{G_F^2}{\pi}m_W^2\frac{s}{s+m_W^2} \end{aligned} \tag{3.23}$$

where we have used the relation (3.19) in the last step. From (3.23), it is obvious that the cross-section tends asymptotically (i.e. for $s \gg m_W^2$) to a constant:

$$\lim_{s\to\infty}\sigma_{\text{IVB}}^{(\nu e)}(s) = \frac{G_F^2}{\pi}m_W^2 \tag{3.24}$$

Thus, we see that the W boson exchange indeed ameliorates the high-energy behaviour observed earlier within the theory of Fermi type — instead of rising rapidly, the considered cross-section is now asymptotically flat. The difference between the two theories is schematically depicted in Fig. 3.5 — the W boson mass obviously plays the role of a natural high-energy "cut-off". (Note that for the antineutrino–electron scattering, the effect of the W exchange leads to a cross-section that vanishes for $s \to \infty$; proving this is left to the reader as an instructive exercise.) Of course, the suppression of a power-like growth of the ν–e cross-section becomes clear immediately, when one observes the effective elimination of the $m_W^{-2}q^\rho q^\sigma$ term from the W boson propagator: as there remain no uncompensated constant factors with the dimension of a negative power of mass, the relevant scattering

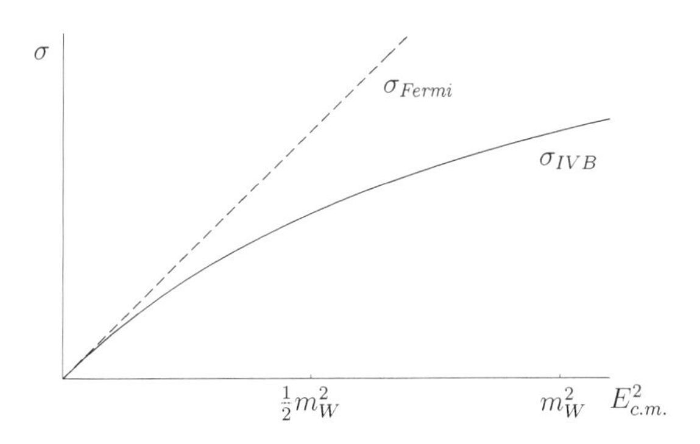

Figure 3.5. Energy dependence of the ν–e elastic scattering cross-section within a Fermi-type model (dashed line) and the IVB model (solid line).

matrix element can behave at most as $O(1)$ for $E \to \infty$ (at a fixed scattering angle)[4] and the formula for the cross-section includes an additional factor of $1/s$.

Let us now discuss the problem of unitarity bound. We will only summarize here briefly the main results; more technical details can be found, e.g., in [Hor]. The relevant high-energy scattering amplitude (corresponding to negative-helicity leptons) that can be guessed from (3.22) has the form

$$\mathcal{M}_{\text{IVB}}^{(\nu e)} = 2g^2 \frac{1}{1 + \cos\vartheta_{\text{c.m.}} + 2m_W^2/s} \tag{3.25}$$

The non-trivial angular dependence in the denominator (which of course is due to the W propagator) implies that now there is an infinite number of partial waves contributing to the expansion of (3.25) (one expands in Legendre polynomials in the considered case). Up to a normalization factor, the amplitude of the lowest ($j = 0$) partial wave is obtained by integrating (3.25) over the $\cos\vartheta_{\text{c.m.}}$ from -1 to 1; the result is

$$\mathcal{M}_{\text{IVB}}^{(0)}(s) = \frac{g^2}{16\pi} \ln\left(\frac{s}{m_W^2} + 1\right) \tag{3.26}$$

Obviously, such a logarithmic dependence on s/m_W^2 is due to the singularity occurring in (3.25) at $\vartheta_{\text{c.m.}} = 180°$ for $s \to \infty$ (or, equivalently, for $m_W = 0$). Note that an analogous result holds for higher partial waves as well. Taking into account that the dimensionless coupling constant g is rather small (say, of the order of electromagnetic coupling constant e), the slow logarithmic rise of a partial-wave amplitude with energy means that a conflict with unitarity may only occur at an astronomically high energy. Indeed, the critical value $s^\star$ for which (3.26) saturates the bound (3.11) is (with a very good accuracy) given by

$$s^\star = m_W^2 \exp\left(\frac{16\pi}{g^2}\right) \tag{3.27}$$

[4]More precisely, the result (3.22) makes it clear that our scattering amplitude is asymptotically flat for any $\theta_{\text{c.m.}} \neq 180°$; on the other hand, for $\theta_{\text{c.m.}} = 180°$ (i.e. for backward scattering) it rises as s/m_W^2. It is not difficult to realize that such an isolated singularity is in fact responsible for the non-zero limit in (3.24); if the matrix element $\mathcal{M}_{\text{IVB}}^{(\nu e)}$ were bounded uniformly, the angular integration in (3.23) would obviously yield a cross-section decreasing as $1/s$ for $s \to \infty$.

which amounts to $E^{\star}_{\text{c.m.}} \approx 10^{29}$ GeV if one employs the present-day values of the relevant parameters, $m_W \doteq 80$ GeV and $g \doteq 0.6$. For other processes, the situation may be even better — in particular, for the $\bar{\nu} - e$ scattering, there is no logarithmic term in the relevant partial-wave amplitude and the bound (3.11) is not violated even at $s \to \infty$. In any case, from the above discussion, it should be clear that the W-exchange mechanism suppressing a rapid (power-like) violation of unitarity is rather general, in the sense that it must work for any fermion–fermion scattering process.

To summarize the results obtained so far, one may say that we have demonstrated explicitly how the IVB model alleviates the unitarity violation problem encountered earlier within the Fermi-type weak interaction theory. In fact, the logarithmic rise of a partial-wave scattering amplitude with energy (at a fixed order of perturbation expansion) cannot be avoided even within a renormalizable field theory — in this sense, the "logarithmic violation of unitarity" exhibited in (3.26) is the best high-energy behaviour attainable within a variety of perturbative QFT models. Let us stress that according to our criterion formulated in Section 3.1, within a renormalizable QFT model, one can have *at worst* a logarithmic violation of perturbative unitarity since there can be no scattering amplitude rising asymptotically as a positive power of energy.

3.3 Difficulties of the simple IVB model

The progress we have achieved so far is not the whole story of the IVB model. Apart from the four-fermion scattering processes discussed previously, the interaction Lagrangian (3.13) describes, at second order of perturbation expansion, also the production of W^+W^- pairs in fermion–antifermion annihilation. As we shall see, the corresponding tree-level amplitudes can lead, for certain combinations of W boson helicities, to a rapid (power-like) violation of unitarity at high energies. In other words, the history repeats itself: the difficulties characteristic of the Fermi-type theory occur here just for another class of physical processes.

For an explicit illustration of the problems we have in mind, let us start with neutrino–antineutrino annihilation process $\nu\bar{\nu} \to W^-W^+$. Historically, the earliest known reference to this example is probably in [32] (published two years after the formulation of the electroweak standard model!). In the lowest non-trivial order, it is described by the Feynman diagram shown in Fig. 3.6.

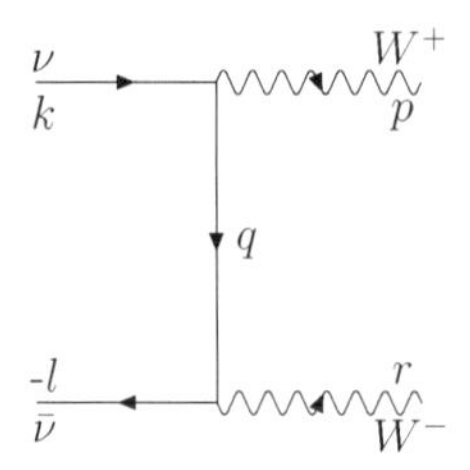

Figure 3.6. Tree-level graph describing the process $\nu\bar{\nu} \to W^+W^-$ within the IVB model (3.13).

The corresponding matrix element can be written as

$$i\mathcal{M}^{(e)}_{\nu\bar{\nu}} = i^3 \left(\frac{g}{2\sqrt{2}}\right)^2 \bar{v}(l)\gamma_\mu(1-\gamma_5)\frac{1}{\not{q} - m_e}\gamma_\nu(1-\gamma_5)u(k)\varepsilon^{*\mu}(r)\varepsilon^{*\nu}(p) \tag{3.28}$$

(note that we have labelled the $\mathcal{M}$ with regard to our later calculations within electroweak standard model). The polarization vectors ε appearing in (3.28) characterize the spin (helicity) states of the final-state vector bosons. Their properties are summarized in Appendix D. There are two possible transverse polarizations corresponding to helicities ± 1 and the longitudinal polarization corresponding to helicity 0. The existence of the zero-helicity state is a characteristic distinguishing feature of a massive vector boson — there is no such thing for massless photon. In fact, it is precisely the W boson longitudinal polarization vector ε_L that will play a crucial role in our subsequent considerations. The important property of the ε_L is that its components grow indefinitely in the high-energy limit as the corresponding four-momentum itself:

$$\varepsilon^\mu_L(p) = \frac{1}{m_W}p^\mu + O\left(\frac{m_W}{p_0}\right) \tag{3.29}$$

The last expression makes it clear, on simple dimensional grounds, why one should worry about the high-energy behaviour of a scattering amplitude involving longitudinally polarized W bosons: the leading contribution from each polarization vector ε_L introduces a factor of inverse mass and one thus expects that a corresponding positive power of energy will be needed to get a dimensionless matrix element (for a binary process). In particular, for (3.28), one expects a quadratic growth with energy when both final-state W's are longitudinally polarized. To make this claim more transparent, let us now work out the corresponding leading asymptotic term (in a form that will be also useful in our later calculations within the standard electroweak model). First of all, from the decomposition (3.29), one can infer easily that

$$\varepsilon_L^{*\mu}(r)\varepsilon_L^{*\nu}(p) = \frac{r^\mu}{m_W}\frac{p^\nu}{m_W} + O(1) \tag{3.30}$$

Consequently, (3.28) can be rewritten as

$$\mathcal{M}_{\nu\bar{\nu}}^{(e)} = -\frac{g^2}{8m_W^2}\bar{v}(l)\not{r}(1-\gamma_5)\frac{1}{\not{q}-m_e}\not{p}(1-\gamma_5)u(k) + O(1) \tag{3.31}$$

Further, one can employ momentum conservation, equations of motion and some simple algebraic manipulations to cancel partially the denominator in (3.31); one thus gets

$$\begin{aligned}\mathcal{M}_{\nu\bar{\nu}}^{(e)} &= -\frac{g^2}{4m_W^2}\bar{v}(l)\not{p}(1-\gamma_5)u(k) \\ &\quad -\frac{g^2}{8m_W^2}m_e\bar{v}(l)(1+\gamma_5)\frac{\not{q}+m_e}{q^2-m_e^2}\not{p}(1-\gamma_5)u(k) + O(1)\end{aligned} \tag{3.32}$$

Clearly, owing to the presence of an otherwise uncompensated factor of m_W^{-2}, the first term in (3.32) embodies the quadratic high-energy divergence.[5] In the second term, only a contribution proportional to m_e^2 survives, which obviously can be absorbed into the asymptotically flat $O(1)$ remainder in (3.32). Thus, the tree-level matrix element for $\nu\bar{\nu} \to W_L^+W_L^-$ can be decomposed as

$$\mathcal{M}_{\nu\bar{\nu}}^{(e)} = -\frac{g^2}{4m_W^2}\bar{v}(l)\not{p}(1-\gamma_5)u(k) + O(1) \tag{3.33}$$

In this result, the leading asymptotic $O(E^2)$ part of the considered matrix element is singled out in a rather simple form. As we shall see later, such a form is in general well suited for a discussion of divergence cancellations among different diagrams contributing within the standard electroweak model. The high-energy divergence in (3.33) cannot vanish identically for an arbitrary scattering angle (unless $g = 0$). Thus, in the corresponding partial-wave expansion, one must necessarily run into the problem with rapid violation of unitarity, completely analogous to that encountered within the old Fermi-type theory. We will not calculate here explicitly the relevant

[5] We are not going to work it out as an explicit function of energy, but it is useful to observe that the u and v spinors behave (within our normalization convention) as $E^{1/2}$ in the high-energy limit; together with the factor of $\not{p}$, this then makes up the quadratic rise with energy anticipated on dimensional grounds.

partial-wave amplitudes (the interested reader is referred to the original paper in [32]). Instead, it may be instructive to see what the high-energy behaviour of the corresponding cross-section is. Using (3.33) and the usual trace techniques, it is straightforward to show that

$$\begin{aligned}\sigma^{(e)}(\nu\bar{\nu} \to W_L^+ W_L^-) &\approx \frac{1}{64\pi^2}\frac{1}{s}\left(\frac{g^2}{4m_W^2}\right)^2 \int \mathrm{Tr}[\not{l}\not{p}(1-\gamma_5)\not{k}\not{p}(1-\gamma_5)]d\Omega_{\text{c.m.}} \\ &\approx \frac{g^4}{512\pi}\frac{s}{m_W^4}\int_{-1}^{1}(1-\cos^2\vartheta_{\text{c.m.}})d(\cos\vartheta_{\text{c.m.}}) \\ &= \frac{G_F^2}{12\pi}s \end{aligned} \tag{3.34}$$

for $s \gg m_W^2$. Note that in the last step, we have reintroduced the Fermi constant G_F through the relation (3.19) in order to stress the close analogy of the considered case with our earlier results. Of course, one would get the same asymptotic behaviour for the unpolarized W boson cross-section calculated directly from (3.28) with the help of the standard formula for the polarization sum

$$\sum_{\lambda=1}^{3}\varepsilon_\mu(k,\lambda)\varepsilon_\nu^*(k,\lambda) = -g_{\mu\nu} + \frac{1}{m_W^2}k_\mu k_\nu \tag{3.35}$$

(the point is that the longitudinally polarized W bosons give dominant contribution in the high-energy limit). There is an important general feature of the above results that should be noticed here. While fermion masses become irrelevant in the high-energy limit, the mass of a physical W boson cannot be generally neglected for $E \to \infty$ simply because this appears in a negative power in the expressions like (3.34). Of course, the source of such an anomalous behaviour is the longitudinal polarization vector (3.29), which is also responsible for the factor of m_W^{-2} in the polarization sum (3.35).[6]

[6]An astute reader might object that the different character of the mass dependence manifested in the spin sums for fermions and vector bosons is due merely to our normalization conventions: we take $\bar{u}(k)u(k) = 2m$ for Dirac spinors (with the corresponding spin sum being $\not{k}+m$), while $\varepsilon(k)\cdot\varepsilon^*(k) = -1$ for vector boson polarization vectors (leading to (3.35)). In fact, these normalization conventions do match each other, for simple dimensional reasons — since the vector and Dirac fields have dimensions of M and $M^{3/2}$ resp., the one-particle states thus become normalized in the same way for both cases and this in turn fits into the general cross-section formula given in Appendix B.

The rapid violation of tree-level unitarity, observed here for the process $\nu\bar{\nu} \to W_L^+ W_L^-$, also indicates — according to the criterion formulated at the end of Section 3.1 — that the model based on the Lagrangian (3.13) is not renormalizable in higher orders of perturbation expansion. Such a claim was indeed proved in the 1960s (see [33]). Thus, the considered IVB model of weak interactions constitutes in fact only a partial improvement of the Fermi-type theory: while some old problems (concerning four-fermion processes) are solved, new difficulties show up due to longitudinally polarized physical W bosons. Such a flaw obviously cannot be removed within the simple model (3.13) itself, and thus, it is clear that a further amelioration of technical properties of the weak interaction theory may only be achieved within a broader theory.

3.4 Electromagnetic interactions of W bosons

One possible extension of the IVB theory is immediately clear. Since the W boson carries electric charge, one should also consider its electromagnetic interactions. This subject has been discussed in considerable detail in [Hor], so we are going to give here only a concise summary of the most important results.

Let us start with the free-field Lagrangian for $W_\mu^\pm$ (understood here as the "matter fields"). It can be written as

$$\mathscr{L}_0 = -\frac{1}{2} W_{\mu\nu}^- W^{+\mu\nu} + m_W^2 W_\mu^- W^{+\mu} \tag{3.36}$$

where we have denoted $W_{\mu\nu}^\pm = \partial_\mu W_\nu^\pm - \partial_\nu W_\mu^\pm$. A usual way of introducing the electromagnetic interaction consists in the "minimal substitution" for the derivatives in the corresponding kinetic term. In this case, this means that (3.36) is replaced by

$$\begin{aligned}\mathscr{L}_{\text{em}}^{(\text{min})} &= -\frac{1}{2}(D_\mu W_\nu^- - D_\nu W_\mu^-)(D^{\mu *} W^{+\nu} - D^{\nu *} W^{+\mu}) \\ &\quad + m_W^2 W_\mu^- W^{+\mu}\end{aligned} \tag{3.37}$$

where $D_\mu = \partial_\mu + ieA_\mu$ and $D_\mu^* = \partial_\mu - ieA_\mu$, with A_μ being the electromagnetic four-potential and e denoting the relevant coupling constant ($e > 0$ and $e^2/4\pi = \alpha$ is the fine-structure constant, $\alpha \doteq 1/137$). Note that (3.37) is built in a straightforward analogy with the familiar electrodynamics of charged Dirac field (which the reader is supposed to know from an introductory field-theory course). One may observe that the Lagrangian (3.37) is invariant under the gradient transformations

$$A'_\mu(x) = A_\mu(x) + \frac{1}{e}\partial_\mu \omega(x) \tag{3.38}$$

accompanied with the corresponding local phase transformations of the charged fields $W_\mu^\pm$

$$\begin{aligned} W_\mu^{-'}(x) &= \mathrm{e}^{-i\omega(x)} W_\mu^-(x) \\ W_\mu^{+'}(x) &= \mathrm{e}^{i\omega(x)} W_\mu^+(x) \end{aligned} \tag{3.39}$$

Using a standard terminology, (3.38) and (3.39) represent local gauge transformations or, simply, gauge transformations. We will discuss the concept of gauge invariance more thoroughly in the following chapter — here, we only stress again the similarity with electrodynamics of a charged Dirac field. The interaction terms descending from (3.37) are

$$\begin{aligned} \mathscr{L}_{\text{int}}^{(\text{min})} &= -ie[(A^\mu W^{-\nu} - A^\nu W^{-\mu})\partial_\mu W_\nu^+ - (A^\mu W^{+\nu} - A^\nu W^{+\mu})\partial_\mu W_\nu^-] \\ &\quad - e^2(A_\mu A^\mu W_\nu^- W^{+\nu} - A^\mu A^\nu W_\mu^- W_\nu^+) \end{aligned} \tag{3.40}$$

In fact, one may consider more general gauge-invariant interaction terms than those contained in (3.37). We restrict ourselves *a priori* to interaction Lagrangians with dimension not greater than 4 in order to avoid a coupling constant with dimension of a negative power of mass (that would lead automatically to the by now familiar difficulties in high-energy limit), and for simplicity, we will also assume the parity invariance. Then there is only one possible addition to (3.40), namely

$$\mathscr{L}'_{\text{int}} = -i\kappa e W_\mu^- W_\nu^+ F^{\mu\nu} \tag{3.41}$$

where, of course, $F_{\mu\nu} = \partial_\mu A_\nu - \partial_\nu A_\mu$ and κ is an arbitrary real parameter (this determines the value of the W boson magnetic moment and electric quadrupole moment — see, e.g., [33, 34] and also [Tay]).[7] Thus, a general

[7] Note that within electrodynamics of spin-$\frac{1}{2}$ fermions, a term analogous to (3.41) would have the form $\bar{\psi}\sigma_{\mu\nu}\psi F^{\mu\nu}$. However, in contrast to (3.41), this is of dimension 5 and spoils perturbative renormalizability.

electromagnetic interaction of W bosons can be written as

$$\begin{aligned}\mathscr{L}_{\text{int}}^{(\text{em})} &= -ie[A^{\mu}(W^{-\nu}\partial_{\mu}W_{\nu}^{+} - \partial_{\mu}W_{\nu}^{-}W^{+\nu}) \\ &\quad + W^{-\mu}(\kappa W^{+\nu}\partial_{\mu}A_{\nu} - \partial_{\mu}W^{+\nu}A_{\nu}) \\ &\quad + W^{+\mu}(A^{\nu}\partial_{\mu}W_{\nu}^{-} - \kappa\partial_{\mu}A^{\nu}W_{\nu}^{-})] \\ &\quad - e^{2}(A_{\mu}A^{\mu}W_{\nu}^{-}W^{+\nu} - A^{\mu}A^{\nu}W_{\mu}^{-}W_{\nu}^{+}) \\ &= \mathscr{L}_{WW\gamma}^{(\kappa)} + \mathscr{L}_{WW\gamma\gamma} \end{aligned} \tag{3.42}$$

where we have marked explicitly the trilinear ($WW\gamma$) and quadrilinear ($WW\gamma\gamma$) parts, respectively (we prefer to label the interaction Lagrangians in terms of the corresponding particle symbols, i.e. $WW\gamma$ instead of WWA, etc.). Note that the $WW\gamma\gamma$ part is independent of κ. The interaction terms appearing in (3.42) are of renormalizable type as they have dimension 4 (consequently, the corresponding coupling constants are dimensionless). On the other hand, we are already well aware of the difficulties associated with zero-helicity states of the physical W bosons. Therefore, one might worry that the rapid violation of unitarity (and the ensuing loss of perturbative renormalizability) could show up for W boson electromagnetic interactions as well in analogy with the weak interaction case discussed previously. The problem is analyzed in detail in [Hor], and it turns out that such expectations are indeed fulfilled. We will summarize here the salient points of such an analysis.

One may start with a particular tree-level electromagnetic process, e.g., with the two-photon annihilation of the W^+W^- pair. The relevant Feynman diagrams of the order $O(e^2)$ in the electromagnetic coupling are depicted in Fig. 3.7. Note that the considered process is particularly interesting for our purpose since the corresponding diagrams involve both external and internal W lines and both potential sources of a "bad"

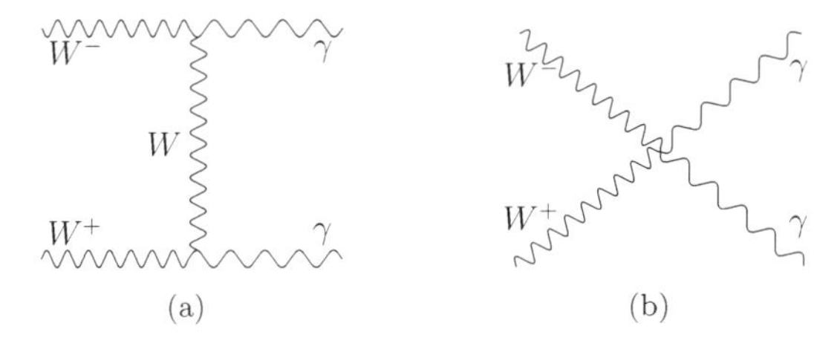

Figure 3.7. Tree-level diagrams contributing to the electromagnetic process $W^+W^- \to \gamma\gamma$.

high-energy behaviour (longitudinal polarization vectors and the W propagator) thus occur here. Since the interaction Lagrangian contains an arbitrary parameter κ, one may also wonder how its value can influence the high-energy asymptotics of the diagrams in question. The corresponding calculations are somewhat tedious, but the conclusion that emerges is rather remarkable. **The tree-level $W^+W^- \to \gamma\gamma$ amplitude is asymptotically flat (i.e. free of power-like divergences) for any combination of external W boson helicities if and only if $\kappa = 1$.** The lesson to be learnt from this example is that the only possible candidate for a renormalizable electrodynamics of W bosons is the model with $\kappa = 1$. Indeed, our little theorem claims that if $\kappa \neq 1$, the tree-level unitarity (which is a necessary condition for renormalizability) would be violated for a particular combination of W boson polarizations in the amplitude of $W^+W^- \to \gamma\gamma$.

Thus, let us adopt the interaction Lagrangian (3.42) with $\kappa = 1$. Its $WW\gamma$ part can then be written as[8]

$$\mathscr{L}^{(\kappa=1)}_{WW\gamma} = -ie(A^\mu W^{-\nu} \overleftrightarrow{\partial}_\mu W^+_\nu + W^{-\mu} W^{+\nu} \overleftrightarrow{\partial}_\mu A_\nu + W^{+\mu} A^\nu \overleftrightarrow{\partial}_\mu W^-_\nu) \tag{3.43}$$

where the symbol $\overleftrightarrow{\partial}$ is defined by $f\overleftrightarrow{\partial}_\mu g = f(\partial_\mu g) - (\partial_\mu f)g$. This in turn leads to the following momentum-space Feynman rule: when each line involved in the corresponding vertex is labelled by a corresponding four-momentum and a Lorentz index as shown in Fig. 3.8, then the contribution of the $WW\gamma$ vertex is given by the function

$$V_{\lambda\mu\nu}(k,p,q) = g_{\lambda\mu}(k-p)_\nu + g_{\mu\nu}(p-q)_\lambda + g_{\lambda\nu}(q-k)_\mu \tag{3.44}$$

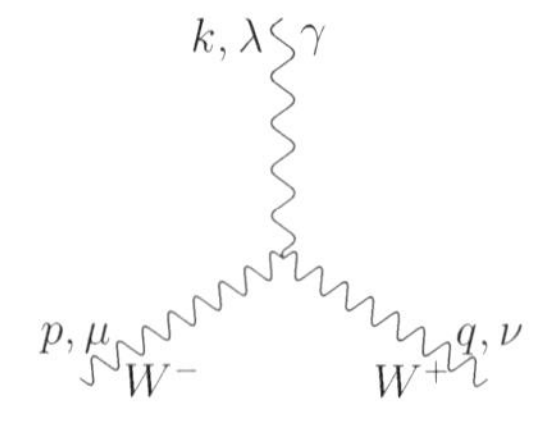

Figure 3.8. An example of the labelling of the $WW\gamma$ vertex. All four-momenta are taken as outgoing.

[8]For reasons that will become clear in the following chapters, the interaction (3.43) may be called the $WW\gamma$ coupling of Yang–Mills type.

multiplied by the coupling constant e. For a vertex involving an incoming line, the corresponding four-momentum in (3.44) is taken with opposite sign. Note also that an incoming $W^{\pm}$ is equivalent to an outgoing $W^{\mp}$.

In view of our previous observations, the form (3.43) (or, equivalently, (3.44)) represents, in a sense, an "optimal choice" for the QED of W bosons. However, it is not difficult to demonstrate that even such an option for the $WW\gamma$ vertex is not able to tame the bad high-energy behaviour for all possible electromagnetic processes. As a pertinent example illustrating this, one may choose the WW elastic scattering. The corresponding lowest-order Feynman graphs are shown in Fig. 3.9. Using (3.44), the contribution of Fig. 3.9(a) can be written as

$$i\mathcal{M}_a = i^3 e^2 \varepsilon^{*\mu}(p)\varepsilon^{\nu}(k)V_{\mu\nu\rho}(p,-k,q)\frac{-g^{\rho\sigma}}{q^2}V_{\sigma\alpha\beta}(-q,r,-l)\varepsilon^{*\alpha}(r)\varepsilon^{\beta}(l) \tag{3.45}$$

where we have also employed the standard form of the photon propagator in Feynman gauge. Of course, the corresponding expression for Fig. 3.9(b) is obtained from (3.45) by interchanging p and r. Now, the worst high-energy behaviour can be expected in the case when all the external W bosons are longitudinally polarized. Taking into account (3.29), one may guess, on simple dimensional grounds, that the leading asymptotic term in (3.45) behaves as $O(E^4/m_W^4)$. This is indeed confirmed by a direct calculation. Substituting into (3.45) the decomposition (3.29) for each polarization vector, one obviously gets an expansion

$$\begin{aligned}\mathcal{M}_a = &-e^2\frac{p^{\mu}}{m_W}\frac{k^{\nu}}{m_W}V_{\mu\nu\rho}(p,-k,q)\frac{-g^{\rho\sigma}}{q^2}V_{\sigma\alpha\beta}(-q,r,-l)\frac{r^{\alpha}}{m_W}\frac{l^{\beta}}{m_W}\\ &+O\left(\frac{E^2}{m_W^2}\right)+O(1)\end{aligned} \tag{3.46}$$

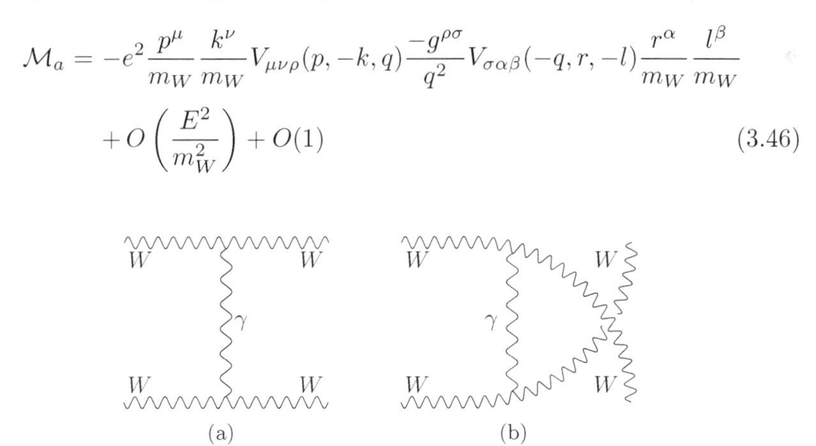

Figure 3.9. Tree-level QED diagrams for the process $WW \to WW$.

To work out the first (leading) term explicitly, one can employ the identity[9]

$$p^{\mu} V_{\lambda\mu\nu}(k, p, q) = (k^2 g_{\lambda\nu} - k_{\lambda} k_{\nu}) - (q^2 g_{\lambda\nu} - q_{\lambda} q_{\nu}) \tag{3.47}$$

(A practically important feature of this formula is that its right-hand side is a difference of two transverse expressions. It should be stressed that the last identity is valid for arbitrary four-momenta satisfying $k + p + q = 0$. A proof of (3.47) is left to the reader as an easy exercise.) After some simple manipulations and taking into account that $k^2 = l^2 = p^2 = r^2 = m_W^2$, the expression (3.46) can then be recast as

$$\mathcal{M}_a = \frac{e^2}{4m_W^4}(t^2 + 2ts) + O\left(\frac{E^2}{m_W^2}\right) + O(1) \tag{3.48}$$

where we have used the standard notation $s = (k + l)^2$ and $t = (k - p)^2$. The contribution of Fig. 3.9(b) is then obtained from (3.48) by the replacement $t \to u$, with $u = (k - r)^2$. Adding the two contributions and using the kinematical identity $s + t + u = 4m_W^2$, the $W_L W_L$ scattering amplitude thus finally becomes

$$\mathcal{M}_{WW}^{(\gamma)} = \frac{e^2}{4m_W^4}(t^2 + u^2 - 2s^2) + O\left(\frac{E^2}{m_W^2}\right) + O(1) \tag{3.49}$$

This result exhibits clearly the quartic growth of the considered amplitude with energy. The leading $O(E^4)$ term depends on the scattering angle through the Mandelstam variables t and u, but obviously it cannot vanish identically (unless $e = 0$). The next-to-leading term $O(E^2)$ has a rather complicated form, but we will not need it now.

Thus, we may conclude that there is no choice of the parameter κ in (3.42), which would eliminate all potential high-energy divergences in the tree-level scattering amplitudes. Consequently, **the quantum electrodynamics of W bosons cannot be renormalizable** in contrast to the "textbook" case of the spinor QED. In any case, the $WW\gamma$ interaction of Yang–Mills type, corresponding to $\kappa = 1$, seems to be the "best" choice for QED of W bosons and we will use it in what follows as an appropriate reference model.

[9]Note that the relation (3.47) is sometimes called 't Hooft identity since it has been probably given first in [35].

3.5 The case for electroweak unification

One can find other examples showing that the IVB model of weak interactions and electrodynamics of W bosons suffer from the same technical difficulties. In particular, there are processes that receive both weak and electromagnetic contributions at the level of tree diagrams. One such example is the process $e^+e^- \to W^+W^-$. The relevant lowest-order Feynman diagrams are depicted in Fig. 3.10. Let us examine the weak and electromagnetic contribution separately. For Fig. 3.10(a), one has

$$i\mathcal{M}^{(\nu)}_{e^+e^-} = i^3 \left(\frac{g}{2\sqrt{2}}\right)^2 \bar{v}(l)\gamma_\mu(1-\gamma_5)\frac{1}{\not{q}}\gamma_\nu(1-\gamma_5)u(k)\varepsilon^{*\mu}(r)\varepsilon^{*\nu}(p) \qquad (3.50)$$

Invoking the usual dimensional arguments, one may guess easily that such a matrix element grows quadratically with energy when both final-state W bosons are longitudinally polarized. To evaluate the leading $O(E^2)$ divergence, one can proceed in analogy with the process $\nu\bar{\nu} \to W_L^+W_L^-$ discussed in Section 3.3. Substituting $\varepsilon_L^\mu(r)\varepsilon_L^\nu(p)$ in the general expression (3.50) and using the decomposition (3.29), one gets, after some manipulations, the result

$$\mathcal{M}^{(\nu)}_{e^+e^-} = -\frac{g^2}{4m_W^2}\bar{v}(l)\not{p}(1-\gamma_5)u(k) + O\left(\frac{m_e}{m_W^2}E\right) + O(1) \qquad (3.51)$$

Note that in contrast to the $\nu\bar{\nu}$ annihilation case, here one also gets a linearly divergent term.

The contribution of Fig. 3.10(b) is given by

$$i\mathcal{M}^{(\gamma)}_{e^+e^-} = -i^3e^2\bar{v}(l)\gamma_\alpha u(k)\frac{-g^{\alpha\nu}}{Q^2}V_{\lambda\mu\nu}(p,r,-Q)\varepsilon^{*\lambda}(p)\varepsilon^{*\mu}(r) \qquad (3.52)$$

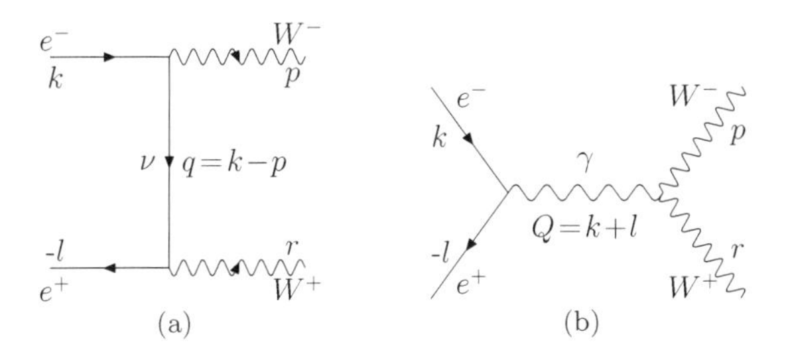

Figure 3.10. Weak and electromagnetic second-order contributions to $e^+e^- \to W^+W^-$.

(note that in writing (3.52), we have taken into account that the coupling factor for the $e^+e^-\gamma$ vertex is $(-e)$). Again, for longitudinally polarized $W^\pm$, one can use (3.29) and the identity (3.47). One then gets after some algebra

$$\mathcal{M}^{(\gamma)}_{e^+e^-} = \frac{e^2}{m_W^2}\bar{v}(l)\not{p}u(k) + O(1) \tag{3.53}$$

Although the considered two diagrams look rather different (Fig. 3.10(a) corresponds to a t-channel fermion exchange, while Fig. 3.10(b) represents a bosonic exchange in the s-channel), the leading divergent terms in (3.51) and (3.53) come out in a similar form. Such a similarity raises a hope that the high-energy divergences of weak and electromagnetic origin might cancel within a broader unified theory if, e.g., the ratio of e and g is chosen appropriately. In fact, if one simply adds (3.51) and (3.53), a complete cancellation of the $O(E^2)$ terms obviously cannot be achieved since the matrix factor of $1-\gamma_5$ occurs in (3.51), while (3.53) can be split into two equal parts involving $1-\gamma_5$ and $1+\gamma_5$. In other words, weak interactions violate parity maximally, while the electromagnetic interactions are parity-conserving — such a deep difference cannot be simply compensated in the two diagrams themselves. Moreover, (3.51) includes another term that diverges linearly for $E \to \infty$, but this is absent in (3.53). Thus, a new particle exchange would be clearly needed to cancel the divergence in the sum of Figs. 3.10(a) and 3.10(b).

The above example — as well as those discussed in preceding sections — make it obvious that a simple addition of the weak and electromagnetic interaction Lagrangians cannot remedy, in general, the technical flaws inherent in these models. Nevertheless, it is in order to remark that there is at least one type of an "electro-weak" process, for which the interactions considered so far do produce a well-behaved scattering amplitude. The simplest example is provided by the reaction $\bar{\nu}e \to W\gamma$, described by the diagrams shown in Fig. 3.11 (a variant of such a process, which is far more realistic from the point of view of present-day experiments, is $\bar{u}d \to W\gamma$, where the d and u are quarks with charges $-1/3$ and $2/3$, respectively). We are not going to perform the corresponding calculation in detail, but one salient point should perhaps be emphasized here. Remembering the usual power-counting dimensional analysis, one might worry that the graph in Fig. 3.11(b) diverges faster than Fig. 3.11(a) in the high-energy limit since the longitudinal term in the W boson propagator introduces an extra factor of m_W^{-2}. In fact, it is easy to show that such a term leads to a contribution

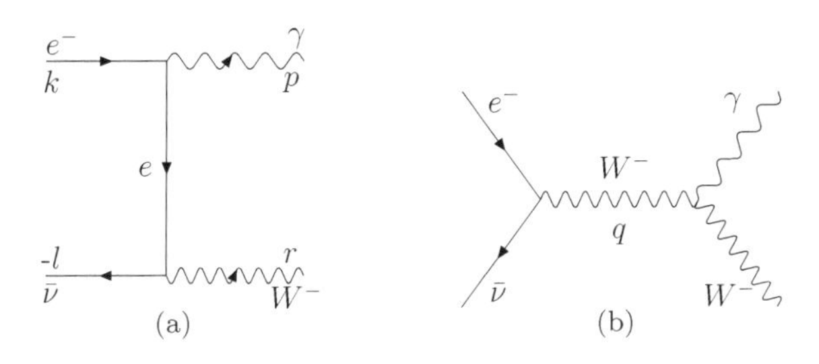

Figure 3.11. Tree-level graphs for the electro-weak process $\bar{\nu}_e e^- \to W^-\gamma$. These two graphs actually represent the full second-order contribution of the current standard model.

to the scattering matrix element that is always asymptotically flat (even for longitudinally polarized external W boson!). To see this, let us consider the expression

$$\bar{v}(l)\gamma_\rho(1-\gamma_5)u(k)q^\rho q^\nu V_{\lambda\mu\nu}(p,r,-q)\varepsilon^{*\lambda}(p)\varepsilon^{*\mu}(r) \tag{3.54}$$

which constitutes the potentially dangerous part of the contribution of Fig. 3.11(b). Using the obvious cyclicity property of the $WW\gamma$ vertex function (3.44) (i.e. $V_{\lambda\mu\nu}(p,r,-q) = V_{\mu\nu\lambda}(r,-q,p)$, etc.), the 't Hooft identity (3.47) and equations of motion (i.e. $p\cdot\varepsilon^*(p) = 0$, $r\cdot\varepsilon^*(r) = 0$, $p^2 = 0$, $r^2 = m_W^2$ and $\bar{v}(l)\not{l} = 0$, $\not{k}u(k) = m_e u(k)$), it becomes

$$-m_e m_W^2 \bar{v}(l)(1+\gamma_5)u(k)\varepsilon^*(p)\cdot\varepsilon^*(r) \tag{3.55}$$

Thus, the expression (3.54) is proportional to a factor of $(\text{mass})^3$ and this is sufficient to swamp completely any negative power of m_W that would arise from the W propagator and a polarization vector of the external W boson (this can be m_W^{-3} at worst). The diagram in Fig. 3.11(b) can therefore only produce a linear high-energy divergence (in case of a longitudinally polarized W boson) and this is exactly compensated by the contribution of Fig. 3.11(a).

The lesson to be learnt from the examples given in this chapter is that there are certainly some *technical* reasons for a non-trivial unification of weak and electromagnetic forces: when the weak interaction theory and the electrodynamics of W bosons are taken separately, one encounters rapid violation of perturbative unitarity at various places and, consequently, the renormalizability is lost. Thus, if one wishes to cancel somehow the high-energy divergences in both theories, the only logical possibility apparently consists in unifying the two interactions. However, as we have seen, the simple addition $\mathscr{L}_{\text{int}}^{(w)} + \mathscr{L}_{\text{int}}^{(\text{em})}$ is not sufficient for such a purpose. Therefore, one obviously has to envisage a broader unification framework, including additional particles and interactions. Since these additional interactions are designed to compensate the high-energy divergences of both weak and electromagnetic origin, they should in a sense "interpolate" between the original two forces, i.e. they must necessarily mix the weak and electromagnetic couplings (in other words, one should expect that the coupling strengths of the additional "compensating" interactions are non-trivial combinations of e and g). Such a theoretical scheme can then be naturally called **electroweak unification**.

Taking into account our previous knowledge, one may envisage the corresponding interaction Lagrangian in a form

$$\mathscr{L}_{\text{int}}^{(\text{ew})} = \mathscr{L}_{CC} + \mathscr{L}_{\text{fermion}}^{(\text{em})} + \mathscr{L}_{WW\gamma} + \mathscr{L}_{WW\gamma\gamma} + \cdots \tag{3.56}$$

where the first term represents the charged-current weak interaction (3.13) and the remaining ones stand for the electromagnetic interactions of fermions (quarks and leptons) and W bosons (note that the Yang–Mills form (3.43) is assumed tacitly for the $\mathscr{L}_{WW\gamma}$). The ellipsis symbolizes the "missing links" of the electroweak unification that should presumably make the theory well behaved in the high-energy limit. One can indeed construct a solution to this problem by adding in (3.56) new interaction terms so as to systematically cancel the high-energy divergences arising within the provisional model. In fact, there are infinitely many solutions that may be obtained in this way, but it turns out that the minimal[10] electroweak theory satisfying the criterion of tree unitarity is just the present-day Standard Model. The construction of renormalizable models of weak and electromagnetic interactions from the high-energy constraints on tree-level Feynman diagrams has been first implemented in the papers [31, 36, 37] and for SM, it is also described in detail in [Hor] (for another pedagogical exposition see the lecture notes [38]). Such a derivation of the electroweak standard model "from scratch" is quite remarkable not only technically but also conceptually: it shows that the whole structure of SM (which admittedly may seem rather complicated to an uninitiated person) is in fact inevitable if one insists on perturbative renormalizability.

[10]The adjective "minimal" refers to the particle content of a considered model.

However, the right solution to the electroweak unification problem has originally been found in a completely different way [39–41]. Instead of going through a tedious diagram analysis, the inventors of the electroweak SM simply had the right inspiration: they employed a rather abstract principle of (broken) gauge symmetry, which in fact has become subsequently the theoretical backbone of the whole modern particle physics. This is precisely the path we are going to follow in the subsequent chapters. The desired cancellations of high-energy divergences must then be verified *a posteriori*, but such a "symmetry construction" does have certain advantage over the aforementioned "engineering approach" — a specific formulation of the scheme of broken gauge symmetry within SM provides a deeper insight into the meaning of the cancellation mechanism.

As a prelude to the discussion of this fundamental method, one may observe that there is in fact a simple *a priori aesthetic* argument in favour of a unified treatment of weak and electromagnetic forces: both interactions are of *vectorial nature* (Lorentz vector or pseudovector currents are involved in both cases) and they are *universal*, i.e. they act between widely different particles (such as quarks and leptons) with equal strength. Moreover, the pure vector part of the weak current belongs to the same isospin multiplet (isotriplet) as the electromagnetic current (cf. the discussion of CVC in Section 2.7). The vectorial character of the two forces means that the corresponding mediators (photon and $W^{\pm}$) have spin 1 and one thus may imagine placing them — at least formally — into a common symmetry multiplet. It turns out that the concept of non-Abelian gauge symmetry [42] (discovered originally without any direct motivation from the side of weak interaction theory) fits, in fact, precisely to this situation.

Thus, as we have seen, **there are both technical and "moral" (aesthetic) arguments in favour of a unification of weak and electromagnetic interactions**. In the following chapters, we will develop the ideas and techniques of broken gauge symmetry that are crucial for the construction of a technically successful (renormalizable) electroweak theory.

Problems

3.1 Calculate the decay width for $W^- \to \ell^- + \bar{\nu}_\ell$ for unpolarized particles. Neglect $m_{\bar{\nu}}$, but keep $m_\ell \neq 0$.

3.2 Calculate longitudinal polarization of charged leptons produced in the decay of an unpolarized W boson. Check correctness of the obtained result by setting there $m_\ell = 0$.

3.3 Calculate the angular distribution of charged leptons produced in the decay of a polarized W boson at rest.

3.4 Determine the asymptotic behaviour of the photon-exchange contribution to the tree-level amplitude for $e^+e^- \to W^+W^-$, assuming that the vertex $WW\gamma$ corresponds to the *minimal* electromagnetic interaction (3.40).
Hint: As a relevant Feynman rule, take the expression (4.13) in [Hor] with $\kappa = 0$.

3.5 Consider the process $e^+e^- \to \gamma_L\gamma_L$ within spinor QED with a *massive* photon. Show that the corresponding tree-level amplitude behaves as $O(1)$ in high-energy limit (i.e. for $E_{\text{c.m.}} \gg m_\gamma$).

3.6 Consider the quark–antiquark annihilation process $d + \bar{u} \to W^- + \gamma$ within the provisional electro-weak theory described by the first four terms in the interaction Lagrangian (3.56). Show that the corresponding tree-level amplitude behaves asymptotically as $O(1)$ for any polarization of the W.

Chapter 4

Gauge Invariance and Yang–Mills Field

4.1 Abelian gauge invariance

Let us consider, e.g., the Lagrangian of a free classical Dirac field

$$\mathscr{L}_0 = i\bar{\psi}\gamma^\mu\partial_\mu\psi - m\bar{\psi}\psi \tag{4.1}$$

where ψ denotes the corresponding bispinor field variable. It is easy to verify that the expression (4.1) is invariant under *global* phase transformations

$$\psi'(x) = \mathrm{e}^{i\omega}\psi(x) \tag{4.2}$$

$$\bar{\psi}'(x) = \mathrm{e}^{-i\omega}\bar{\psi}(x) \tag{4.3}$$

where ω is a constant independent of coordinates (the adjective *global* refers to the x-independence of the transformation parameter). ω can otherwise take on an arbitrary real value and the unitary transformations (4.2) and (4.3) thus form an Abelian (i.e. commutative) group called $U(1)$. Let us recall that such a continuous one-parameter symmetry leads in general to a conserved Noether current, which in the present case has the familiar form

$$J_\mu = \bar{\psi}\gamma_\mu\psi \tag{4.4}$$

One may now ask what happens if we let the parameter ω depend on x, i.e. if we consider *local* transformations

$$\psi'(x) = \mathrm{e}^{i\omega(x)}\psi(x) \tag{4.5}$$

$$\bar{\psi}'(x) = \mathrm{e}^{-i\omega(x)}\bar{\psi}(x) \tag{4.6}$$

When the Lagrangian (4.1) is transformed according to (4.5) and (4.6), one obtains

$$\begin{aligned}\mathscr{L}_0' &= i\bar{\psi}'\gamma^\mu\partial_\mu\psi' - m\bar{\psi}'\psi' = i\mathrm{e}^{-i\omega}\bar{\psi}\gamma^\mu(i\partial_\mu\omega\mathrm{e}^{i\omega}\psi + \mathrm{e}^{i\omega}\partial_\mu\psi) - m\bar{\psi}\psi \\ &= -\bar{\psi}\gamma^\mu\psi\partial_\mu\omega + i\bar{\psi}\gamma^\mu\partial_\mu\psi - m\bar{\psi}\psi = -\bar{\psi}\gamma^\mu\psi\partial_\mu\omega + \mathscr{L}_0 \end{aligned} \tag{4.7}$$

Thus, $\mathscr{L}_0$ is *not* invariant under local phase transformations and its non-invariance (which is obviously due to the derivative involved in the kinetic term) can be represented as a coupling of the gradient of the local phase parameter to the vector current (4.4). Now, one can make a simple observation, which will be of crucial importance for our later considerations. The contribution proportional to $\partial_\mu \omega$, which has shown up in the last expression, can be cancelled by adding to the original free Lagrangian an *interaction term* involving a new vector field (coupled to the current (4.4)), endowed with appropriate transformation properties. In particular, the term to be added may be written as

$$\mathscr{L}_{\text{int}} = g\bar{\psi}\gamma^\mu \psi A_\mu \tag{4.8}$$

where g denotes a coupling constant and the vector field A_μ is required to transform according to

$$A'_\mu(x) = A_\mu(x) + \frac{1}{g}\partial_\mu \omega(x) \tag{4.9}$$

The extended Lagrangian

$$\mathscr{L} = \mathscr{L}_0 + g\bar{\psi}\gamma^\mu \psi A_\mu \tag{4.10}$$

is then invariant under the transformations (4.5), (4.6), and (4.9), as now we have

$$\begin{aligned}\mathscr{L}' &= \mathscr{L}'_0 + g\bar{\psi}'\gamma^\mu \psi' A'_\mu = \mathscr{L}_0 - \bar{\psi}\gamma^\mu \psi \partial_\mu \omega + g\bar{\psi}\gamma^\mu \psi \left(A_\mu + \frac{1}{g}\partial_\mu \omega\right) \\ &= \mathscr{L}\end{aligned} \tag{4.11}$$

In the standard terminology, the relations (4.5), (4.6), (4.9) represent the (Abelian) **gauge transformations** and the vector field A_μ is called accordingly the Abelian **gauge field**. The Lagrangian (4.10) can be recast as

$$\begin{aligned}\mathscr{L} &= i\bar{\psi}\gamma^\mu \partial_\mu \psi + g\bar{\psi}\gamma^\mu \psi A_\mu - m\bar{\psi}\psi \\ &= i\bar{\psi}\gamma^\mu(\partial_\mu - igA_\mu)\psi - m\bar{\psi}\psi = i\bar{\psi}\not{D}\psi - m\bar{\psi}\psi\end{aligned} \tag{4.12}$$

where we have introduced a usual symbol D_μ denoting the **covariant derivative**

$$D_\mu = \partial_\mu - igA_\mu \tag{4.13}$$

(this is another piece of the standard gauge-theory vocabulary).

Let us now pause here to discuss briefly the meaning of the preceding manipulations. Of course, in (4.12), one may easily recognize the "minimal electromagnetic coupling" well known from classical electrodynamics, and the electromagnetic vector potential thus can serve as an obvious example of an Abelian gauge field. Historically, the classical Maxwell electrodynamics has been deduced from the wealth of known experimental data, and its gauge (or "gradient") invariance shows up as an additional mathematical property of the relevant system of equations (this should presumably be familiar to everybody who followed a corresponding introductory course). Here, however, we have proceeded in a reverse direction: starting with a free field Lagrangian (which violates the local gauge symmetry), we have subsequently extended it by including a particular interaction (of an "electromagnetic" type) to meet the requirement of local gauge invariance. This is actually the most important lesson to be learnt from the preceding discussion, so let us formulate it once again in a more concise form. **Promoting the global phase invariance of a free matter-field Lagrangian to the local gauge symmetry, one is forced to introduce an interaction involving a vector (gauge) field with rather specific properties.** Such a simple observation is in fact the core of all modern gauge theories, which are based on the non-Abelian generalization of the concept of local symmetry (to be discussed in the next section). The *a priori* requirement of local gauge symmetry, though in a sense natural, is a rather abstract mathematical principle and its physical meaning is not immediately obvious. Nevertheless, it has proved to be an immensely successful heuristic principle in modern particle theory, as it led to the formulation of the present-day standard model of fundamental interactions (incorporating the quantum chromodynamics (QCD) for strong interactions and the Glashow–Weinberg–Salam theory of electroweak unification).

Let us now return to the technical development of the gauge theory ideas. First, let us observe that the gradient transformation of the gauge field (4.9) corresponds to the transformation of the covariant derivative

$$D'_\mu = \mathrm{e}^{i\omega} D_\mu \mathrm{e}^{-i\omega} \tag{4.14}$$

(which in fact justifies the adjective "covariant"). The last relation is easy to prove; letting the relevant differential operator act on an arbitrary test function f, one gets, on the one hand,

$$\begin{aligned}(\partial_\mu - igA'_\mu)f &= \partial_\mu f - ig\left(A_\mu + \frac{1}{g}\partial_\mu\omega\right)f \\ &= \partial_\mu f - igA_\mu f - i\partial_\mu\omega f \end{aligned} \tag{4.15}$$

and on the other hand,

$$\begin{aligned}
\mathrm{e}^{i\omega}(\partial_\mu - igA_\mu)\mathrm{e}^{-i\omega} f &= \mathrm{e}^{i\omega}\partial_\mu(\mathrm{e}^{-i\omega} f) - igA_\mu f \\
&= \mathrm{e}^{i\omega}(-i\partial_\mu\omega\mathrm{e}^{-i\omega} f + \mathrm{e}^{-i\omega}\partial_\mu f) - igA_\mu f \\
&= -i\partial_\mu\omega f + \partial_\mu f - igA_\mu f
\end{aligned} \tag{4.16}$$

so comparing the results (4.15) and (4.16), the identity (4.14) is seen to be valid. Note also that the transformation property (4.14) now makes the gauge invariance of the Lagrangian (4.12) transparent.

Once we have introduced a new field A_μ, we should add a corresponding kinetic term (i.e. a term involving the A_μ derivatives) as well in order to arrive at non-trivial Euler–Lagrange equations of motion for A_μ. If one wants to maintain gauge invariance, one has to invoke the familiar antisymmetric electromagnetic field tensor

$$F_{\mu\nu} = \partial_\mu A_\nu - \partial_\nu A_\mu \tag{4.17}$$

which is manifestly invariant under (4.9). The Lagrangian (4.10) may now be completed by adding a term quadratic in the $F_{\mu\nu}$ to get finally

$$\mathscr{L}_{\text{g.inv.}} = -\frac{1}{4}F_{\mu\nu}F^{\mu\nu} + i\bar{\psi}\not{D}\psi - m\bar{\psi}\psi \tag{4.18}$$

where the relevant coefficient has been fixed so as to correctly reproduce the standard Maxwell–Dirac equations.

When elaborating on the gauge theory formalism, it is important to realize that the gauge field tensor $F_{\mu\nu}$ can in fact be expressed in terms of the commutator of covariant derivatives, namely

$$-igF_{\mu\nu} = [D_\mu, D_\nu] \tag{4.19}$$

Indeed, let the commutator act on an arbitrary test function; one readily gets

$$\begin{aligned}
[D_\mu, D_\nu]f &= (\partial_\mu - igA_\mu)(\partial_\nu - igA_\nu)f - (\mu \leftrightarrow \nu) \\
&= \partial_\mu\partial_\nu f - ig\partial_\mu A_\nu f - igA_\nu\partial_\mu f - igA_\mu\partial_\nu f - g^2 A_\mu A_\nu f - (\mu \leftrightarrow \nu) \\
&= -ig(\partial_\mu A_\nu - \partial_\nu A_\mu)f = -igF_{\mu\nu} f
\end{aligned} \tag{4.20}$$

Note that the identity (4.19) makes the gauge invariance of the $F_{\mu\nu}$ obvious (taking into account the transformation properties of the covariant derivative shown in (4.14)). Of course, in the Abelian case, we know a right form of the gauge field kinetic term anyway, so the identity (4.19) is actually not of vital importance here (and essentially the same can be said about the transformation law of the covariant derivative (4.14)). However, the knowledge of such identities (which in the Abelian case can be viewed merely as an elegant reformulation of some familiar elementary relations) is extremely useful for the successful generalization of the gauge theory concepts to the non-Abelian case. This crucial development is the subject of the next section.

4.2 Non-Abelian gauge invariance

The ideas and techniques of the preceding section can be extended in a non-trivial way to the field theory models involving non-Abelian (i.e. non-commutative) internal symmetries, such as isospin. This extension is due to Yang and Mills [42], and it has become a true conceptual foundation of modern particle theory, fully recognized since the early 1970s. The famous Yang–Mills construction can be described in the following way. Let us consider again a free-field Lagrangian

$$\mathscr{L}_0 = i\bar{\Psi}\gamma^\mu\partial_\mu\Psi - m\bar{\Psi}\Psi \tag{4.21}$$

where Ψ now means a doublet of Dirac spinors

$$\Psi = \begin{pmatrix} \psi_1 \\ \psi_2 \end{pmatrix} \tag{4.22}$$

(the Ψ is thus in fact an eight-component object). The individual spinor fields ψ_1 and ψ_2 may be viewed as corresponding, e.g., to proton and neutron or neutrino and electron (or any other natural "isotopic doublet" in a generalized sense), but our considerations in this section will be in fact purely methodical and stay on a rather abstract level. Note also that in (4.21), one obviously has

$$m\bar{\Psi}\Psi = m(\bar{\psi}_1\psi_1 + \bar{\psi}_2\psi_2) \tag{4.23}$$

so that the components of the doublet are degenerate in mass. It is easy to realize that the Lagrangian (4.21) is invariant under matrix transformations

$$\begin{aligned} \Psi'(x) &= U\Psi(x) \\ \bar{\Psi}'(x) &= \bar{\Psi}(x)U^\dagger \end{aligned} \tag{4.24}$$

where U is a 2×2 unitary matrix (i.e. $U^\dagger = U^{-1}$) with constant elements. The U can otherwise be arbitrary, so the transformations (4.24) constitute the group $U(2)$. In what follows, we shall restrict ourselves to matrices with unit determinant, i.e. we consider only the special unitary group $SU(2)$, which is the lowest-dimensional non-Abelian group suitable for our discussion. When imposing such a restriction, one actually does not lose any essential feature connected with the non-Abelian nature of the general transformations (4.24), since any $U(2)$ matrix can be written as a $SU(2)$ matrix multiplied by a $U(1)$ phase factor. In other words, the $U(2)$ group is actually factorized as $SU(2) \times U(1)$ and the Abelian factor $U(1)$ can be treated separately in the manner already described in the preceding section. Thus, we will examine symmetry properties of the Lagrangian (4.21) with respect to the transformations

$$\begin{aligned} \Psi' &= S\Psi \\ \bar{\Psi}' &= \bar{\Psi}S^{-1} \end{aligned} \tag{4.25}$$

with $S^{-1} = S^\dagger$, $\det S = 1$. Any $SU(2)$ matrix can be described in terms of three independent real parameters; in particular, the S may be conveniently written in exponential form as

$$S = \exp(i\omega^a T^a) \tag{4.26}$$

with $T^a = \frac{1}{2}\tau^a$, where τ^a, $a = 1, 2, 3$ denote the Pauli matrices, and ω^a are the relevant parameters. The $SU(2)$ matrix (4.26) represents a rotation in an abstract internal-symmetry (isospin) space. Note that in a more general context, the exponential form (4.26) reflects the fact that $SU(2)$ is a particular example of a Lie group, with generators T^a satisfying commutation relations of the corresponding Lie algebra

$$[T^a, T^b] = if^{abc}T^c \tag{4.27}$$

where the symbol f^{abc} denotes generally the relevant structure constants; in the particular $SU(2)$ case, $f^{abc} = \epsilon^{abc}$ with ϵ^{abc} being the totally antisymmetric (three-dimensional) Levi-Civita symbol.

In analogy with the previously discussed Abelian case, let us now consider *local* $SU(2)$ transformations, i.e. let the parameters ω^a in (4.26) depend on x. As before, the derivative kinetic term in the Lagrangian (4.21) obviously violates such a local symmetry, and the corresponding non-invariance can now be expressed in terms of gradients of the three parameters ω^a. Invoking the ideas developed in the preceding section, one may therefore try to compensate the "local isospin" non-invariance by introducing an appropriate number of vector fields (three in the present case) endowed with suitable transformation properties. With the identity (4.14) in mind, it is not difficult to guess how such a procedure can be implemented technically: one can introduce the relevant compensation term by means of a covariant derivative (in analogy with (4.12), (4.13)) required to obey a transformation law which would represent a straightforward generalization of (4.14). The relevant transformation properties of the vector fields can then be deduced from the rule for the covariant derivative. Thus, we will introduce a triplet of vector fields A_μ^a, $a = 1, 2, 3$ (corresponding to the three "phases" $\omega^a(x)$) which can equivalently be described in terms of the matrix

$$A_\mu(x) = A_\mu^a(x) T^a \tag{4.28}$$

In the original free-field Lagrangian (4.21), we replace the ordinary derivative by the covariant one, i.e. extend (4.21) to the form

$$\begin{aligned} \mathscr{L} &= i\bar{\Psi}\gamma^\mu D_\mu \Psi - m\bar{\Psi}\Psi \\ &= i\bar{\Psi}\gamma^\mu(\partial_\mu - igA_\mu)\Psi - m\bar{\Psi}\Psi \end{aligned} \tag{4.29}$$

As we have stated above, D_μ should transform "covariantly" under the local $SU(2)$, i.e.

$$D'_\mu = S D_\mu S^{-1} \tag{4.30}$$

where $D'_\mu = \partial_\mu - igA'_\mu$. From (4.30), the corresponding transformation law for the matrix field A_μ (see (4.28)) can be deduced easily. Indeed, using (4.30) for an arbitrary test function (two-component column vector) f, one gets

$$\begin{aligned} (\partial_\mu - igA'_\mu)f &= S(\partial_\mu - igA_\mu)S^{-1}f \\ &= S(\partial_\mu S^{-1} f + S^{-1}\partial_\mu f - igA_\mu S^{-1} f) \\ &= S\partial_\mu S^{-1} f + \partial_\mu f - igSA_\mu S^{-1} f \end{aligned} \tag{4.31}$$

and from (4.31) then immediately follows

$$A'_\mu = SA_\mu S^{-1} + \frac{i}{g} S\partial_\mu S^{-1} \tag{4.32}$$

We should now make sure that the local $SU(2)$ transformation (4.32) is compatible with the structure (4.28), namely that the transformed matrix field A'_μ can be decomposed in terms of the $SU(2)$ generators in accordance with (4.28) (in other words, A'_μ should also be equivalent to a triplet of components A'^a_μ). Having in mind that a basis in the space of 2×2 matrices can be taken as consisting of the three Pauli matrices (which are traceless) and the unit matrix, it is clear that the problem reduces to showing that $\mathrm{Tr}\ A'_\mu = 0$. The first term on the right-hand side of (4.32) is manifestly traceless as a consequence of $\mathrm{Tr}\ A_\mu = 0$. As for the second term, vanishing of its trace is not immediately obvious, but the proof can be accomplished in an elementary way. Indeed, for any matrix $M(x) = \exp \Omega(x)$, one can show that

$$\mathrm{Tr}(M^{-1}\partial_\mu M) = \mathrm{Tr}(\partial_\mu \Omega) \quad (= \mathrm{Tr}(\partial_\mu \ln M)) \tag{4.33}$$

(this can be done by means of a straightforward power-series expansion of the relevant exponentials and by employing the cyclic property of the trace — of course, the trace symbol in (4.33) is absolutely essential for the validity of such an identity). From (4.33), the desired result

$$\mathrm{Tr}(S\partial_\mu S^{-1}) = 0 \tag{4.34}$$

follows immediately if one takes into account (4.26). Let us remark that an alternative proof of (4.34) can be accomplished by invoking an elegant general formula for differentiating a matrix exponential, namely

$$\partial_\mu \mathrm{e}^{\Omega(x)} = \int_0^1 dt\ \mathrm{e}^{t\Omega(x)} \partial_\mu \Omega(x) \mathrm{e}^{(1-t)\Omega(x)} \tag{4.35}$$

It is clear that the knowledge of the last identity already makes the proof of (4.34) trivial. The formula (4.35) is also highly useful in other field-theory applications; we leave its proof to the interested reader as an instructive exercise.

The matrix field A_μ (or an individual component A^a_μ of the corresponding "isomultiplet") obeying the local transformation law (4.32) is called the **non-Abelian gauge field** or **Yang–Mills field** corresponding to the gauge group $SU(2)$. Of course, the preceding construction can be generalized in a straightforward way, e.g., to any unitary group $SU(n)$. There we would have a traceless $n \times n$ matrix field, equivalent to a multiplet of $n^2 - 1$ Yang–Mills components; in particular, for $n = 3$, a relevant set of generators is represented by the well-known Gell-Mann matrices. The rule (4.32) represents a non-trivial generalization of the original gradient transformation (4.9); it is easy to check that in the Abelian case, i.e. when the S is taken simply as exp $(i\omega)$, the form (4.32) is indeed reduced to (4.9):

$$A'_\mu = \mathrm{e}^{i\omega} A_\mu \mathrm{e}^{-i\omega} + \frac{i}{g}\mathrm{e}^{i\omega}(-i\partial_\mu\omega)\mathrm{e}^{-i\omega} = A_\mu + \frac{1}{g}\partial_\mu\omega \tag{4.36}$$

There is still one point concerning the non-Abelian transformation (4.32) that should be clarified here. Once we have shown that A'_μ can be written as

$$A'_\mu = A^{a\prime}_\mu T^a \tag{4.37}$$

one may wonder whether the relation (4.32) could be recast in terms of the isotriplet components. For finite gauge transformations, it is not possible to obtain a transformation relation for the Yang–Mills components in a closed form (technically, this is precluded by complications stemming from the multiplication of matrix exponentials). However, for *infinitesimal* gauge transformations, one can get a simple and intuitively transparent result, which we are going to derive now. To this end, let us write the transformation matrices S and S^{-1} in the form

$$\begin{aligned} S(x) &= 1 + i\epsilon^a(x)T^a \\ S^{-1}(x) &= 1 - i\epsilon^a(x)T^a \end{aligned} \tag{4.38}$$

where $\epsilon^a(x)$ is an infinitesimal local parameter. Substituting (4.38) into (4.32), neglecting systematically terms of the order $O(\epsilon^2)$ and employing the commutation relation (4.27), one gets the desired result:

$$A^{a\prime}_\mu = A^a_\mu - f^{abc}\epsilon^b A^c_\mu + \frac{1}{g}\partial_\mu\epsilon^a \tag{4.39}$$

which is a standard (and in fact most frequently used) form of the gauge transformation of a Yang–Mills field (the rule (4.39) is of course quite general, not restricted to the $SU(2)$ case we have started with). Note that the second term on the right-hand side of (4.39) clearly reflects the non-Abelian nature of the considered transformation, and it is non-vanishing even for the parameters ϵ^a independent of the space–time coordinates (i.e. for *global* transformations), while the last term is simply an infinitesimal gradient transformation analogous to the Abelian case.

Now, in analogy with the Abelian case, we should look for an appropriate kinetic term for the Yang–Mills field. If one attempts to introduce simply a term quadratic in the first derivatives of A^a_μ, one finds that in the non-Abelian case, there is no straightforward way of doing it in a gauge invariant way. In particular, the simplest expression that would come first to one's mind, namely

$$\mathscr{L}_{\text{kin}} = -\frac{1}{4} A_{\mu\nu}^{a} A^{a\mu\nu} \tag{4.40}$$

with $A_{\mu\nu}^{a} = \partial_\mu A_\nu^a - \partial_\nu A_\mu^a$ is not gauge invariant (the reader is recommended to check this statement explicitly using the transformation rule (4.39)). At this point, one may invoke the identity (4.19), which provides a crucial inspiration. Indeed, let us define the quantity $F_{\mu\nu}$ in terms of the matrix covariant derivatives as

$$-igF_{\mu\nu} = [D_\mu, D_\nu] \tag{4.41}$$

Then one has

$$\begin{aligned}
[D_\mu, D_\nu]f &= (\partial_\mu - igA_\mu)(\partial_\nu - igA_\nu)f - (\mu \leftrightarrow \nu) \\
&= \partial_\mu \partial_\nu f - ig\partial_\mu A_\nu f - igA_\nu \partial_\mu f - igA_\mu \partial_\nu f \\
&\quad - g^2 A_\mu A_\nu f - (\mu \leftrightarrow \nu) \\
&= -ig(\partial_\mu A_\nu - \partial_\nu A_\mu)f - g^2[A_\mu, A_\nu]f \\
&= -ig(\partial_\mu A_\nu - \partial_\nu A_\mu - ig[A_\mu, A_\nu])f
\end{aligned} \tag{4.42}$$

since, in contrast with the Abelian case, the commutator $[A_\mu, A_\nu]$ is now non-zero. The $F_{\mu\nu}$ defined by (4.41) thus becomes

$$F_{\mu\nu} = \partial_\mu A_\nu - \partial_\nu A_\mu - ig[A_\mu, A_\nu] \tag{4.43}$$

For components defined by $F_{\mu\nu} = F_{\mu\nu}^{a} T^a$, one then gets, using the commutation relation (4.27),

$$F_{\mu\nu}^{a} = \partial_\mu A_\nu^a - \partial_\nu A_\mu^a + gf^{abc} A_\mu^b A_\nu^c \tag{4.44}$$

Obviously, the meaning of the construction (4.41) is that $F_{\mu\nu}$ now transforms covariantly under (4.32), i.e. in the same way as the covariant derivative:

$$F'_{\mu\nu} = SF_{\mu\nu}S^{-1} \tag{4.45}$$

Note that (4.45) also means that the change of $F_{\mu\nu}$ under local and global transformations is the same. For infinitesimal transformations one gets

$$F_{\mu\nu}^{a\prime} = F_{\mu\nu}^{a} - f^{abc}\epsilon^b F_{\mu\nu}^{c} \tag{4.46}$$

Thus, while the $F_{\mu\nu}$ is *not* gauge invariant (in contrast to the Abelian case), it is gauge *covariant*, and therefore, it can be used to construct readily a quadratic invariant, namely

$$\mathscr{L}^{(2)} = c\mathrm{Tr}(F_{\mu\nu}F^{\mu\nu}) \tag{4.47}$$

where c is an arbitrary constant. Taking into account that the generators T^a are normalized as

$$\mathrm{Tr}(T^aT^b) = \frac{1}{2}\delta^{ab} \tag{4.48}$$

we can recast (4.47) as

$$\mathscr{L}^{(2)} = \frac{1}{2}cF^a_{\mu\nu}F^{a\mu\nu} \tag{4.49}$$

For convenience, we will fix the overall coefficient in (4.49) in analogy with the Abelian (Maxwell) case (cf. (4.18)); then, putting together the gauge invariant pieces (4.29) and (4.49), the full Yang–Mills Lagrangian can be written as

$$\mathscr{L}_{\mathrm{YM}} = -\frac{1}{4}F^a_{\mu\nu}F^{a\mu\nu} + i\bar{\Psi}\not{D}\Psi - m\bar{\Psi}\Psi \tag{4.50}$$

The most remarkable contribution is contained in the first term, made entirely of the gauge fields. Let us denote it as $\mathscr{L}_{\mathrm{gauge}}$; using for the $F^a_{\mu\nu}$ the expression (4.44), one gets, after a simple manipulation,

$$\begin{aligned}\mathscr{L}_{\mathrm{gauge}} = -\frac{1}{4}F^a_{\mu\nu}F^{a\mu\nu} &= -\frac{1}{4}A^a_{\mu\nu}A^{a\mu\nu} - \frac{1}{2}gf^{abc}(\partial_\mu A^a_\nu - \partial_\nu A^a_\mu)A^{b\mu}A^{c\nu}\\ &\quad - \frac{1}{4}g^2 f^{abc}f^{ajk}A^b_\mu A^c_\nu A^{j\mu}A^{k\nu}\end{aligned} \tag{4.51}$$

Thus, the $\mathscr{L}_{\mathrm{gauge}}$ is seen to contain the desired kinetic terms, but in addition, we have earned some new cubic and quartic terms, i.e. contributions corresponding to *self-interactions* of the Yang–Mills fields, which have no analogue in the Abelian case (note that a term quartic in the electromagnetic field would describe classical light-by-light scattering, which of course does not exist in Maxwell electrodynamics). Obviously, the form of the Yang–Mills interaction terms is severely constrained by the gauge symmetry — note, e.g., that the coupling constant at the quartic term is the square of a relevant factor corresponding to the triple gauge field interaction, and the polynomial structure of both terms is also determined completely. Let us emphasize, however, that we have restricted ourselves to the terms with lowest dimension (equal to four). In principle, we could introduce higher powers of the $F_{\mu\nu}$ as well; however, at the quantum level, such contributions would spoil perturbative renormalizability, which is a desired technical aspect of the theory of electroweak interactions and in fact has been a primary goal of the inventors of the Standard Model in the late 1960s.

One more remark is perhaps in order here, concerning the gauge coupling constant appearing in the covariant derivative. Although we have been working with a multiplet ($SU(2)$ triplet) of vector Yang–Mills fields, we are allowed to introduce only a single coupling constant g. This is due to the fact that the gauge group under consideration (let us say $SU(n)$ in general) is simple (mathematically, this means that the corresponding Lie algebra does not contain any non-trivial invariant subalgebra); in other words, introducing more coupling constants in the covariant derivative would not be compatible with the commutation relations (4.27). The standard model of electroweak interactions is based on the gauge group $SU(2) \times U(1) = U(2)$ (which, from the mathematical point of view, is not even semi-simple because of the Abelian factor $U(1)$) and one then has to introduce, in general, two independent coupling constants g and g'. Some models of the so-called grand unification (unifying all forces except gravity) are based on simple groups (e.g., $SU(5)$ or $SO(10)$) and this leads to an interrelation between the coupling strengths of the electromagnetic, weak and strong interactions.

In the preceding discussion, we have not included any mass term for the Yang–Mills fields. The reason why we did not do so is that a mass term of the vector fields violates gauge invariance. In fact, from our pragmatic point of view, it is the renormalizability which is of interest to us, rather than a symmetry. As we shall see later, the bad news is that a naive mass term for the Yang–Mills field would in general spoil the renormalizability as well. Massless gauge fields do describe at least a part of our physical world: the modern theory of strong interaction (QCD) is based on the idea of exact gauge invariance under $SU(3)_{\text{colour}}$ and the corresponding (eight) gauge fields represent massless gluons interacting with coloured quarks (and with themselves). On the other hand, for the weak interaction theory, we need (a multiplet of) massive vector bosons to reproduce the familiar phenomenology of the beta decay, muon decay, etc. We will defer the subtle issue of mass generation in gauge theories to Chapter 6. In the next chapter, we will show how far one can get if the concept of Yang–Mills field is applied to the unification of weak and electromagnetic interactions, assuming for the moment that the relevant mass terms are added simply by hand. In other words, we will first discuss the theory proposed by Glashow in 1961, which now constitutes a well-established part of the present-day standard electroweak model and leads by itself to remarkable valid predictions.

Problems

4.1 Prove the identity (4.35).

4.2 We have seen that the term $\mathrm{Tr}(F_{\mu\nu}F^{\mu\nu})$ is gauge invariant by construction and has dimension *four*. Is there any other gauge invariant expression made of Yang–Mills fields A_μ only and carrying the same dimension?

4.3 Consider the $SU(2)$ gauge theory involving Yang–Mills fields coupled to a doublet of fermions. Write down the relevant equations of motion. What are the conserved Noether currents corresponding to the global $SU(2)$ symmetry?

4.4 Find an appropriate set of generators for the group $SU(2) \times SU(2)$.

4.5 What are generators of the group $SU(3)$, satisfying the normalization condition $\mathrm{Tr}(T^aT^b) = \frac{1}{2}\delta^{ab}$ (c.f. (4.48))? Find such a set of generators for $SU(4)$ and $SU(5)$.

Chapter 5

Electroweak Unification and Gauge Symmetry

5.1 $SU(2) \times U(1)$ gauge theory for leptons

In Chapter 3, we have emphasized that new particles and new interactions must be added to the old theory of weak and electromagnetic forces if one wants to tame divergent high-energy behaviour of the tree-level S-matrix elements, and we have argued that accomplishing this goal would also bear on the issue of perturbative renormalizability at higher orders. From a purely theoretical point of view, any such scenario should introduce either a new (neutral) massive vector boson or "exotic" fermions (such as heavy leptons). Of course, a combination of both schemes would be possible as well. At the same time, looking back in history, the technical experience gained by various people from the early studies of Yang–Mills theories suggested that non-Abelian gauge symmetry might control at least a part of the desirable divergence cancellations (one of the pioneering personalities in this direction was Veltman). A rigorous proof of perturbative renormalizability of a broad class of non-Abelian gauge models (incorporating the by now famous Higgs mechanism for the mass generation) was finally invented by 't Hooft in 1971 and this triggered the boom of "gauge model building" in the early 1970s. Various options were discussed, but the crucial moment was the experimental discovery of weak neutral currents in 1973 that pointed rather clearly towards a model involving a neutral vector boson as a viable candidate for the realistic description of the physical world. This development ultimately led to the recognition of a "minimal" gauge theory model, proposed by Weinberg and Salam in the late 1960s (who followed, in a sense, the earlier Glashow's attempt), as the "standard model" of electroweak interactions. The theory passed many stringent experimental tests in subsequent years and today represents one of the most successful physical theories of the 20th century.

The only essential missing link of the standard model is represented by the Higgs scalar boson, which emerges in the theory as a leftover of the electroweak symmetry breaking mechanism.

Thus, from now on, we will follow a path leading to the standard electroweak model. In this chapter, we will discuss the gauge structure of the model, leaving aside, for the moment, the subtle issue of the mass generation via Higgs mechanism. We will restrict ourselves to the leptonic sector of the elementary fermion spectrum, since most of the important aspects of the electroweak gauge symmetry can be displayed within such a reduced framework. Thus, in this chapter, we will stay essentially within the 1961 Glashow model, which constitutes a part of the present-day Standard Model.

The idea of unifying weak and electromagnetic interactions on the basis of a non-Abelian gauge symmetry is in fact rather appealing *a priori*. Both forces are universal and involve vector (or axial-vector) currents that can be coupled naturally to vector fields, and these may constitute a Yang–Mills multiplet (note that a pioneering work in this direction is due to Schwinger (1957)). Using then our previous considerations as a technical guide, one may guess that in a "minimal variant" of the electroweak unification, *four* gauge fields are actually needed, corresponding to the W^+, W^-, photon and a new neutral vector boson. Thus, an appropriate gauge group is $SU(2) \times U(1)$. Later on, we shall see that a fourth vector boson is indeed necessary for a successful electroweak gauge unification (involving only conventional leptons) even for purely "algebraic" reasons, i.e. without making any further reference to the high-energy behaviour of Feynman diagrams. Obviously, an important conceptual problem is how to accommodate in the envisaged unified theory both the vectorial (parity-conserving) electromagnetic current and the left-handed weak charged current manifesting maximum parity violation. At first sight, this dramatic difference between the two forces might seem to be a major obstacle to their unification, but as we shall see, the distinct chiral structure of the relevant currents can in fact be incorporated quite easily. The right idea is to consider the chiral components of the fermion fields as independent "building blocks" and assign different transformation properties to the left-handed and right-handed components when writing down the $SU(2) \times U(1)$ gauge invariant Lagrangian. In particular, the left-handed fermion fields are placed in $SU(2)$ doublets, while the right-handed fermions are taken to be singlets — as we shall see, this is precisely the choice leading automatically to the desired $V - A$ structure of the charged weak current.

Thus, let us now proceed to construct the relevant Lagrangian invariant under the local $SU(2) \times U(1)$. Needless to say, as in the preceding chapter, we start at the level of classical field theory; we will comment on the quantization later on. To begin with, we are going to consider leptons of the electron type. The left-handed components $\nu_L = \frac{1}{2}(1 - \gamma_5)\nu$ and $e_L = \frac{1}{2}(1 - \gamma_5)e$ form an $SU(2)$ doublet

$$L^{(e)} = \begin{pmatrix} \nu_L \\ e_L \end{pmatrix} \tag{5.1}$$

(as usual, we denote the individual fields by letters labelling normally the corresponding particles and write for the moment ν instead of ν_e; in what follows, we will also drop the superscript on the L for brevity). The right-handed fields $e_R = \frac{1}{2}(1 + \gamma_5)e$ and $\nu_R = \frac{1}{2}(1 + \gamma_5)\nu$ are $SU(2)$ singlets. Note that we have included the right-handed component of the neutrino field in addition to the mandatory ν_L (by doing it, we keep an open mind about a possibility of non-vanishing neutrino mass). We should also specify the transformation properties of lepton fields under the Abelian subgroup $U(1)$ and clarify the form of the relevant covariant derivatives acting on the lepton fields. To this end, the following simple observation will be helpful: if an Abelian gauge field B_μ transforms as

$$B'_\mu = B_\mu + \frac{1}{g}\partial_\mu \omega \tag{5.2}$$

and a Dirac field Ψ is transformed according to

$$\Psi' = \mathrm{e}^{iY\omega}\Psi \tag{5.3}$$

with Y being a real number (note that Ψ may in general mean a multiplet of fields), then the expression

$$\bar{\Psi}\gamma^\mu D_\mu \Psi \tag{5.4}$$

is invariant under the local $U(1)$ if the covariant derivative D_μ has the form

$$D_\mu = \partial_\mu - igYB_\mu \tag{5.5}$$

(the proof of this statement is left to the reader as a trivial exercise).

The meaning of (5.3) consists in pointing out the existence of infinitely many (inequivalent) representations of the Abelian group $U(1)$, labelled here by an arbitrary real parameter Y. We may now use this freedom to assign different (in general arbitrary) values of the real parameter — called usually "weak hypercharge" — to the doublet L and singlets e_R,

ν_R, to characterize their transformation properties with respect to the Abelian factor $U(1)$ of the considered gauge group. The $SU(2)\times U(1)$ gauge invariant Lagrangian involving lepton interactions thus can be written as

$$\begin{aligned}\mathscr{L}_{\text{lepton}} &= i\bar{L}\gamma^\mu(\partial_\mu - igA_\mu^a\frac{\tau^a}{2} - ig'Y_L B_\mu)L \\ &\quad + i\bar{e}_R\gamma^\mu(\partial_\mu - ig'Y_R^{(e)}B_\mu)e_R + i\bar{\nu}_R\gamma^\mu(\partial_\mu - ig'Y_R^{(\nu)}B_\mu)\nu_R\end{aligned} \tag{5.6}$$

where A_μ^a, $a = 1, 2, 3$ denote the triplet of Yang–Mills fields corresponding to the "weak isospin" subgroup $SU(2)$, and the B_μ is the gauge field associated with the weak hypercharge subgroup $U(1)$. Of course, in the covariant derivatives acting on the e_R and ν_R, the non-Abelian part is absent since the $SU(2)$ generators are trivial in the singlet representation.

In (5.6), we have introduced arbitrary values of the weak hypercharges for the L, e_R and ν_R. It is important to employ such a general parametrization at this initial stage; we shall see later that the relevant values of weak hypercharge are constrained non-trivially by the requirement of recovering — within the considered unified theory — a standard electromagnetic interaction of leptons carrying the usual charges. Looking ahead, let us state already here that a rule providing automatically the right values of Y reads

$$Q = T_3 + Y \tag{5.7}$$

where Q is the relevant charge (in units of the positron charge, so, $Q_e = -1$, etc.) and T_3 is the value of weak isospin, defined as the eigenvalue of the corresponding $SU(2)$ generator; in particular, for the considered $SU(2)$ doublet and singlets, respectively, one has

$$T_3(\nu_L) = +\frac{1}{2}, \quad T_3(e_L) = -\frac{1}{2}, \quad T_3(\nu_R) = 0, \quad T_3(e_R) = 0 \tag{5.8}$$

From (5.7) and (5.8), one then gets

$$Y_L = -\frac{1}{2}, \quad Y_R^{(e)} = -1, \quad Y_R^{(\nu)} = 0 \tag{5.9}$$

The textbook expositions of the standard electroweak model usually start immediately with the relation (5.7) yielding the "physical values" (5.9). Here, we will keep the general parametrization (5.6) and derive the rule (5.7) yielding the values (5.9) in Section 5.3, where the electromagnetic interaction will be discussed in detail.

Note that in (5.6) we have included two independent coupling constants g and g'. This, of course, is related to the fact that the considered gauge group is not simple (cf. the discussion at the end of the previous chapter). Such a dichotomy represents an obvious aesthetic flaw of the envisaged electroweak unification — one would certainly prefer a unified picture of the two interactions based on a single common coupling constant. However, one cannot arbitrarily set $g = g'$, since such a relation would be violated by renormalization effects at the quantum level (and, as we know today, it would also contradict experimental facts). Nevertheless, in the course of the subsequent discussion, it will become clear that the term "electroweak unification" does match with the $SU(2) \times U(1)$ model (the discussion of the weak neutral currents in Section 5.4 is particularly instructive in this respect). On the other hand, it may well be that a simple group of "grand unification" of the electroweak and strong interactions lies ahead in our future. The goal of the "incomplete" electroweak model is much more modest: it unifies the electrodynamics with the low-energy $V - A$ theory of weak interactions, and does provide a realistic and highly accurate description of the electroweak forces at the currently accessible energies. In this sense, the standard electroweak model can be viewed as an effective approximation (at relatively low energies) of a deeper theory whose contours we may now only guess.

Coming back to the structure of the Lagrangian (5.6), it should be stressed that the gauge fields A^a_μ and B_μ need not (and in fact do not) have any direct physical meaning. The physical vector fields will emerge as their linear combinations, displaying thus a characteristic feature of the electroweak unification. The physical contents of the leptonic Lagrangian will be discussed in subsequent sections, but before proceeding to this fundamental task, we should add to (5.6) a gauge invariant contribution involving the kinetic term of the vector fields. In the spirit of the general Yang–Mills construction described in the preceding chapter, we may write

$$\mathscr{L}_{\text{gauge}} = -\frac{1}{4} F^a_{\mu\nu} F^{a\mu\nu} - \frac{1}{4} B_{\mu\nu} B^{\mu\nu} \tag{5.10}$$

where

$$F^a_{\mu\nu} = \partial_\mu A^a_\nu - \partial_\nu A^a_\mu + g\epsilon^{abc} A^b_\mu A^c_\nu \tag{5.11}$$

and

$$B_{\mu\nu} = \partial_\mu B_\nu - \partial_\nu B_\mu \tag{5.12}$$

In the remainder of this chapter, we will thus analyze the form

$$\mathscr{L}_1 = \mathscr{L}_{\text{gauge}} + \mathscr{L}_{\text{lepton}} \tag{5.13}$$

which in fact constitutes the first part of gauge invariant Glashow–Weinberg–Salam (GWS) electroweak Lagrangian.

5.2 Charged current weak interaction

Let us consider the lepton Lagrangian (5.6). We should check that it contains, among other things, the conventional weak interaction of the left-handed (i.e. $V - A$) charged current (made of the neutrino and electron fields) with a charged intermediate vector boson. It is not difficult to guess that such a term could originate from the non-Abelian part of the covariant derivative in (5.6), in particular from the two terms involving the antidiagonal Pauli matrices τ^1 and τ^2. Indeed, the interaction part of (5.6) reads

$$\begin{aligned}\mathscr{L}^{(\text{int})}_{\text{lepton}} &= g\bar{L}\gamma^\mu \frac{\tau^a}{2} L A^a_\mu + g' Y_L \bar{L}\gamma^\mu L B_\mu \\ &\quad + g' Y_R^{(e)} \bar{e}_R \gamma^\mu e_R B_\mu + g' Y_R^{(\nu)} \bar{\nu}_R \gamma^\mu \nu_R B_\mu \end{aligned} \tag{5.14}$$

This can be recast as

$$\begin{aligned}\mathscr{L}^{(\text{int})}_{\text{lepton}} &= g\left(\bar{L}\gamma^\mu \frac{\tau^+}{2} L A^-_\mu + \bar{L}\gamma^\mu \frac{\tau^-}{2} L A^+_\mu + \bar{L}\gamma^\mu \frac{\tau^3}{2} L A^3_\mu\right) \\ &\quad + g' Y_L \bar{L}\gamma^\mu L B_\mu + g' Y_R^{(e)} \bar{e}_R \gamma^\mu e_R B_\mu + g' Y_R^{(\nu)} \bar{\nu}_R \gamma^\mu \nu_R B_\mu \end{aligned} \tag{5.15}$$

where $\tau^\pm = \frac{1}{\sqrt{2}}(\tau^1 \pm i\tau^2)$ and $A^\pm_\mu = \frac{1}{\sqrt{2}}(A^1_\mu \pm iA^2_\mu)$, i.e.

$$\frac{1}{2}\tau^+ = \frac{1}{\sqrt{2}}\begin{pmatrix} 0 & 1 \\ 0 & 0 \end{pmatrix}, \quad \frac{1}{2}\tau^- = \frac{1}{\sqrt{2}}\begin{pmatrix} 0 & 0 \\ 1 & 0 \end{pmatrix} \tag{5.16}$$

Using (5.16) in (5.15) and working out the simple matrix products in the first two terms, one readily gets

$$\mathscr{L}^{(\text{int})}_{\text{lepton}} = \frac{g}{\sqrt{2}}(\bar{\nu}_L \gamma^\mu e_L W^+_\mu + \bar{e}_L \gamma^\mu \nu_L W^-_\mu) + \mathscr{L}_{\text{diag.}} \tag{5.17}$$

where we have denoted $W^\pm_\mu = A^\mp_\mu$ and under the symbol $\mathscr{L}_{\text{diag.}}$, we have collected all the remaining terms from (5.15), i.e. those involving diagonal 2×2 matrices (τ^3 or $\mathbb{1}$). We have passed from $A^\pm_\mu$ to $W^\mp_\mu$ so as to recover precisely the structure of the old charged current weak interaction discussed in the previous chapters. Indeed, the first two terms in (5.17) can be obviously written in the form

$$\mathscr{L}_{CC} = \frac{g}{2\sqrt{2}}\bar{\nu}\gamma^\mu(1-\gamma_5)eW^+_\mu + \text{h.c.} \tag{5.18}$$

which is seen to coincide with the electron part of the weak interaction Lagrangian (3.13) and also explains the convention used for the definition of the weak coupling constant in the old theory.

Of course, recovering the charged current leptonic weak interaction within the framework of the considered gauge theory should not come as a surprise — we have actually "ordered" this result by imposing different $SU(2)$ transformation properties for the left-handed and right-handed fields, respectively. Now, it is also clear that introducing a doublet of right-handed leptons along with the L in (5.6) would produce a purely vector-like charged weak current (which would be a phenomenological disaster). Thus, the assignment of the $SU(2)$ transformation properties to the chiral components of lepton fields that we have chosen here simply means that we have "translated" the requirement of the $V-A$ structure of weak charged currents into the gauge theory language — by specifying the representation contents of the matter (lepton) fields.

5.3 Electromagnetic interaction

Let us now proceed to analyze the "diagonal" part of the leptonic interaction Lagrangian (5.17), i.e. the contribution

$$\begin{aligned}\mathscr{L}_{\text{diag}} &= \frac{1}{2}g\bar{L}\gamma^\mu\tau^3 LA^3_\mu + g'Y_L\bar{L}\gamma^\mu LB_\mu \\ &\quad + g'Y_R^{(e)}\bar{e}_R\gamma^\mu e_R B_\mu + g'Y_R^{(\nu)}\bar{\nu}_R\gamma^\mu\nu_R B_\mu\end{aligned} \tag{5.19}$$

Taking into account that

$$\tau^3 = \begin{pmatrix}1 & 0\\ 0 & -1\end{pmatrix} \tag{5.20}$$

and working out the matrix multiplication in (5.19), one readily gets

$$\begin{aligned}\mathscr{L}_{\text{diag.}} &= \frac{1}{2}g\bar{\nu}_L\gamma^\mu\nu_L A^3_\mu - \frac{1}{2}g\bar{e}_L\gamma^\mu e_L A^3_\mu + g'Y_L\bar{\nu}_L\gamma^\mu\nu_L B_\mu \\ &\quad + g'Y_L\bar{e}_L\gamma^\mu e_L B_\mu + g'Y_R^{(e)}\bar{e}_R\gamma^\mu e_R B_\mu + g'Y_R^{(\nu)}\bar{\nu}_R\gamma^\mu\nu_R B_\mu\end{aligned} \tag{5.21}$$

From the last expression, it is obvious that neither A^3_μ nor B_μ can be identified directly with the electromagnetic field; more precisely, there is no choice of the weak hypercharges that would enable one to make such an identification — both these fields are generally coupled to the neutrino and none of them has a purely vectorial coupling to the electron. Now, it is also clear why an electroweak unification based on the simple group $SU(2)$ (and involving only the ordinary leptons) would not work: discarding the field B_μ, one is left with the A^3_μ couplings only, which certainly are not of an electromagnetic type. In other words, the $SU(2)$ gauge theory would not be able to accommodate both the left-handed weak charged currents and the vectorial electromagnetic current made of the conventional leptons. We thus arrive at an independent, "purely algebraic" argument in favour of the $SU(2) \times U(1)$ electroweak unification without invoking an analysis of the high-energy behaviour of Feynman diagrams. It should be stressed, however, that an $SU(2)$ unification does work if one introduces some extra leptons of the electron type; such a theoretical scenario was developed by Georgi and Glashow in 1972, but no exotic (heavy) leptons demanded by that theory have been observed so far.

Although the gauge fields A^3_μ and B_μ have no direct physical interpretation, one may try to produce physical fields by making appropriate linear combinations of the former. In particular, we are going to consider an orthogonal transformation

$$\begin{aligned} A^3_\mu &= \cos\theta_W Z_\mu + \sin\theta_W A_\mu \\ B_\mu &= -\sin\theta_W Z_\mu + \cos\theta_W A_\mu \end{aligned} \tag{5.22}$$

where A_μ will be required to have properties of the electromagnetic field and Z_μ represents a new neutral vector field. θ_W is an arbitrary angle at the present moment, but it will be expressed through the other parameters of the theory after imposing the necessary physical requirements. It is usually called the "Weinberg angle" or "weak mixing angle". We should perhaps explain here the reason why we have chosen an *orthogonal* transformation. The orthogonality is in fact necessary for preserving the diagonal structure of kinetic terms of the vector fields, as one can see easily. For the original gauge fields A^a_μ and B_μ, one has

$$\mathscr{L}^{(\text{kin})}_{\text{gauge}} = -\frac{1}{4} A^a_{\mu\nu} A^{a\mu\nu} - \frac{1}{4} B_{\mu\nu} B^{\mu\nu} \tag{5.23}$$

where $A^a_{\mu\nu} = \partial_\mu A^a_\nu - \partial_\nu A^a_\mu$ and $B_{\mu\nu} = \partial_\mu B_\nu - \partial_\nu B_\mu$ (cf. (5.10)). In the preceding section, we have already passed from A^1_μ and A^2_μ to the charged vector fields $W^\pm_\mu$ through another (complex) orthogonal transformation; the expression (5.23) can thus be recast as

$$\mathscr{L}^{(\text{kin})}_{\text{gauge}} = -\frac{1}{2} W^-_{\mu\nu} W^{+\mu\nu} - \frac{1}{4} A^3_{\mu\nu} A^{3\mu\nu} - \frac{1}{4} B_{\mu\nu} B^{\mu\nu} \tag{5.24}$$

Using (5.22) in (5.24), one finally gets

$$\mathscr{L}^{(\text{kin})}_{\text{gauge}} = -\frac{1}{2} W^-_{\mu\nu} W^{+\mu\nu} - \frac{1}{4} A_{\mu\nu} A^{\mu\nu} - \frac{1}{4} Z_{\mu\nu} Z^{\mu\nu} \tag{5.25}$$

The orthogonality of (5.22) thus prevents any $A - Z$ mixing terms from appearing in (5.25).

After these explanatory remarks, let us now substitute the transformation (5.22) into (5.21). One gets

$$\mathscr{L}_{\text{diag}} = \mathscr{L}^{(A)}_{\text{diag.}} + \mathscr{L}^{(Z)}_{\text{diag.}} \tag{5.26}$$

where

$$\begin{aligned}
\mathscr{L}^{(A)}_{\text{diag.}} = \Bigg[& \frac{1}{2} g \sin\theta_W \bar{\nu}_L \gamma^\mu \nu_L + Y_L g' \cos\theta_W \bar{\nu}_L \gamma^\mu \nu_L + Y^{(\nu)}_R g' \cos\theta_W \bar{\nu}_R \gamma^\mu \nu_R \\
& - \frac{1}{2} g \sin\theta_W \bar{e}_L \gamma^\mu e_L + Y_L g' \cos\theta_W \bar{e}_L \gamma^\mu e_L \\
& + Y^{(e)}_R g' \cos\theta_W \bar{e}_R \gamma^\mu e_R \Bigg] A_\mu
\end{aligned} \tag{5.27}$$

and $\mathscr{L}^{(Z)}_{\text{diag.}}$ denotes the part involving the Z field; this term will be discussed in detail later on. We would like to interpret (5.27) as the standard electromagnetic interaction of leptons, so it should have the corresponding familiar properties: in particular, the neutrino fields should be absent from (5.27) and the right-handed and left-handed components of the electron field should interact with A_μ with an equal strength (in other words, the electromagnetic current must be pure vector). Of course, to meet these requirements, we can use the freedom we still have in the assignments of the weak hypercharge values. The first requirement thus leads to the conditions

$$Y^{(\nu)}_R = 0 \tag{5.28}$$

and

$$\frac{1}{2} g \sin\theta_W + Y_L g' \cos\theta_W = 0 \tag{5.29}$$

Similarly, the requirement of vectorial (i.e. parity-conserving) nature of the electromagnetic current leads to

$$-\frac{1}{2}g\sin\theta_W + Y_L g'\cos\theta_W = Y_R^{(e)} g'\cos\theta_W \tag{5.30}$$

Combining (5.29) with (5.30), we obtain immediately

$$Y_R^{(e)} = 2Y_L \tag{5.31}$$

and from (5.29), the weak mixing angle can be expressed as

$$\tan\theta_W = -2Y_L\frac{g'}{g} \tag{5.32}$$

Writing now the electromagnetic interaction conventionally as

$$\mathscr{L}_{\text{lepton}}^{(\text{EM})} = -e\ \bar{e}\gamma^\mu e A_\mu \tag{5.33}$$

the relevant coupling constant e is given by one of the equivalent expressions in (5.30) with the negative sign; using (5.31) and (5.32), one then obtains a remarkably simple relation

$$e = g\sin\theta_W \tag{5.34}$$

or in terms of g, g' and Y_L

$$e = -2Y_L\frac{gg'}{\sqrt{g^2 + 4Y_L^2 g'^2}} \tag{5.35}$$

Note that (5.34) means

$$e < g \tag{5.36}$$

From the previous discussion, it is clear that the *strict inequality* must hold indeed, as the $\cos\theta_W$ must not be zero. The relation (5.36) (or (5.34), respectively) is usually called the **unification condition** as it relates the coupling strengths of the old weak interaction and electromagnetism, unified within the $SU(2) \times U(1)$ gauge theory; note that before the unification, the ratio of e and g was completely unconstrained. We will discuss an important physical consequence of the unification condition in the next section.

We have seen that the weak hypercharge values are essentially fixed by the requirement of internal consistency of the electroweak unification, up to Y_L, which can be arbitrary (but non-zero). We believe that it may be instructive for the reader, in particular for a beginner in the field, to fully realize such a freedom of parametrization before adopting the conventional values of Y mentioned in Section 3.1 (cf. (5.7)–(5.9)) — this is why we have devoted a relatively large space to this general discussion.

In fact, it is natural to expect that the electric charge should be a linear combination of the weak isospin and weak hypercharge simply because the generators T^3 and Y are both represented by diagonal matrices. Summarizing now our previous knowledge, one may note that the Y values indeed satisfy a relation

$$Q = T_3 + cY \tag{5.37}$$

where c is a real coefficient (its value being fixed, e.g., by an arbitrarily chosen Y_L). The conventional choice corresponds to $c = 1$: this leads to $Y_L = -1/2$ and the formula (5.32) for the weak mixing angle is thus simplified to

$$\tan\theta_W = \frac{g'}{g} \tag{5.38}$$

i.e. the $\sin\theta_W$ and $\cos\theta_W$ are expressed by the aesthetically pleasing formulae

$$\cos\theta_W = \frac{g}{\sqrt{g^2 + g'^2}}, \quad \sin\theta_W = \frac{g'}{\sqrt{g^2 + g'^2}} \tag{5.39}$$

The upshot of all this is that in further study of the standard electroweak model, the reader can, for convenience, use the weak hypercharge values determined by the rule

$$Q = T_3 + Y \tag{5.40}$$

which automatically lead to the correct structure of the electromagnetic current and to a simple relation for the mixing angle θ_W. Nevertheless, we will come back to the general parametrization in Section 5.5 (and also in the next chapter in connection with a mass formula for the W and Z fields) to show that physical results do not depend on the choice of the non-zero value of the Y_L.

5.4 Unification condition and *W* boson mass

Let us now return to the unification condition (5.36). It gives a simple lower bound for the weak coupling constant g (at least at the level of the classical Lagrangian, i.e. at the tree level within quantum theory) and one might therefore employ it to obtain useful constraints on the physical quantities, expressed in terms of g. In particular, in this section, we will discuss a lower bound for the W boson mass which follows from (5.36). We have not introduced any mass terms for the vector fields so far, but the good old W boson model should of course be fully reproduced within the $SU(2) \times U(1)$ unification scheme. Thus, at the present stage, we may simply put a mass term for the $W^{\pm}$ fields by hand (the Z boson should also become massive in order to avoid a new long-range force different from the electromagnetism) and the photon will remain massless. We may then adopt, for the moment, a standard (perturbative) canonical quantization procedure and consider the corresponding Feynman graphs — such a program can be successfully carried out at least at the tree level. As we have seen in Chapter 3, in a model involving charged intermediate vector boson, there is a relation between the weak coupling constant g, the W boson mass and the Fermi coupling G_{F}

$$\frac{G_{\mathrm{F}}}{\sqrt{2}} = \frac{g^2}{8m_W^2} \tag{5.41}$$

which tells us that in the low-energy limit, a Fermi-type model represents a good effective weak interaction theory. The relation (5.41) must then also be valid within the $SU(2) \times U(1)$ unified theory and using the unification condition (5.34) in (5.41), one gets a formula for the W boson mass:

$$m_W = \left(\frac{\pi\alpha}{G_F\sqrt{2}}\right)^{1/2} \frac{1}{\sin\theta_W} \tag{5.42}$$

where we have introduced the fine structure constant $\alpha = e^2/(4\pi)$. The weak mixing angle θ_W is a free parameter of the considered model of electroweak unification, which must be measured independently (θ_W can be traded for other physical parameters, but its numerical value is not predicted by the standard electroweak model — such a prediction can only be accomplished within an appropriate grand unification scheme). θ_W can be measured, e.g., in neutrino scattering processes (see Section 5.6 for an explicit example), so that the formula (5.42) (corrected by including higher-order quantum effects within the full standard model) did provide a *prediction* for the W boson mass before its actual discovery in 1983. Using in (5.42) the current experimental value $\sin^2\theta_W \doteq 0.23$, taking $\alpha \doteq 1/137$ and $G_F \doteq 1.166 \times 10^{-5}\ \mathrm{GeV}^{-2}$, one gets $m_W \doteq 77.7$ GeV. For the corrected value, one then obtains roughly 80 GeV (the relevant corrections were calculated first by Veltman in 1980), which is in agreement with the current experimental value $m_W = (80.377 \pm 0.012)$ GeV (cf. [6]). Note that the main effect of these higher-order corrections on the W mass can be reproduced by replacing the traditional low-energy value of the fine structure constant by the "running electromagnetic coupling" at the W mass scale, i.e. $\alpha(m_W^2) \doteq 1/128$.

In any case, (5.42) obviously implies a *lower bound* for the W mass, namely

$$m_W > \left(\frac{\pi\alpha}{G_F\sqrt{2}} \right)^{1/2} \tag{5.43}$$

(of course, the bound (5.43) is an immediate consequence of (5.41) and the unification condition written as the inequality $g > e$). For $\alpha = 1/137$, one thus gets roughly

$$m_W > 37 \text{ GeV} \tag{5.44}$$

Let us also remark that at this stage, the Z boson mass can be entirely arbitrary; we will touch the problem of the Z mass determination from the low-energy scattering experiments in Section 5.6. To get a prediction for the Z mass, one has to settle the subtle issue of mass generation in gauge theories (which we have trivialized for the moment). This will be the subject of the next chapter.

Finally, let us stress that the lower bound for the W boson mass (5.43) is not a universal feature of any electroweak unification — the condition (5.36) is indeed intimately connected with the particular $SU(2) \times U(1)$ unification scheme. For example, in the $SU(2)$ (or $O(3)$) model of Georgi and Glashow [69] involving heavy leptons, the unification condition reads

$$g \leq e\sqrt{2} \tag{5.45}$$

which implies an *upper* bound for the W mass, namely

$$m_W \leq \left(\frac{\pi\alpha\sqrt{2}}{G_F} \right)^{1/2} \doteq 53 \text{ GeV} \tag{5.46}$$

The currently known experimental value of the m_W thus certainly excludes the minimal scenario based on heavy leptons, but this still remains to be of methodical interest as a construction of electroweak unification alternative to the standard model. For details of the heavy lepton scheme, the interested reader is referred to the literature.

5.5 Weak neutral currents

Let us now examine the interactions of leptons with the Z boson, i.e. the term denoted by $\mathscr{L}_{\text{diag.}}^{(Z)}$ in (5.26). To keep the discussion as general as possible, we will maintain an arbitrary value of the weak hypercharge Y_L — of course, and at the same time we will utilize the relations $Y_R^{(e)} = 2Y_L$ and $Y_R^{(\nu)} = 0$ established in Section 5.3. We will show that physical results do not depend on Y_L. According to (5.21) (where the substitution (5.22) is performed), we have, grouping together the interactions of the individual chiral components of lepton fields,

$$\begin{aligned}\mathscr{L}^{(Z)}_{\text{diag.}} = &\left(\frac{1}{2}g\cos\theta_W - Y_L g'\sin\theta_W\right)\bar{\nu}_L\gamma^\mu\nu_L Z_\mu \\ &+ \left(-\frac{1}{2}g\cos\theta_W - Y_L g'\sin\theta_W\right)\bar{e}_L\gamma^\mu e_L Z_\mu \\ &- 2Y_L g'\sin\theta_W\bar{e}_R\gamma^\mu e_R Z_\mu \end{aligned} \tag{5.47}$$

The last expression can be conveniently recast as

$$\begin{aligned}\mathscr{L}^{(Z)}_{\text{diag.}} = &\frac{g}{\cos\theta_W}\left(\frac{1}{2}\cos^2\theta_W - Y_L\frac{g'}{g}\sin\theta_W\cos\theta_W\right)\bar{\nu}_L\gamma^\mu\nu_L Z_\mu \\ &+ \frac{g}{\cos\theta_W}\left(-\frac{1}{2}\cos^2\theta_W - Y_L\frac{g'}{g}\sin\theta_W\cos\theta_W\right)\bar{e}_L\gamma^\mu e_L Z_\mu \\ &+ \frac{g}{\cos\theta_W}\left(-2Y_L\frac{g'}{g}\sin\theta_W\cos\theta_W\right)\bar{e}_R\gamma^\mu e_R Z_\mu \end{aligned} \tag{5.48}$$

and employing the relation $\tan\theta_W = -2Y_L g'/g$ (see (5.32)), one gets, after a simple manipulation,

$$\begin{aligned}\mathscr{L}^{(Z)}_{\text{diag.}} = &\frac{g}{\cos\theta_W}\left[\frac{1}{2}\bar{\nu}_L\gamma^\mu\nu_L + \left(-\frac{1}{2} + \sin^2\theta_W\right)\bar{e}_L\gamma^\mu e_L \right. \\ &\left. + \sin^2\theta_W\bar{e}_R\gamma^\mu e_R\right] Z_\mu \end{aligned} \tag{5.49}$$

The form (5.49) represents an interaction of the Z boson field with "weak neutral leptonic currents". Note that the adjective "neutral" in the present context means that the corresponding current is composed of fermion fields carrying the same charge — in this sense, the electromagnetic current is neutral as well. We see that any possible dependence on Y_L drops out, and the neutral current (NC) interaction is fully parametrized in terms of the CC coupling strength g and the weak mixing angle θ_W. Let us stress that the form of the NC interaction (5.49) is a non-trivial *prediction* of the considered electroweak unification — having fixed the values of the free parameters (weak hypercharges) so as to recover the standard electromagnetic interaction, the weak NC interactions are fully determined. From now on, we will denote the term (5.49) by the symbol $\mathscr{L}_{NC}^{(e)}$ (referring explicitly to leptons of electron type) and introduce a frequently used notation for the relevant coupling strengths by writing

$$\mathscr{L}_{NC}^{(e)} = \frac{g}{\cos\theta_W} \sum_{f=\nu,e} (\varepsilon_L^{(f)} \bar{f}_L \gamma^\mu f_L + \varepsilon_R^{(f)} \bar{f}_R \gamma^\mu f_R) Z_\mu \tag{5.50}$$

where

$$\varepsilon_L^{(\nu)} = \frac{1}{2}, \quad \varepsilon_R^{(\nu)} = 0, \quad \varepsilon_L^{(e)} = -\frac{1}{2} + \sin^2\theta_W, \quad \varepsilon_R^{(e)} = \sin^2\theta_W \tag{5.51}$$

This description of the neutral current structure in terms of the parameter $\sin^2\theta_W$ exhibits a famous rule characteristic of the $SU(2) \times U(1)$ standard electroweak model, namely

$$\varepsilon_{L,R}^{(f)} = T_{3L,R}^{(f)} - Q^{(f)} \sin^2\theta_W \tag{5.52}$$

The reader can easily verify that (5.51) is indeed reproduced when one uses in (5.52) the relevant values of the electric charge and weak isospin for the chiral components of lepton fields (cf. (5.8)).

We could also express the coupling constants for NC interactions in terms of g and e — in other words, in terms of the parameters of the "old physics" (weak and electromagnetic interactions before the gauge unification). Introducing an alternative notation for the NC couplings (5.50)

$$\mathscr{L}_{NC}^{(e)} = g_L^{(\nu)} \bar{\nu}_L \gamma^\mu \nu_L Z_\mu + g_L^{(e)} \bar{e}_L \gamma^\mu e_L Z_\mu + g_R^{(e)} \bar{e}_R \gamma^\mu e_R Z_\mu \tag{5.53}$$

and using the relation

$$\sin\theta_W = \frac{e}{g} \tag{5.54}$$

(which is valid independently of the Y_L value — cf. (5.34)), one obtains

$$g_L^{(\nu)} = \frac{g^2}{2\sqrt{g^2 - e^2}}, \quad g_L^{(e)} = \frac{-\frac{1}{2}g^2 + e^2}{\sqrt{g^2 - e^2}}, \quad g_R^{(e)} = \frac{e^2}{\sqrt{g^2 - e^2}} \tag{5.55}$$

From (5.55), it is particularly clear that the term "electroweak unification" is indeed justified in connection with the considered $SU(2) \times U(1)$ gauge model: the NC couplings are non-trivial functions of e and g and "interpolate" thus between the electromagnetic and weak interactions. We will see more examples of such a functional dependence of the electroweak couplings (and mass ratios) in other sectors of the standard model — in particular, in the sector of vector bosons discussed in detail in Section 5.7. In any case, the weak mixing angle, which can be expressed in terms of the ratio e/g, is an arbitrary parameter of the electroweak unification and must be measured independently. In the next section, we will show in an example how the parameter $\sin^2\theta_W$ can be determined from the low-energy neutrino scattering processes mediated by the weak neutral currents.

5.6 Low-energy neutrino–electron scattering

Scattering of muon neutrino or antineutrino on the electron is a typical process which goes via neutral currents; in the old Feynman–Gell-Mann theory, it can only occur at one-loop (and higher) level (the reader is recommended to draw a one-loop diagram describing this process in the old weak interaction theory). Before proceeding to a detailed discussion of the $\overset{(-)}{\nu}_\mu - e$ scattering within the $SU(2) \times U(1)$ electroweak model, we have to incorporate leptons of muon type into this framework. In fact, this can be done in an almost trivial way. In complete analogy with the scheme explained in Section 5.1, one introduces left-handed doublet and right-handed singlets for the second (muonic) generation

$$L^{(\mu)} = \begin{pmatrix} \nu_{\mu L} \\ \mu_L \end{pmatrix}, \quad \mu_R, \quad \nu_{\mu R} \tag{5.56}$$

with weak hypercharges following the pattern of the electron-type leptons. Then the muonic contributions to the weak charged current and electromagnetic current have the right form and the corresponding weak neutral currents obviously repeat precisely the structure shown in (5.49). Let us stress that in writing (5.56) along with (5.1), we neglect *a priori* a possible mixing between the two lepton generations (and exclude thus phenomena like the neutrino oscillations) — we will comment on this issue later on, in the context of the full standard electroweak model, including the mechanism for generating masses. It is also clear that the $SU(2) \times U(1)$ electroweak model can be extended in this way to an arbitrary number of lepton generations; as we know now, in our physical world, there are precisely three generations of leptons (labelled as e, μ, τ) involving light neutrinos.

With the above remarks in mind, we are ready to write down the part of the neutral-current interaction Lagrangian relevant for the description of the $\overset{(-)}{\nu}_\mu - e$ scattering processes. This can be written as

$$\mathscr{L}_{NC}^{(\nu_\mu e)} = \frac{g}{2\cos\theta_W}\left[\frac{1}{2}\bar{\nu}_\mu\gamma^\alpha(1-\gamma_5)\nu_\mu + \bar{e}\gamma^\alpha(v - a\gamma_5)e\right] Z_\alpha \tag{5.57}$$

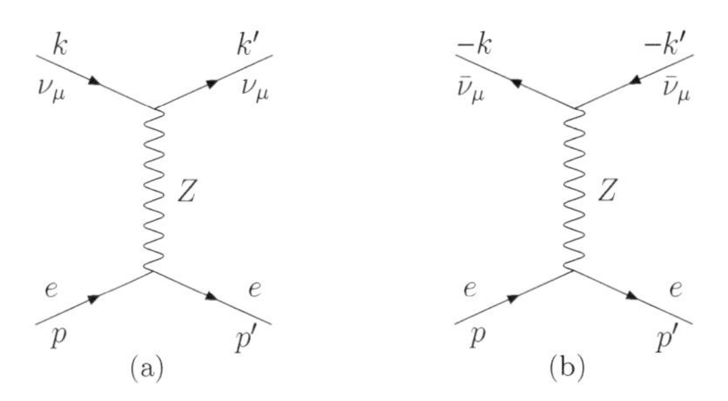

Figure 5.1. Tree-level Feynman graphs for the $\nu_\mu - e$ scattering (a) and $\bar{\nu}_\mu - e$ scattering (b).

where the axial-vector and vector NC couplings for the electron are, according to the results of the preceding section,

$$\begin{aligned} v &= \varepsilon_L + \varepsilon_R = -\frac{1}{2} + 2\sin^2\theta_W \\ a &= \varepsilon_L - \varepsilon_R = -\frac{1}{2} \end{aligned} \tag{5.58}$$

(cf. (5.50) and (5.51)). The lowest-order Feynman graphs for the considered processes are shown in Fig. 5.1. Let us start with the neutrino process. The Lorentz invariant matrix element corresponding to Fig. 5.1(a) is, following (5.57),

$$\begin{aligned} \mathcal{M}_a = &-\frac{g^2}{8\cos^2\theta_W}[\bar{u}(k')\gamma_\alpha(1-\gamma_5)u(k)][\bar{u}(p')\gamma_\beta(v-a\gamma_5)u(p)] \\ &\times \frac{-g^{\alpha\beta} + m_Z^{-2}q^\alpha q^\beta}{q^2 - m_Z^2} \end{aligned} \tag{5.59}$$

The kinematical conditions are assumed to be such that

$$m_e^2 \ll s \ll m_Z^2 \tag{5.60}$$

where $s = (k+p)^2$, so we will neglect the electron mass in what follows, and q^2 in the Z boson propagator can be neglected as well. Of course, the contribution of the longitudinal term in the numerator of the Z propagator is also strongly suppressed (it vanishes exactly for a massless neutrino). Thus, (5.59) is approximately equal to

$$\mathcal{M}_a \doteq -\frac{g^2}{8\cos^2\theta_W}\frac{1}{m_Z^2}[\bar{u}(k')\gamma_\alpha(1-\gamma_5)u(k)][\bar{u}(p')\gamma^\alpha(v-a\gamma_5)u(p)] \tag{5.61}$$

The last expression may be conveniently recast as

$$\mathcal{M}_a = -\frac{G_F}{\sqrt{2}}\rho[\bar{u}(k')\gamma_\alpha(1-\gamma_5)u(k)][\bar{u}(p')\gamma^\alpha(v-a\gamma_5)u(p)] \tag{5.62}$$

where we have introduced the Fermi coupling constant through the relation $G_F/\sqrt{2} = g^2/8m_W^2$ and ρ denotes the ratio

$$\rho = \frac{m_W^2}{m_Z^2\cos^2\theta_W} \tag{5.63}$$

As we shall see in the next chapter, the Weinberg–Salam standard model predicts classically (i.e. at the tree level) the value $\rho = 1$ as a consequence of the specific realization of the Higgs mechanism generating the vector boson masses — a prediction that has indeed turned out to be phenomenologically successful. At the present stage of our discussion, the ρ value is essentially arbitrary, but we will see shortly that it may be determined experimentally (along with the weak mixing angle) when both neutrino and antineutrino low energy cross-sections are measured.

In calculation of the cross-section for the neutrino process, we will assume that the electron is unpolarized; the spin-averaged square of the matrix element (5.62) then becomes

$$\begin{aligned}\overline{|\mathcal{M}_a|^2} &= \frac{1}{2}\sum_{\text{pol.}}|\mathcal{M}_a|^2 \\ &= \frac{1}{4}G_F^2\rho^2\mathrm{Tr}[\not{k}'\gamma_\alpha(1-\gamma_5)\not{k}\gamma_\beta(1-\gamma_5)] \\ &\quad\cdot\mathrm{Tr}[\not{p}'\gamma^\alpha(v-a\gamma_5)\not{p}\gamma^\beta(v-a\gamma_5)]\end{aligned} \tag{5.64}$$

After some simple manipulations and using the formulae (A.51), one gets from (5.64)

$$\begin{aligned}\overline{|\mathcal{M}_a|^2} &= \frac{1}{2}G_F^2\rho^2[(v^2+a^2)\mathrm{Tr}(\not{k}'\gamma_\alpha\not{k}\gamma_\beta)\cdot\mathrm{Tr}(\not{p}'\gamma^\alpha\not{p}\gamma^\beta) \\ &\quad + 2va\mathrm{Tr}(\not{k}'\gamma_\alpha\not{k}\gamma_\beta\gamma_5)\cdot\mathrm{Tr}(\not{p}'\gamma^\alpha\not{p}\gamma^\beta\gamma_5)] \\ &= 16G_F^2\rho^2[(v^2+a^2)\left((k\cdot p)(k'\cdot p')+(k\cdot p')(k'\cdot p)\right) \\ &\quad + 2va\left((k\cdot p)(k'\cdot p')-(k\cdot p')(k'\cdot p)\right)] \\ &= 16G_F^2\rho^2[(v+a)^2(k\cdot p)(k'\cdot p')+(v-a)^2(k\cdot p')(k'\cdot p)]\end{aligned} \tag{5.65}$$

The last expression can be rewritten in terms of the Mandelstam invariants $s = (k+p)^2$ and $u = (k-p')^2$ as

$$\overline{|\mathcal{M}_a|^2} = 4G_F^2\rho^2[(v+a)^2s^2 + (v-a)^2u^2] \tag{5.66}$$

(let us stress again that we neglect systematically the electron mass). Alternatively, one may introduce the dimensionless invariant $y = p\cdot q/p\cdot k$ which in the considered massless case satisfies a simple relation $u = -s(1-y)$ (cf. Appendix B); one thus obtains

$$\overline{|\mathcal{M}_a|^2} = 4G_F^2\rho^2s^2[(v+a)^2 + (v-a)^2(1-y)^2] \tag{5.67}$$

Similarly, one can calculate a corresponding quantity for the antineutrino process described by the graph in Fig. 5.1(b). One gets

$$\overline{|\mathcal{M}_b|^2} = 4G_F^2\rho^2s^2[(v+a)^2(1-y)^2 + (v-a)^2] \tag{5.68}$$

(of course, the result (5.68) can also be obtained directly from (5.66) by employing the crossing symmetry, i.e. interchanging the s and u variables). Using now a standard formula for the differential cross-section (cf. Appendix B), one gets

$$\begin{aligned}\frac{d\sigma}{dy}(\nu_\mu e \to \nu_\mu e) &= \frac{G_F^2 s}{\pi}\rho^2[\varepsilon_L^2 + \varepsilon_R^2(1-y)^2]\\ \frac{d\sigma}{dy}(\bar{\nu}_\mu e \to \bar{\nu}_\mu e) &= \frac{G_F^2 s}{\pi}\rho^2[\varepsilon_L^2(1-y)^2 + \varepsilon_R^2]\end{aligned} \tag{5.69}$$

where we have retrieved the original "chiral" parameters for the electron neutral current (cf. (5.58)). Integrating the expressions (5.69) over y from 0 to 1, one obtains the total cross-sections

$$\begin{aligned}\sigma(\nu_\mu e \to \nu_\mu e) &= \frac{G_F^2 s}{\pi}\rho^2\left(\varepsilon_L^2 + \frac{1}{3}\varepsilon_R^2\right)\\ \sigma(\bar{\nu}_\mu e \to \bar{\nu}_\mu e) &= \frac{G_F^2 s}{\pi}\rho^2\left(\frac{1}{3}\varepsilon_L^2 + \varepsilon_R^2\right)\end{aligned} \tag{5.70}$$

Now, it is clear that a measurement of the cross-sections (5.69) or (5.70) leads to the determination of both the weak mixing angle and the parameter ρ. In particular, taking the ratio of the neutrino and antineutrino total cross-sections (5.70) and using $\varepsilon_L = -\frac{1}{2} + \sin^2\theta_W, \varepsilon_R = \sin^2\theta_W$, one gets

$$R_{\nu/\bar{\nu}} = \frac{\sigma(\nu_\mu e \to \nu_\mu e)}{\sigma(\bar{\nu}_\mu e \to \bar{\nu}_\mu e)} = \frac{3 - 12\sin^2\theta_W + 16\sin^4\theta_W}{1 - 4\sin^2\theta_W + 16\sin^4\theta_W} \tag{5.71}$$

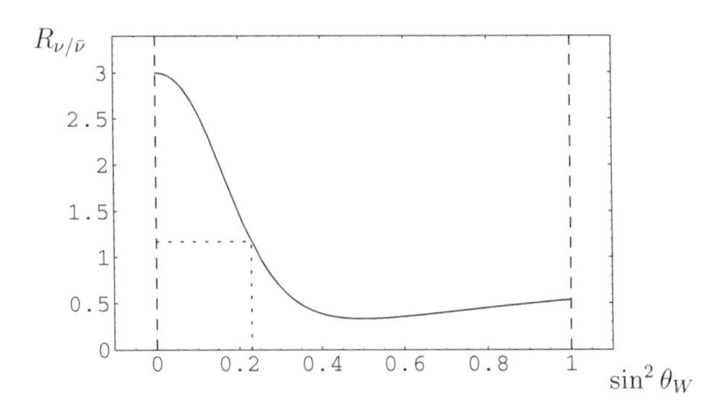

Figure 5.2. The dependence of the ratio of neutrino and antineutrino cross-sections on the parameter $\sin^2\theta_W$.

The functional dependence (5.71) is graphically depicted in Fig. 5.2. The relevant experimental value is roughly around $R = 1.2$, where the slope of the curve in Fig. 5.2 is rather favourable for a reasonably accurate determination of the $\sin^2\theta_W$. From an absolute value of one of the cross-sections (5.70), one can then determine the parameter ρ. Currently, the best data for the considered scattering processes are provided by the collaboration CHARM II. An analysis of the data accumulated till 1991 led to the following results:

$$\begin{aligned} \sin^2\theta_W &= 0.232 \pm 0.008 \\ \rho &= 1.006 \pm 0.047 \end{aligned} \tag{5.72}$$

(see 5.90). Of course, the knowledge of the parameters ρ and $\sin^2\theta_W$ together with the result for the W mass (see (5.42)) enables one to determine the Z mass. In particular, taking $\rho = 1$, $\sin^2\theta_W = 0.23$ and $m_W = 80$ GeV, one obtains $m_Z \doteq 90$ GeV (the current experimental value is $m_Z \doteq 91.19$ GeV). Let us stress again that the Z boson mass — obtained here from an experimental value of the parameter ρ — is in fact a successful prediction of the full GWS standard electroweak model; this will be made clear in the next chapter. The cross-sections (5.70) are very small; using the values (5.72), one has roughly

$$\begin{aligned} \sigma(\nu_\mu e \to \nu_\mu e) &\doteq 1.5\, E_{\text{lab.}}(\text{GeV}) \times 10^{-42}\text{cm}^2 \\ \sigma(\bar{\nu}_\mu e \to \bar{\nu}_\mu e) &\doteq 1.3\, E_{\text{lab.}}(\text{GeV}) \times 10^{-42}\text{cm}^2 \end{aligned} \tag{5.73}$$

The corresponding experimental measurement therefore represents a formidable task — on the other hand, these purely leptonic processes are theoretically clean and provide a simple and instructive example of a calculation involving the neutral current interactions. Needless to say, there are other more accurate determinations of the relevant NC parameters from processes with higher statistics (such as lepton-nucleon scattering, electron-positron annihilation, etc.). Nevertheless, from the historical point of view, the $\nu_\mu - e$ scattering was actually the first neutral current process observed (in 1973) and thus provided a decisive experimental support to the $SU(2) \times U(1)$ gauge theory of weak and electromagnetic interactions.

5.7 Interactions of vector bosons

Let us now turn to the investigation of the term $\mathscr{L}_{\text{gauge}}$ in the Lagrangian (5.13), which contains the interactions of the gauge fields with themselves. According to (4.51) and (5.10)–(5.12), the relevant interaction term can be written as

$$\begin{aligned}\mathscr{L}_{\text{gauge}}^{\text{int}} = &-\frac{1}{2}g\epsilon^{abc}(\partial_\mu A_\nu^a - \partial_\nu A_\mu^a)A^{b\mu}A^{c\nu} \\ &-\frac{1}{4}g^2\epsilon^{abc}\epsilon^{ajk}A_\mu^b A_\nu^c A^{j\mu}A^{k\nu}\end{aligned} \tag{5.74}$$

Working out explicitly the first term in (5.74) and employing the identity $\epsilon^{abc}\epsilon^{ajk} = \delta^{bj}\delta^{ck} - \delta^{bk}\delta^{cj}$ in the second term, one first gets

$$\begin{aligned}\mathscr{L}_{\text{gauge}}^{\text{int.}} = &- g[(\partial_\mu A_\nu^1 - \partial_\nu A_\mu^1)A^{2\mu}A^{3\nu} + \text{ cycl. perm. } (123)] \\ &-\frac{1}{4}g^2[(A_\mu^a A^{a\mu})(A_\nu^b A^{b\nu}) - (A_\mu^a A_\nu^a)(A^{b\mu}A^{b\nu})]\end{aligned} \tag{5.75}$$

which can be further recast as

$$\begin{aligned}\mathscr{L}_{\text{gauge}}^{(\text{int.})} = &- g(A_\mu^1 A_\nu^2 \overleftrightarrow{\partial}^\mu A^{3\nu} + A_\mu^2 A_\nu^3 \overleftrightarrow{\partial}^\mu A^{1\nu} + A_\mu^3 A_\nu^1 \overleftrightarrow{\partial}^\mu A^{2\nu}) \\ &-\frac{1}{4}g^2[(\vec{A}_\mu \cdot \vec{A}^\mu)^2 - (\vec{A}_\mu \cdot \vec{A}_\nu)(\vec{A}^\mu \cdot \vec{A}^\nu)]\end{aligned} \tag{5.76}$$

where the symbol $\overleftrightarrow{\partial}$ is defined by $f\overleftrightarrow{\partial}g = f(\partial g) - (\partial f)g$ and we have used the standard shorthand notation, $\vec{A}_\mu \cdot \vec{A}^\mu = A_\mu^a A^{a\mu}$ etc. Replacing the Yang–Mills fields A_μ^1 and A_μ^2 by the physical charged vector fields $W_\mu^\pm$ according to

$$A_\mu^1 = \frac{1}{\sqrt{2}}(W_\mu^+ + W_\mu^-), \quad A_\mu^2 = \frac{i}{\sqrt{2}}(W_\mu^+ - W_\mu^-)$$

(cf. (5.17)), the form (5.76) becomes

$$\begin{aligned}\mathscr{L}^{(\text{int.})}_{\text{gauge}} = &- ig(W^0_\mu W^-_\nu \overleftrightarrow{\partial}^\mu W^{+\nu} + W^-_\mu W^+_\nu \overleftrightarrow{\partial}^\mu W^{0\nu} + W^+_\mu W^0_\nu \overleftrightarrow{\partial}^\mu W^{-\nu}) \\ &- g^2 \Big[\frac{1}{2}(W^+_\mu W^{-\mu})^2 - \frac{1}{2}(W^+_\mu W^{+\mu})(W^-_\nu W^{-\nu}) \\ &+ (W^0_\mu W^{0\mu})(W^+_\nu W^{-\nu}) - (W^-_\mu W^+_\nu)(W^{0\mu} W^{0\nu})\Big] \end{aligned} \tag{5.77}$$

where the W^0_μ denotes the linear combination

$$W^0_\mu = \cos\theta_W Z_\mu + \sin\theta_W A_\mu \tag{5.78}$$

(W^0_μ is simply a different name for the original Yang–Mills field A^3_μ, cf. (5.22)).

The expression (5.77) is seen to contain trilinear and quadrilinear interactions of the vector fields; when it is worked out by employing (5.78), one can identify two trilinear and four quadrilinear couplings of the W, Z and photon, namely

$$\begin{aligned}\mathscr{L}_{WW\gamma} = &-ie(A_\mu W^-_\nu \overleftrightarrow{\partial}^\mu W^{+\nu} + W^-_\mu W^+_\nu \overleftrightarrow{\partial}^\mu A^\nu \\ &+ W^+_\mu A_\nu \overleftrightarrow{\partial}^\mu W^{-\nu})\end{aligned} \tag{5.79}$$

$$\begin{aligned}\mathscr{L}_{WWZ} = &-ig\cos\theta_W(Z_\mu W^-_\nu \overleftrightarrow{\partial}^\mu W^{+\nu} + W^-_\mu W^+_\nu \overleftrightarrow{\partial}^\mu Z^\nu \\ &+ W^+_\mu Z_\nu \overleftrightarrow{\partial}^\mu W^{-\nu})\end{aligned} \tag{5.80}$$

$$\mathscr{L}_{WW\gamma\gamma} = -e^2(W^-_\mu W^{+\mu} A_\nu A^\nu - W^-_\mu A^\mu W^+_\nu A^\nu) \tag{5.81}$$

$$\mathscr{L}_{WWWW} = \frac{1}{2}g^2(W^-_\mu W^{-\mu} W^+_\nu W^{+\nu} - W^-_\mu W^{+\mu} W^-_\nu W^{+\nu}) \tag{5.82}$$

$$\mathscr{L}_{WWZZ} = -g^2\cos^2\theta_W(W^-_\mu W^{+\mu} Z_\nu Z^\nu - W^-_\mu Z^\mu W^+_\nu Z^\nu) \tag{5.83}$$

$$\begin{aligned}\mathscr{L}_{WWZ\gamma} = &\, g^2\sin\theta_W\cos\theta_W(-2W^-_\mu W^{+\mu} A_\nu Z^\nu + W^-_\mu Z^\mu W^+_\nu A^\nu \\ &+ W^-_\mu A^\mu W^+_\nu Z^\nu)\end{aligned} \tag{5.84}$$

Note that in the electromagnetic interactions of W bosons, i.e. in (5.79) and (5.81), we have used the unification condition $e = g\sin\theta_W$ (see (5.34)). One may observe that the triple coupling (5.79) corresponds to the value $\kappa = 1$ in the Lagrangian (3.42) discussed in Chapter 3. In other words, the $SU(2) \times U(1)$ electroweak gauge model automatically predicts a very specific *non-minimal* electromagnetic interaction of the W bosons.

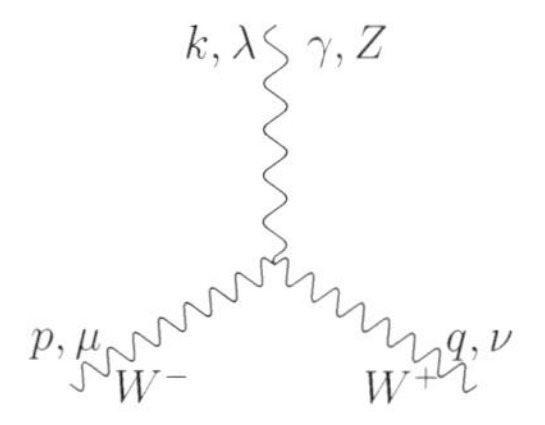

Figure 5.3. Feynman-graph vertex for the triple vector boson interactions $WW\gamma$ or WWZ.

In general, the coupling constants of vector boson interactions appearing in (5.79)–(5.84) can obviously be expressed in terms of e and g when the unification condition is employed, e.g.,

$$g_{WWZ} = g\cos\theta_W = \sqrt{g^2 - e^2} \tag{5.85}$$

This is again an explicit illustration of the characteristic feature of the $SU(2) \times U(1)$ electroweak unification, mentioned earlier in this chapter: the new interactions stemming from the Yang–Mills construction involve coupling constants that are non-trivial functions of the parameters of the old theory of weak interactions and electromagnetism.

In quantum theory (i.e. at the level of Feynman diagrams), the interaction Lagrangians (5.79) or (5.80), respectively, lead to the vertex shown in Fig. 5.3. The relevant Feynman rule is given by the function

$$V_{\lambda\mu\nu}(k,p,q) = (k-p)_\nu g_{\lambda\mu} + (p-q)_\lambda g_{\mu\nu} + (q-k)_\mu g_{\lambda\nu} \tag{5.86}$$

(multiplied by an appropriate coupling constant). The function (5.86) is obviously invariant under cyclic permutations

$$V_{\lambda\mu\nu}(k,p,q) = V_{\mu\nu\lambda}(p,q,k) = V_{\nu\lambda\mu}(q,k,p) \tag{5.87}$$

and also satisfies a highly useful relation

$$p^\mu V_{\lambda\mu\nu}(k,p,q) = (k^2 g_{\lambda\nu} - k_\lambda k_\nu) - (q^2 g_{\lambda\nu} - q_\lambda q_\nu) \tag{5.88}$$

(called the 't Hooft identity, cf. (3.47)). A proof of the relation (5.88) (which is valid for any four-momenta satisfying $k+p+q=0$) is left to the reader as an easy exercise. The form (5.88) can be used in Feynman diagrams for different configurations of the outgoing and incoming particles — one has to remember that an incoming $W^\pm$ line is equivalent to the outgoing $W^\mp$ line carrying opposite momentum. As for the Feynman rules for the quartic interactions, these can be read off rather easily from the Lagrangians (5.81)–(5.84); some illustrations will be provided in the subsequent calculations.

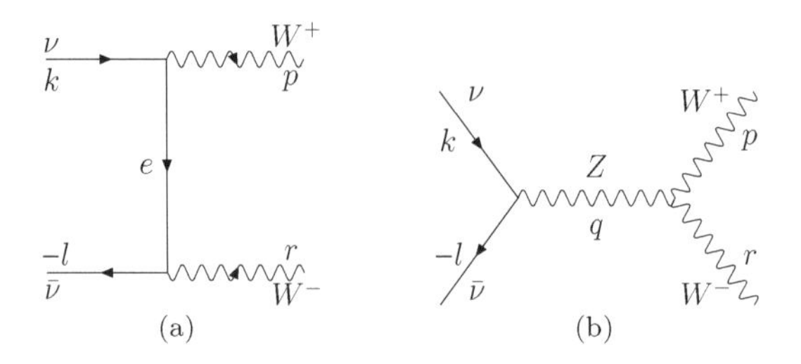

Figure 5.4. Tree-level diagrams contributing to the process $\nu\bar{\nu} \to W^+W^-$ within the $SU(2) \times U(1)$ gauge model of electroweak interactions. No further graphs arise in the full standard electroweak model if the neutrino mass is neglected.

The Yang–Mills structure of the vector boson interactions and the non-trivial relations among the relevant coupling parameters have dramatic consequences for the cancellation of high-energy divergences in tree-level scattering amplitudes for various electroweak processes. Some examples of such "gauge cancellations" are discussed in the next section.

5.8 Cancellation of leading divergences

To begin with, let us consider the process $\nu\bar{\nu} \to W^+W^-$, which we have already mentioned briefly in Chapter 3. Within our $SU(2) \times U(1)$ electroweak model, this is described, at the tree level, by the two Feynman graphs shown in Fig. 5.4. We will examine the case of longitudinally polarized W bosons, where one can expect the worst high-energy behaviour. The matrix element corresponding to the diagram in Fig. 5.4(a) can then be written, according to the results of Chapter 3, as

$$\mathcal{M}_{\nu\bar{\nu}}^{(e)} = -\frac{g^2}{4m_W^2}\bar{v}(l)\not{p}(1-\gamma_5)u(k) + O(1) \tag{5.89}$$

The quadratic divergence occurring in (5.89) for $E \to \infty$ is embodied in the first term. Now, we are going to show that this divergence is cancelled by a corresponding contribution coming from the diagram in Fig. 5.4(b).

According to the standard Feynman rules, the matrix element for the graph in Fig. 5.4(b) can be written (for arbitrary polarizations of the external W bosons) as

$$\begin{aligned}\mathcal{M}_{\nu\bar{\nu}}^{(Z)} = &-\frac{1}{4}\frac{g}{\cos\theta_W} \cdot g\cos\theta_W \bar{v}(l)\gamma_\rho(1-\gamma_5)u(k) \\ &\times \frac{-g^{\rho\nu} + m_Z^{-2}q^\rho q^\nu}{q^2 - m_Z^2} V_{\nu\mu\lambda}(q,r,p)\varepsilon^{*\mu}(r)\varepsilon^{*\lambda}(p)\end{aligned} \tag{5.90}$$

where we have employed the earlier results for the neutral-current $\nu\nu Z$ vertex (see (5.49)) and the Yang–Mills WWZ vertex (see (5.80) and (5.86)). By naive power counting, one might expect that the leading high-energy divergence associated with this graph could be more severe than that occurring in Fig. 5.4(a), owing to the extra factor m_Z^{-2} in the longitudinal part of the Z boson propagator. However, using the cyclicity property (5.87) and the 't Hooft identity (5.88) for the Yang–Mills vertex, along with the familiar properties of the polarization vectors, it is not difficult to show that the potentially dangerous contribution proportional to m_Z^{-2} vanishes exactly for any combination of the external polarizations (the proof is left to the reader as a simple exercise). For longitudinally polarized external W bosons, one then gets, using the usual high-energy decomposition of the polarization vectors,

$$\mathcal{M}_{\nu\bar{\nu}}^{(Z)} = \frac{1}{4} g^2 \bar{v}(l) \gamma^\nu (1-\gamma_5) u(k) \frac{1}{q^2 - m_Z^2} V_{\nu\mu\lambda}(q,r,p) \frac{r^\mu}{m_W} \frac{p^\lambda}{m_W} + O(1) \tag{5.91}$$

where we have singled out explicitly the diverging part of the whole contribution — now it is obvious that we are left with only a quadratic divergence, similar to the graph in Fig. 5.4(a). Employing once more the 't Hooft identity, as well as the equations of motion for the Dirac spinors, the expression (5.91) can be finally recast, after some simple algebraic manipulations, as

$$\mathcal{M}_{\nu\bar{\nu}}^{(Z)} = \frac{g^2}{4m_W^2} \bar{v}(l) \not{p} (1-\gamma_5) u(k) + O(1) \tag{5.92}$$

Comparing this with (5.89), one can see that for the considered process, we have indeed achieved the desired compensation of the original high-energy divergence lurking in the old W boson weak interaction model — it was the gauge structure of the electroweak $SU(2) \times U(1)$ theory, manifested in the combination of the two relevant graphs, which played an important role in the cancellation mechanism. Let us add that the above treatment remains unaltered even within the full GWS standard electroweak model, at least if the neutrino is taken to be massless. In fact, there are several other processes of similar type (involving a fermion pair along with a pair of vector bosons), where complete cancellation of the high-energy divergences is achieved already at the level of the $SU(2) \times U(1)$ gauge theory (i.e. without invoking the Higgs mechanism of the full GWS standard model); finding some relevant examples is left as a challenge for the reader.

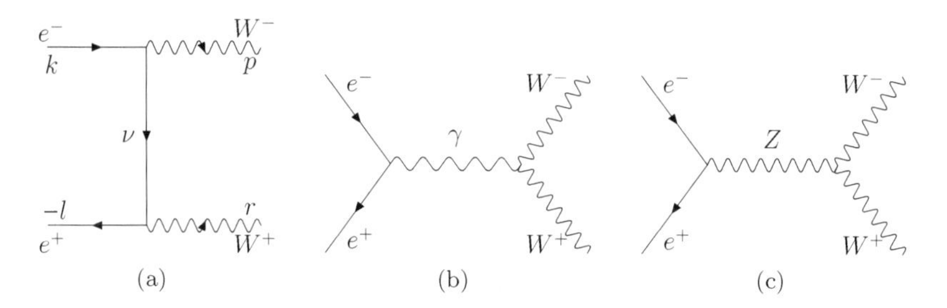

Figure 5.5. Tree-level diagrams for the process $e^+e^- \to W^+W^-$ in the $SU(2) \times U(1)$ electroweak theory, including the t-channel neutrino exchange and s-channel photon or Z boson exchange.

Next, let us turn to the process $e^+e^- \to W^+W^-$ for the longitudinally polarized external W bosons. Within our $SU(2) \times U(1)$ electroweak theory, there are now three Feynman graphs that contribute at the tree level, namely those depicted in Fig. 5.5. The graphs Figs. 5.5(a) and 5.5(b) were discussed in Chapter 3 and the results can be written as

$$\mathcal{M}^{(\nu)}_{e^+e^-} = -\frac{g^2}{4m_W^2}\bar{v}(l)\not{p}(1-\gamma_5)u(k) + O\left(\frac{m_e}{m_W^2}E\right) + O(1) \tag{5.93}$$

for the weak contribution (i.e. the neutrino exchange in Fig. 5.5(a)) and

$$\mathcal{M}^{(\gamma)}_{e^+e^-} = \frac{e^2}{m_W^2}\bar{v}(l)\not{p}u(k) + O(1) \tag{5.94}$$

for the electromagnetic contribution in Fig. 5.5(b) — let us stress that the last expression corresponds to the $WW\gamma$ vertex of the Yang–Mills type (cf. (5.79)). The calculation of the Z-exchange graph in Fig. 5.5(c) proceeds essentially along the same lines as in the neutrino–antineutrino case described earlier. Again, the longitudinal part of the Z propagator does not contribute, and the final result can be written as

$$\begin{aligned}\mathcal{M}^{(Z)}_{e^+e^-} = &-\frac{1}{2m_W^2}g\cos\theta_W\frac{g}{\cos\theta_W}\left(-\frac{1}{2}+\sin^2\theta_W\right)\bar{v}(l)\not{p}(1-\gamma_5)u(k)\\ &-\frac{1}{2m_W^2}g\cos\theta_W\frac{g}{\cos\theta_W}\sin^2\theta_W\,\bar{v}(l)\not{p}(1+\gamma_5)u(k)\\ &+O\left(\frac{m_e}{m_W^2}E\right)+O(1)\end{aligned} \tag{5.95}$$

where we have singled out a term involving the leading (quadratic) high-energy divergences and used the neutral current parametrization (5.49).

An explicit form of the non-leading (linear) divergences can be found in [Hor]. We will ignore these terms for the moment, but we will return to them in the next chapter. Adding now the expressions (5.93), (5.94), and (5.95) and using the familiar relation $\sin\theta_W = e/g$, it is seen that the quadratic divergences indeed cancel, but — in contrast to the preceding example — a residual linear divergence still persists in the sum of the graphs (a), (b), and (c). Thus, in the present case, the electroweak gauge structure alone cannot ensure a complete cancellation of the high-energy divergences — one may observe that, technically, this is related to the non-vanishing electron mass. Nevertheless, as before, the gauge couplings do control the *leading* divergences — this is the most important lesson to be learnt from the present example.

Last but not least, let us reconsider the process of WW scattering, which we have been able to describe only via photon exchange in the naive W boson model. Within the $SU(2) \times U(1)$ electroweak gauge theory, there are three types of tree-level diagrams representing such a process, as shown in Fig. 5.6 (of course, for graphs in Figs. 5.6(a) and 5.6(b) one must also take into account the relevant crossing of the external lines). We will consider again the case of longitudinally polarized external W bosons. For the photon exchange contribution, we have obtained earlier (cf. Chapter 3) the following result:

$$\mathcal{M}_{WW}^{(\gamma)} = \frac{e^2}{4m_W^4}(t^2 + u^2 - 2s^2) + O(E^2) + O(1) \tag{5.96}$$

(where s, t, u are the standard Mandelstam variables), which exhibits the quartic high-energy divergence expected on dimensional grounds. As for the Z boson exchange, the longitudinal part of the propagator does not contribute, owing to the Yang–Mills structure of the WWZ vertex. This in turn means that the worst divergent behaviour to be expected for this

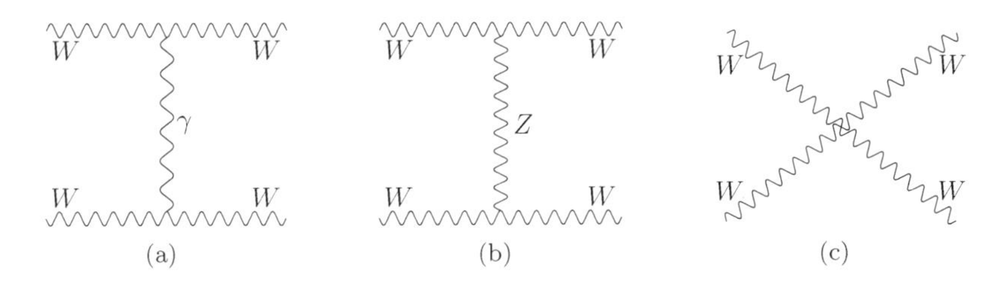

Figure 5.6. Tree-level Feynman graphs for the process $WW \to WW$, including the photon and Z boson exchange and the direct coupling of four W bosons.

graph is the same as for the photon exchange. Further, the Z mass cannot play any role in the coefficient of the leading divergence, so one may readily write for the sum of the graphs in Figs. 5.6(a) and 5.6(b)

$$\mathcal{M}_{WW}^{(Z,\gamma)} = (e^2 + g_{WWZ}^2)\frac{1}{4m_W^4}(t^2 + u^2 - 2s^2) + O(E^2) + O(1) \qquad (5.97)$$

Using the relations $e = g\sin\theta_W$ and $g_{WWZ} = g\cos\theta_W$ (see (5.34) and (5.80)), the coupling factor in (5.97) is reduced to

$$e^2 + g_{WWZ}^2 = g^2 \qquad (5.98)$$

The contribution of the diagram in Fig. 5.6(c) can be calculated in a straightforward way from the interaction term (5.82); for the leading divergences, one gets, after some simple algebraic manipulations,

$$\mathcal{M}_{WW}^{(\text{direct})} = -g^2\frac{1}{4m_W^4}(t^2 + u^2) + g^2\frac{1}{2m_W^4}s^2 + O(E^2) + O(1) \qquad (5.99)$$

Let us remark that the non-leading terms (quadratic in energy) have, in general, a rather complicated form for the individual diagrams; similar to the preceding example, we relegate their treatment to the next chapter. From (5.97), (5.98), and (5.99), it is clear that the quartic divergences are cancelled in the sum of the three considered graphs — again, the Yang–Mills structure of the vector boson sector (manifested in the interplay of the three-boson and four-boson couplings) is responsible for a "miraculous" cancellation of the leading high-energy divergences.

The examples discussed in this section illustrate nicely some remarkable technical consequences of the non-Abelian gauge invariance in the theory of electroweak unification. It turns out that within such a theory, the scattering amplitudes involving massive vector bosons have much softer high-energy behaviour than one might naively guess on simple dimensional grounds; in particular, the leading high-energy divergences are cancelled owing to the gauge structure of the relevant interactions. After such cancellations, there are still some residual non-leading divergences and their ultimate elimination within the full GWS standard model is related to the subtle issue of the mass generation in gauge theories, which is the subject of the next chapter.

Problems

5.1 What is the effective four-fermion Lagrangian describing low-energy neutral current interactions?

5.2 Calculate the partial decay width $\Gamma(Z \to \ell^+\ell^-)$ for unpolarized Z and $\ell^\pm$. First, set $m_\ell = 0$ for the sake of simplicity (of course, this is expected to be a very good approximation, since $m_\ell^2 \ll m_Z^2$ for any $\ell = e, \mu, \tau$). How is the result changed (numerically) when the effects of $m_\ell \neq 0$ are taken into account? In particular, make such a comparison for the heaviest known lepton, τ. Next, calculate the decay width $\Gamma(Z \to \nu_\ell \bar{\nu}_\ell)$.

5.3 Evaluate longitudinal polarization of a lepton ℓ produced in the decay of an unpolarized Z boson at rest. The degree of polarization in question is defined as

$$P = \frac{w_R - w_L}{w_R + w_L}$$

with w_L and w_R denoting the probability of the production of left-handed and right-handed lepton, respectively. First of all, set $m_\ell = 0$ for simplicity. Actually, in such a massless case, the outcome can be guessed quite easily. An astute expert should then anticipate the result (to be verified by an explicit calculation)

$$P = \frac{\varepsilon_R^2 - \varepsilon_L^2}{\varepsilon_R^2 + \varepsilon_L^2} = -\frac{2va}{v^2 + a^2}$$

where the lepton coupling factors are given by (5.58) (these are independent of the lepton species). For $m_\ell \neq 0$, the calculation is algebraically more complicated (and its result cannot be guessed so easily). Anyway, any hard-working reader is encouraged to derive the formula

$$P = -\frac{2va\sqrt{1 - \frac{4m_\ell^2}{m_Z^2}}}{(v^2 + a^2)\left(1 - \frac{m_\ell^2}{m_Z^2}\right) + 3(v^2 - a^2)\frac{m_\ell^2}{m_Z^2}}$$

5.4 Calculate the angular distribution of electrons produced in decays of a polarized Z boson at rest (again, work in the approximation $m_\ell = 0$). As a follow-up, evaluate the up-down asymmetry for electrons produced in decays of a Z with spin "up" (i.e. directed along positive z axis). The asymmetry in question is defined as $(\Gamma_+ - \Gamma_-)/(\Gamma_+ + \Gamma_-)$, where Γ_+ denotes the angular distribution integrated over the upper hemisphere (with the azimuthal angle ϑ lying between 0 and $\frac{\pi}{2}$) and Γ_- has an analogous meaning with respect to the lower hemisphere ($\vartheta \in (0, \frac{\pi}{2})$).

5.5 Consider the process $e^+e^- \to \mu^+\mu^-$ in the c.m. system and at a sufficiently high energy ($E_{\text{c.m.}} \gg m_\mu$), so that the lepton masses can be safely neglected. In lowest order, the relevant matrix element can be written as $\mathcal{M}_\gamma + \mathcal{M}_Z$, with $\mathcal{M}_\gamma$ and $\mathcal{M}_Z$ corresponding to the exchange of photon and Z boson, respectively. Let us denote the cross-sections obtained from $|\mathcal{M}_\gamma|^2$ and $|\mathcal{M}_Z|^2$ as σ_γ and σ_Z. What is the numerical value of the ratio σ_Z/σ_γ for $E_{\text{c.m.}} = 1\,\text{GeV}$, and $20\,\text{GeV}$, and $200\,\text{GeV}$? Next, calculate the full cross-section involving both γ and Z exchange. For the above-mentioned energies, determine a relative magnitude of the interference term $\sigma_{\gamma Z}$, descending from $|\mathcal{M}_\gamma + \mathcal{M}_Z|^2 = |\mathcal{M}_\gamma|^2 + |\mathcal{M}_Z|^2 + \mathcal{M}_\gamma\mathcal{M}_Z^* + \mathcal{M}_\gamma^*\mathcal{M}_Z$.

5.6 For the process $e^+e^- \to \mu^+\mu^-$, evaluate also the forward–backward (or front–back) asymmetry A_{FB} of the cross-section. The A_{FB} is defined in close analogy with the quantity considered in Problem 5.4, namely

$$A_{\text{FB}} = \frac{\sigma_{\text{F}} - \sigma_{\text{B}}}{\sigma_{\text{F}} + \sigma_{\text{B}}}$$

where the σ_{F} and σ_{B} are cross-sections obtained by integrating over the front and back hemisphere, respectively, i.e. over the azimuthal angle ϑ lying in the interval $(0, \frac{\pi}{2})$ and $(\frac{\pi}{2}, \pi)$, respectively (note that ϑ is conventionally chosen as the angle between the momentum of μ^- and that of the incident electron). Is a non-vanishing value of the A_{FB} related to the $\mathcal{C}$ or $\mathcal{P}$ violation in the weak neutral current interaction?

5.7 Show that the tree-level amplitude for $e^+e^- \to Z_L\gamma$ behaves as $O(1)$ in the high-energy limit. What kind of asymptotic behaviour one gets for the process $e^+e^- \to Z_LZ_L$?

5.8 Compute the limiting value of the cross-section $\sigma(\nu_\mu e \to \nu_\mu e)$ for $s \to \infty$. Neglect lepton masses throughout the calculation.

5.9 Consider the annihilation process $e^-e^+ \to \nu_\mu\bar{\nu}_\mu$. At the tree level, evaluate its cross-section as a function of the collision energy $E_{\text{c.m.}} = s^{1/2}$, in the kinematic region $m_e \ll E_{\text{c.m.}} \ll m_Z$. Throughout the calculation, neglect everything that may be safely neglected. Further, find the limit of the relevant cross-section for $s \to \infty$.

5.10 Calculate cross-sections for processes $\nu_e e \to \nu_e e$ and $\bar{\nu}_e e \to \bar{\nu}_e e$ in a low-energy domain $m_e^2 \ll s \ll m_W^2$, taking into account both CC and NC contributions. Compare the results with those obtained within the old IVB model involving the $W^{\pm}$ only. Examine these processes also in the high-energy region, where $s \gg m_Z^2$. What is the asymptotic value of the cross-section ratio $\sigma(\bar{\nu}_e e \to \bar{\nu}_e e)/\sigma(\nu_e e \to \nu_e e)$?

5.11 An instructive illustration of the "miraculous" cancellations of high-energy divergences within SM is provided by the process $W^- W^+ \to Z\gamma$. Show that its amplitude satisfies the condition of tree-level unitarity.
Hint: Any attentive reader may guess immediately that the cancellation in question is due to the interplay of the $WW\gamma$, WWZ and $WWZ\gamma$ Yang–Mills couplings shown in (5.79), (5.80) and (5.84).

5.12 As another example of the divergence cancellation mechanism due to the Yang–Mills structure of the vector boson sector of SM, one may consider the reaction $W_L W_L \to Z_L Z_L$. Demonstrate a compensation of the leading divergences occurring in contributions of the relevant tree-level diagrams. Clearly, for this purpose, one has to invoke just the couplings WWZ and $WWZZ$ shown in (5.80) and (5.83). Note that in this case some residual quadratic divergences persist, which are ultimately eliminated by means of the exchange of the Higgs scalar boson (to be introduced in the next chapter).

Chapter 6

Higgs Mechanism for Masses

6.1 Residual divergences: Need for scalar bosons

The bulk of this chapter is devoted to the so-called Higgs mechanism. This is a tool for generating particle masses in gauge theories through specific interactions involving scalar fields without spoiling perturbative renormalizability. Historically, the first field-theory models exhibiting such a mechanism [47] were developed independent of the program of electroweak unification — only a few years after its discovery, the magic Higgs trick has been applied successfully by Weinberg and Salam to Glashow's $SU(2) \times U(1)$ gauge model of weak and electromagnetic interactions. In most of the current textbooks on particle theory, the Higgs mechanism is usually introduced immediately when formulating the Standard Model. However, the corresponding construction might seem, at first sight, somewhat bizarre to an uninitiated reader and, subsequently, the beginner in the field could wonder whether the electroweak SM *must* indeed be built precisely as it is — in particular, whether the Higgs scalars are necessary or not. Of course, since 2012, we know that a spin-0 particle, which resembles the Higgs scalar closely, indeed exists (though we still cannot be quite sure that the observed scalar is just *the* Higgs boson of SM), and this may dispel possible doubts of a skeptical reader. Nevertheless, it may be instructive to explain the role of a scalar boson in the electroweak theory independent of the current experimental data. So, we will describe the idea of the Higgs construction (and its realization within the GWS standard model) later in this chapter, and in this section, we will start by showing first a rather straightforward motivation for a scalar boson in the electroweak gauge theory in connection with the issue of divergence cancellations investigated in the preceding chapter. Of course, such a preliminary discussion cannot provide us with a detailed insight into the subtle aspects of the Higgs mechanism, but will at least indicate that a scalar boson is a *necessary* ingredient for achieving the tree-level unitarity (which in turn is necessary for renormalizability) in a gauge model incorporating mass terms for vector bosons and fermions. Moreover, it will become clear that the relevant scalar boson couplings must be intimately related to the particle masses.

First, let us come back to the $W_L W_L$ scattering process. In Section 5.8, we have found that the leading (quartic) high-energy divergences cancel in the sum of the diagrams shown in Fig. 5.6. A detailed (somewhat tedious) calculation reveals that the remaining quadratically divergent contribution has a remarkably simple form

$$\mathcal{M}_{WW}^{(\gamma)} + \mathcal{M}_{WW}^{(Z)} + \mathcal{M}_{WW}^{(\text{direct})} = -g^2 \frac{s}{4m_W^2} + O(1) \tag{6.1}$$

(*A technical remark*: Along with the quartic divergences, some ugly-looking $O(E^2)$ terms from the individual graphs are cancelled as well and one is thus happily left with the result (6.1); details of the calculation can be found in Appendix J of [Hor].) Since the coupling factor occurring in this expression is definitely non-zero, there is obviously no way how the divergent term in (6.1) could be eliminated without introducing a new particle and a corresponding new interaction. The crucial observation is that the quadratic divergence in (6.1) can be cancelled in the most natural way by means of an additional diagram involving the exchange of a scalar boson (in fact, one can hardly imagine any other option that would be feasible). An interaction of a pair of W's with a single neutral scalar field σ has an essentially unique form if it is required to be of renormalizable type (i.e. have dimension not greater than four), namely

$$\mathscr{L}_{WW\sigma} = g_{WW\sigma} W_\mu^- W^{+\mu} \sigma \tag{6.2}$$

The relevant coupling constant then obviously has a dimension of mass. The σ-exchange diagrams contributing to the WW scattering in the lowest

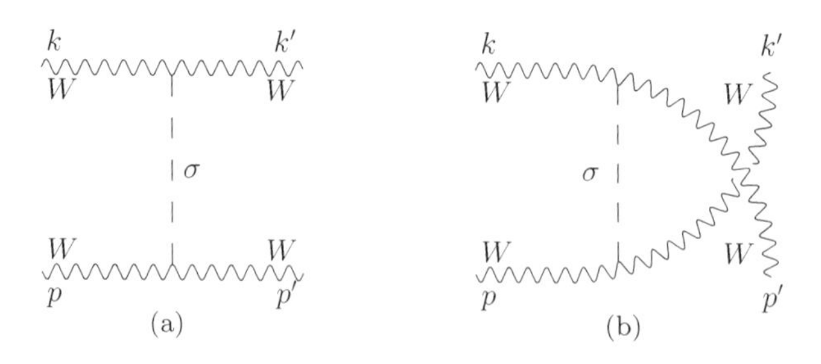

Figure 6.1. Neutral scalar exchange graphs for the WW scattering.

order are shown in Fig. 6.1. The corresponding matrix element can be written as

$$i\mathcal{M}_{WW}^{(\sigma)} = i^3 g_{WW\sigma}^2 \varepsilon^{\alpha}(k)\varepsilon_{\alpha}^{*}(k')\frac{1}{(k-k')^2 - m_{\sigma}^2}\varepsilon^{\beta}(p)\varepsilon_{\beta}^{*}(p') + \{k' \leftrightarrow p'\} \tag{6.3}$$

The leading divergence associated with the graphs of Fig. 6.1 for longitudinally polarized W's is obtained easily from the last expression by replacing the $\varepsilon_L^{\alpha}(k)$ by k^{α}/m_W, etc. After some simple algebraic manipulations, one thus gets

$$\mathcal{M}_{WW}^{(\sigma)} = -g_{WW\sigma}^2 \frac{1}{4m_W^4}(t+u) + O(1) \tag{6.4}$$

As we could have anticipated on simple dimensional grounds (keeping in mind the dimensionality of the coupling constant $g_{WW\sigma}$), the high-energy divergence embodied in (6.3) is indeed quadratic. Using now the kinematical identity $s + t + u = 4m_W^2$, the result (6.3) can be recast as

$$\mathcal{M}_{WW}^{(\sigma)} = g_{WW\sigma}^2 \frac{s}{4m_W^4} + O(1) \tag{6.5}$$

It is obvious that the divergent terms in (6.1) and (6.5) cancel each other if and only if

$$g_{WW\sigma} = g m_W \tag{6.6}$$

We thus see that the extra interaction of W bosons with a neutral scalar field σ, introduced in a rather *ad hoc* way in the context of the WW scattering process, does provide a remedy for the residual non-leading divergence (6.1). At the same time, the result (6.6) reveals a remarkable connection of such a "compensating" coupling with the W boson mass.

Let us work out one more example displaying the above-mentioned feature, i.e. a link between a primordial mass term and a scalar field coupling necessary for achieving the tree-level unitarity within the electroweak gauge theory. The case we have in mind is the process $e^+e^- \to W^+W^-$. In Section 5.8, we have observed that the sum of the three diagrams in Fig. 5.5 is already free of the leading quadratic divergences, but a linear divergence may still persist. Indeed, an explicit calculation yields the result

$$\mathcal{M}_{e^+e^-}^{(\nu)} + \mathcal{M}_{e^+e^-}^{(\gamma)} + \mathcal{M}_{e^+e^-}^{(Z)} = -\frac{g^2}{4m_W^2} m_e \bar{v}(l) u(k) + O(1) \tag{6.7}$$

The formula (6.7) exhibits the residual (linear) high-energy divergence, which obviously cannot be eliminated without an additional diagram.

As before, the scalar boson exchange offers a possible way out. Of course, we will try to utilize the σ field introduced in the previous example. We already know the precise form of the $WW\sigma$ coupling, so one only has to add an interaction of the scalar with leptons. The simple matrix structure of the linearly divergent term in (6.7) makes it obvious that one has to postulate a Yukawa coupling

$$\mathscr{L}_{ee\sigma} = g_{ee\sigma}\bar{e}e\sigma \tag{6.8}$$

The σ-exchange graph designed to cancel the divergent behaviour of (6.7) is shown in Fig. 6.2.

The matrix element corresponding to Fig. 6.2 reads

$$i\mathcal{M}^{(\sigma)}_{e^+e^-} = i^3 g m_W g_{ee\sigma}\bar{v}(l)u(k)\frac{1}{q^2 - m_\sigma^2}\varepsilon^*_\mu(p)\varepsilon^{*\mu}(r) \tag{6.9}$$

(note that we have already taken into account the result (6.6)). On simple dimensional grounds, one may guess that the matrix element (6.9) will be at most linearly divergent for longitudinal external W's. An explicit expression is easily calculated; proceeding in the usual way, one gets

$$\mathcal{M}^{(\sigma)}_{e^+e^-} = -\frac{1}{2m_W^2}g_{ee\sigma}g m_W\bar{v}(l)u(k) + O(1) \tag{6.10}$$

Comparing now (6.10) with (6.7), it is obvious that the divergent parts are cancelled if and only if

$$g_{ee\sigma} = -\frac{g}{2}\frac{m_e}{m_W} \tag{6.11}$$

Thus, we have identified another σ boson coupling that implements successfully the desired divergence cancellation; similar to the previous case, the corresponding coupling constant is proportional to a bare mass. Recall that masses introduced into the electroweak Lagrangian simply by hand break the $SU(2) \times U(1)$ gauge symmetry, so one may also say that the

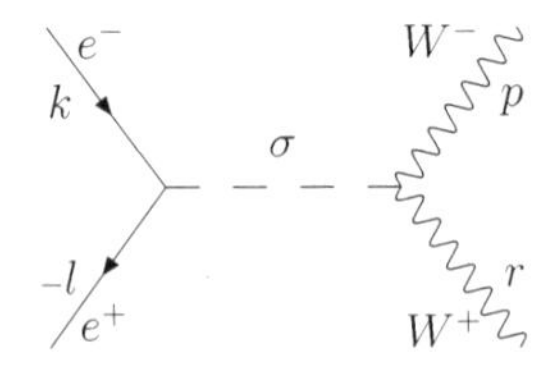

Figure 6.2. Scalar boson exchange contribution to the process $e^+e^- \to W^+W^-$.

interactions of the scalar σ field compensate the effects of the symmetry-breaking terms in the electroweak Lagrangian.

The examples discussed above show that the mass terms incorporated in the electroweak Lagrangian must be tightly correlated with couplings of a newly postulated neutral scalar boson if one wants to accomplish the delicate divergence cancellations necessary for perturbative renormalizability. In fact, this heuristic discussion seems to offer an important clue for building renormalizable electroweak models: instead of introducing the phenomenologically needed mass terms directly, one should perhaps try *to generate masses through appropriate interactions involving scalar fields*. Such a vague statement can indeed be given a more precise meaning within some particular field-theory models developed (though in a slightly different context) in the early 1960s. These field-theoretic constructions will be described in the following sections and finally employed in completing the construction of the full standard model of electroweak interactions.

6.2 Goldstone model

A basic ingredient of the Higgs mechanism is the so-called "Goldstone phenomenon" associated with "spontaneous symmetry breakdown". We shall start our discussion with a simple model of classical scalar field theory (invented originally by Goldstone [43]) that illustrates these concepts. The model we have in mind is described by the Lagrangian density of the type

$$\mathscr{L} = \partial_\mu \varphi \partial^\mu \varphi^* - V(\varphi) \tag{6.12}$$

with

$$V(\varphi) = -\mu^2 \varphi\varphi^* + \lambda(\varphi\varphi^*)^2 \tag{6.13}$$

where φ is a complex scalar field, μ is a real parameter with dimension of mass and λ is a (dimensionless) coupling constant.[1] In what follows, the function (6.13) will sometimes be called the "potential" (though, of course, it has nothing to do with the potential of a classical force). It is depicted schematically in Fig. 6.3. The essential feature of the considered Lagrangian is the "wrong sign" of the mass term in (6.13). Indeed, discarding temporarily the $\lambda\varphi^4$ term, one is left with a quadratic form that leads to the equation of motion $(\Box - \mu^2)\varphi = 0$, i.e. to the Klein–Gordon

[1]We assume $\lambda > 0$ in order that the energy density corresponding to (6.12) be bounded from below.

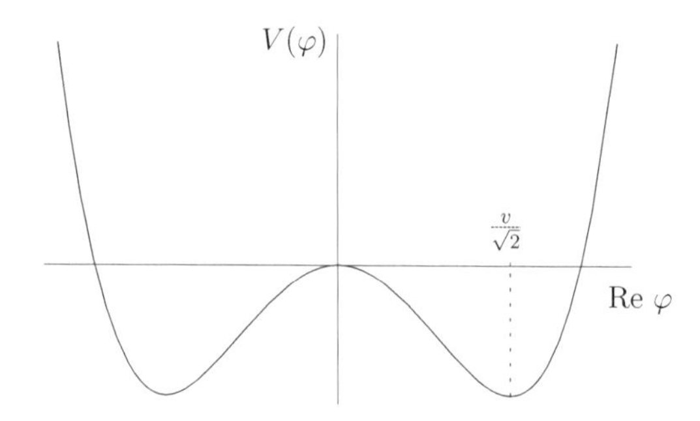

Figure 6.3. A visualization of the Goldstone potential given by (6.13). In fact, the full picture would consist of a surface formed by rotating this curve around the ordinate axis.

equation with reversed sign of mass squared. Thus, it is not immediately clear how the classical model described by (6.12) and (6.13) should be quantized — the quadratic part of the Lagrangian cannot be simply interpreted in terms of free particles and a straightforward perturbative treatment thus becomes inapplicable.

In order to guess a plausible interpretation of the considered model, it is instructive to calculate the corresponding energy (Hamiltonian) density. This is given by the component $\mathscr{T}_{00}$ of the canonical energy–momentum tensor, namely

$$\mathscr{H} = \mathscr{T}_{00} = \frac{\partial \mathscr{L}}{\partial(\partial_0 \varphi)}\partial_0\varphi + \frac{\partial \mathscr{L}}{\partial(\partial_0 \varphi^*)}\partial_0\varphi^* - \mathscr{L} \tag{6.14}$$

(of course, we take φ and φ^* as independent dynamical variables). Using (6.12) in (6.14), one thus gets, after some simple manipulations,

$$\mathscr{H} = \partial_0\varphi\partial_0\varphi^* + \vec{\nabla}\varphi\vec{\nabla}\varphi^* + V(\varphi) \tag{6.15}$$

One may now ask what is the field configuration $\varphi(x)$ corresponding to a *minimum* of the energy density. Obviously, the derivative terms in (6.15) always give a positive contribution for φ that is not a space–time constant. Thus, one should consider a constant φ and find a minimum of the potential V. This can be done easily. The V in fact depends only on one real variable ρ defined as $\rho^2 = \varphi\varphi^*$ so that instead of (6.13), one could write

$$V(\rho) = -\mu^2\rho^2 + \lambda\rho^4 \tag{6.16}$$

The first derivative $V'(\rho)$ vanishes for $\rho = 0$ and for $\rho^2 = \mu^2/2\lambda$. The value $\rho = 0$ corresponds to a local maximum, while for $\rho = \mu/\sqrt{2\lambda}$, one has an absolute minimum of the V. In terms of the original variable φ, it means that the minimum of the energy density corresponds to a one-parametric set of constant values

$$\varphi_0 = \frac{v}{\sqrt{2}} \mathrm{e}^{i\alpha} \tag{6.17}$$

where α is an arbitrary real number and we have denoted[2]

$$v = \frac{\mu}{\sqrt{\lambda}} \tag{6.18}$$

In other words, the φ_0 values that minimize the energy density lie on a circle in the complex plane with radius $v/\sqrt{2}$ and the energy minimum is thus infinitely (continuously) degenerate. Such a finding, namely the observation that the ground state of the considered system is described by a non-zero constant field, leads to the following simple idea: instead of φ, one should perhaps use its deviation from the "vacuum value" (6.17) as a true dynamical variable. Indeed, it seems to be more promising to study small oscillations around a stable ground state with $|\varphi| = v/\sqrt{2}$ rather than take as a reference point the value $\varphi = 0$ corresponding to an unstable state. This idea can be implemented mathematically in a rather elegant way if the original Lagrangian (6.12) is first rewritten in terms of radial and angular field variables defined by

$$\varphi(x) = \rho(x) \exp\left(i\frac{\pi(x)}{v}\right) \tag{6.19}$$

(note that we have introduced the factor of $1/v$ in the exponent in order to get the angular field $\pi(x)$ with the right dimension of mass). Using (6.19) in (6.12), one gets easily

$$\mathscr{L} = \partial_\mu \rho \partial^\mu \rho + \frac{1}{v^2} \rho^2 \partial_\mu \pi \partial^\mu \pi - V(\rho) \tag{6.20}$$

[2]The symbol v introduced in (6.18) stands for "vacuum" — this refers to the fact that the value $|\varphi| = v/\sqrt{2}$ corresponds to the ground state of the considered field system. The term "vacuum" for the ground state would be more appropriate at quantum level, but such a loose terminology is quite customary even in the context of classical field theory. In a systematic quantum treatment of this problem, one is led to the notion of "effective potential" (see, e.g., [Hua]) and v then represents a non-zero vacuum expectation value of the quantum scalar field. The relation (6.18) fixes a notation that has become standard for electroweak theories involving Higgs mechanism.

For further discussion, it is now also useful to recast the potential in a slightly different form; in particular, from (6.16), one immediately gets

$$\begin{aligned} V(\rho) &= \lambda\left[\left(\rho^2 - \frac{\mu^2}{2\lambda}\right)^2 - \left(\frac{\mu^2}{2\lambda}\right)^2\right] \\ &= \lambda\left(\rho^2 - \frac{v^2}{2}\right)^2 - \frac{1}{4}\lambda v^4 \end{aligned} \tag{6.21}$$

Obviously, the additive constant appearing in the last line can be dropped without changing anything essential — only the energy density thus becomes automatically non-negative. In what follows, we shall therefore replace the Lagrangian (6.20) by the equivalent form

$$\mathscr{L} = \partial_\mu \rho \partial^\mu \rho + \frac{1}{v^2}\rho^2 \partial_\mu \pi \partial^\mu \pi - \lambda\left(\rho^2 - \frac{v^2}{2}\right)^2 \tag{6.22}$$

(of course, we could have started with such a positively definite potential from the very beginning, but we find the expression (6.13) to be a more natural starting point since the "wrong-sign mass term" is explicitly singled out there). Let us now perform the shift of the field variable suggested above. The ρ may be written as

$$\rho = \frac{1}{\sqrt{2}}(\sigma + v) \tag{6.23}$$

where the variable σ involves the rescaling factor of $1/\sqrt{2}$ so as to get a proper normalization of its kinetic term. Using (6.23) in (6.22), one gets easily

$$\begin{aligned} \mathscr{L} = &\frac{1}{2}\partial_\mu \sigma \partial^\mu \sigma + \frac{1}{2}\partial_\mu \pi \partial^\mu \pi - \frac{1}{4}\lambda(\sigma^2 + 2v\sigma)^2 \\ &+ \frac{1}{2v^2}\sigma^2 \partial_\mu \pi \partial^\mu \pi + \frac{1}{v}\sigma \partial_\mu \pi \partial^\mu \pi \end{aligned} \tag{6.24}$$

i.e.

$$\mathscr{L} = \frac{1}{2}\partial_\mu \sigma \partial^\mu \sigma + \frac{1}{2}\partial_\mu \pi \partial^\mu \pi - \lambda v^2 \sigma^2 + \text{interactions} \tag{6.25}$$

where all terms higher than quadratic have been generically denoted as "interactions". The important point is that the σ field now has a mass term with the *right sign*, while π came out to be *massless*. In particular, the σ mass value that can be read off from (6.25) is given by $\frac{1}{2}m_\sigma^2 = \lambda v^2$, i.e. $m_\sigma^2 = 2\mu^2$ in view of (6.18). In fact, the appearance of a mass term with correct sign should not be surprising. Our redefinition (shift) of the radial field variable actually means that we perform Taylor expansion around a local minimum of the potential, where its second derivative is of course positive. However, this second derivative determines the coefficient of the term quadratic in the relevant field, which is precisely the mass term in the Lagrangian.

Thus, in the above simple exercise, we have seen that the model (6.12) describes in fact two real scalar fields σ and π, where

$$m_\sigma = \mu\sqrt{2}, \quad m_\pi = 0 \tag{6.26}$$

although such an interpretation is completely hidden in the original form of the Lagrangian written in terms of the variables φ and φ^*. The most remarkable feature of the considered model is the appearance of the massless field π, since this provides an illustration of the so-called Goldstone phenomenon alluded to earlier in this section. To explain this point, we have to make a brief digression here and recall first some important general concepts concerning the problem of symmetry breaking in field-theory models of particle physics.

A familiar manifestation of an (approximate) internal symmetry of such a model is the existence of multiplets of particles with (nearly) degenerate masses. The multiplets correspond to irreducible representations of the relevant symmetry group and become truly degenerate in the limit of exact symmetry, while the observed deviations from degeneracy within multiplets are attributed to small symmetry-violating terms in the Hamiltonian — in this context, the term "explicit symmetry breaking" is used (a good example of such an approximate symmetry is the isospin in strong interaction physics). This pattern corresponds to what is usually called the **Wigner–Weyl realization** of symmetry (cf., e.g., [Mar]); another well-known aspect of a symmetry realized in the Wigner–Weyl mode is the existence of certain selection rules for transition matrix elements with respect to the relevant quantum numbers (an illustration of this is in fact provided by the calculation of pion beta decay in Chapter 2). To put it briefly, in the Wigner–Weyl mode, the physical states transform according to the symmetry group representations; in particular, the vacuum can be taken as invariant.

On the other hand, there is a radically different possibility for a symmetry realization, which corresponds to the case of an invariant Hamiltonian or Lagrangian possessing non-invariant ground state (vacuum). Such a mode is indeed relevant for a wide variety of systems with infinite number of degrees of freedom, both relativistic and non-relativistic.[3] This scheme means that the symmetry is not realized on physical states in the usual way and the structure of quasi-degenerate multiplets as well as the selection rules typical for the Wigner–Weyl realization are completely lost. In current parlance, the term **spontaneous symmetry breaking** [44] is usually used for such a situation ("symmetry breaking" because the symmetry is no longer manifest in the physical spectrum and "spontaneous" because one may imagine that the system occupies spontaneously a non-invariant lowest-energy state, e.g. under the influence of an arbitrarily small asymmetric perturbation that picks a particular ground state). One may note that such a term is slightly deceptive as the symmetry is in fact only *hidden* — it is still present at the level of the Hamiltonian or Lagrangian (cf., e.g., [Col]). The most important aspect of spontaneous symmetry breakdown is that it has a generic signature described by the celebrated **Goldstone theorem** [45] (see also, e.g., the textbook [Wei]): If the symmetry of the considered Hamiltonian or Lagrangian is continuous, the non-invariance of its ground state (which is then necessarily continuously degenerate) implies the existence of a *massless bosonic excitation* (Goldstone boson) in the physical spectrum of the system. In particular, in the context of relativistic quantum field theory, the Goldstone boson is a spin-0 massless particle (its spinless nature is related to the requirement of Lorentz invariance of the vacuum state, but it can be both scalar and pseudoscalar). A familiar example of an (approximate) Goldstone boson in particle physics is the pion: many experimental facts in low-energy hadron phenomenology are naturally explained in terms of an effective theory where pions $\pi^{\pm}$ and π^0 are massless in the limit of exact chiral symmetry $SU(2)_L \times SU(2)_R$ of the strong interaction Lagrangian; their masses as observed in the real world are assumed to be due to an additional explicit symmetry breaking. Historically, it was probably Nambu [46] who came up first with this idea (for more details, see also, e.g., [Wei], [ChL] or [Geo]). To close this general digression, the last terminological remark is perhaps in order here. For reasons that should be obvious from the above discussion, the term **Nambu–Goldstone realization** of a symmetry (or simply Goldstone realization) is also frequently used instead of "spontaneous symmetry breakdown" (in fact, it is even more appropriate), but the

[3]Of course, for a system with finite number of degrees of freedom, one can also have the ground state that does not share a symmetry of the corresponding Hamiltonian, but in such a case, this has no further dramatic consequences.

latter name has certainly become more popular in the present-day particle physics.

Now, it is easy to see how the scalar field model discussed before illustrates the Goldstone phenomenon associated with spontaneous symmetry breakdown. The Lagrangian (6.12) is invariant under global phase transformations

$$\begin{aligned} \varphi'(x) &= \mathrm{e}^{i\omega}\varphi(x) \\ \varphi^{*\prime}(x) &= \mathrm{e}^{-i\omega}\varphi^*(x) \end{aligned} \tag{6.27}$$

where ω is a constant parameter (an arbitrary real number, independent of x). In other words, the symmetry group of our model is $U(1)$ (which is isomorphic to $O(2)$ — the rotation group of two-dimensional plane). The ground state φ_0 shown in (6.17) is obviously not invariant under such transformations (by applying (6.27), one moves around the circle in the complex plane corresponding to (6.17)) and the ground state energy is thus continuously degenerate. The massless field π then may be understood as corresponding to a Goldstone boson (note, however, that we are staying at the *classical* level!). It would be a highly non-trivial task to reformulate this simple Goldstone model for quantum fields, but the manipulations that led to (6.24) and (6.26) are nevertheless quite instructive — a difficult part of the discussion has been done for classical fields, with a result that is in accordance with the general Goldstone theorem. The Lagrangian (6.24) can then be quantized in the usual perturbative way. Note that this of course retains the original symmetry of (6.12), but in terms of the variables ρ and π, the transformation law (6.27) is recast as

$$\begin{aligned} \sigma'(x) &= \sigma(x) \\ \pi'(x) &= \pi(x) + v\omega \end{aligned} \tag{6.28}$$

It is also easy to guess how one can get, in the present context, a classical picture of a Wigner–Weyl realization of symmetry. Clearly, this would correspond to the Lagrangian of the type (6.12), with the opposite sign of μ^2 in the potential, i.e. with $V(\varphi) = \mu^2\varphi\varphi^* + \lambda(\varphi\varphi^*)^2$ instead of (6.13). In such a case, the ground state is unique and corresponds to $\varphi = 0$. The model then can be interpreted, e.g., as a system of two real scalar fields φ_1 and φ_2 (with $\varphi = \varphi_1 + i\varphi_2$) corresponding to particles with equal mass μ, i.e. with the $O(2)$ symmetry manifested directly in the particle spectrum.

As we noted before, the discussion of the simple Goldstone model carried out here is only a necessary prerequisite for the formulation of the mass-generation mechanism in the electroweak theory. Actually, the physics of massless scalar bosons is not of primary interest to us. The truly important thing, from our point of view, happens when the interaction with an Abelian gauge field is switched on in the Lagrangian (6.12). As we shall see in the next section, the magic Higgs trick then works, which means that the Goldstone boson becomes unphysical and one gets a mass term for the vector field.

6.3 Abelian Higgs model

Let us now introduce, following Higgs [47], the interaction with an Abelian gauge field into the Goldstone model considered in preceding section. As we know from Chapter 4, there is a standard way of doing that: ordinary derivatives in the kinetic term in (6.12) are replaced by the covariant ones and the usual kinetic term for the gauge field is added. One thus gets, formally, the scalar electrodynamics incorporating also quartic self-coupling of the complex scalar field and its mass term with the wrong sign. The corresponding Lagrangian can be written as

$$\mathscr{L}_{\text{Higgs}} = -\frac{1}{4}F_{\mu\nu}F^{\mu\nu} + (\partial_\mu - igA_\mu)\varphi(\partial^\mu + igA^\mu)\varphi^* - \lambda\left(\varphi\varphi^* - \frac{v^2}{2}\right)^2 \tag{6.29}$$

where, of course, $F_{\mu\nu} = \partial_\mu A_\nu - \partial_\nu A_\mu$ and g denotes the gauge coupling constant. Note that in (6.29), we have used the form (6.22) for the Goldstone potential, with $v = \mu/\sqrt{\lambda}$ (see (6.18)). By construction, the Lagrangian (6.29) is invariant under local gauge transformations

$$\begin{aligned}
\varphi'(x) &= \mathrm{e}^{i\omega(x)}\varphi(x) \\
\varphi^{*\prime}(x) &= \mathrm{e}^{-i\omega(x)}\varphi^*(x) \\
A'_\mu(x) &= A_\mu(x) + \frac{1}{g}\partial_\mu\omega(x)
\end{aligned} \tag{6.30}$$

In analogy with the discussion of previous section, one may now trade φ and φ^* for the corresponding radial and angular variables and shift the radial field according to (6.23); in other words, the complex field φ is reparametrized as

$$\varphi(x) = \rho(x)\exp\left(i\frac{\pi(x)}{v}\right) \tag{6.31}$$

In terms of the variables ρ and π, the gauge transformations (6.30) are recast as

$$\begin{aligned} \rho'(x) &= \rho(x) \\ \pi'(x) &= \pi(x) + v\omega(x) \\ A'_\mu(x) &= A_\mu(x) + \frac{1}{g}\partial_\mu\omega(x) \end{aligned} \tag{6.32}$$

The gauge invariance means that a field configuration described by some functions $\rho(x), \pi(x)$ and $A_\mu(x)$ (solutions of the corresponding equations of motion) is equivalent to the set $\rho'(x), \pi'(x), A'_\mu(x)$ obtained by the transformation (6.32) (the equivalence is to be understood in the sense that any physical quantity can be calculated either from ρ, π, A_μ or from ρ', π', A'_μ, with the same result). In particular, for a given set ρ, π, A_μ, one can choose $\omega = -\pi/v$ in (6.32) and eliminate thus completely the angular field variable π; in other words, the original field configuration is equivalent, up to a gauge transformation, to that described by

$$\begin{aligned} \rho'(x) &= \rho(x) \\ \pi'(x) &= 0 \\ A'_\mu(x) &= A_\mu(x) - \frac{1}{gv}\partial_\mu\pi(x) \end{aligned} \tag{6.33}$$

One can thus also say that — owing to the local gauge invariance — the angular field π (i.e. the erstwhile Goldstone boson) becomes unphysical within the Higgs model, since it can be eliminated by an appropriate choice of gauge. The gauge fixed by the condition (6.33), i.e. by the requirement $\pi \equiv 0$, is usually called unitary gauge (U-gauge).[4] Now, it is clear that the equations of motion for the U-gauge dynamical variables can be obtained directly from the Lagrangian (6.29) where one fixes the gauge by setting simply $\pi = 0$, i.e. $\varphi = \varphi^* = \rho$. (The reason is obvious: the constraint $\pi = 0$ is implemented via a special gauge transformation and the original

[4]The adjective "unitary" may seem totally obscure at the present moment, but this in fact refers to the envisaged quantum version of the considered model: it is well known that, in general, the S-matrix unitarity becomes transparent in a theory that does not involve any auxiliary unphysical fields. The label "physical gauge" (which would be perhaps most appropriate in the present context) is also sometimes used, but the term "unitary gauge" has become standard in modern electroweak theories.

Lagrangian (6.29) is gauge invariant.) Further, in full analogy with our previous analysis of the Goldstone model, the radial field ρ should be shifted as

$$\rho = \frac{1}{\sqrt{2}}(\sigma + v) \tag{6.34}$$

(see (6.23)) in order to get rid of the wrong-sign scalar mass term (obviously, the resulting mass term for σ must be the same as in the case of the Goldstone model since it is fully determined by the scalar field potential V). Thus, the U-gauge Higgs Lagrangian (6.29) can be written as

$$\begin{aligned}\mathscr{L}_{\text{Higgs}}^{(U)} &= -\frac{1}{4}G_{\mu\nu}G^{\mu\nu} + \frac{1}{2}(\partial_\mu - igB_\mu)(\sigma + v)(\partial^\mu + igB^\mu)(\sigma + v) \\ &\quad - \frac{1}{4}\lambda[(\sigma + v)^2 - v^2]^2\end{aligned} \tag{6.35}$$

where we have introduced, for definiteness, an extra symbol B_μ for the U-gauge value of the vector field (cf. (6.33)), and $G_{\mu\nu} = \partial_\mu B_\nu - \partial_\nu B_\mu$. The form (6.35) can be easily worked out as

$$\begin{aligned}\mathscr{L}_{\text{Higgs}}^{(U)} &= -\frac{1}{4}G_{\mu\nu}G^{\mu\nu} - \frac{1}{4}\lambda(\sigma^2 + 2v\sigma)^2 \\ &\quad + \frac{1}{2}(\partial_\mu\sigma - ig\sigma B_\mu - igvB_\mu)(\partial^\mu\sigma + ig\sigma B^\mu + igvB^\mu) \\ &= -\frac{1}{4}G_{\mu\nu}G^{\mu\nu} + \frac{1}{2}\partial_\mu\sigma\partial^\mu\sigma - \frac{1}{4}\lambda(\sigma^2 + 2v\sigma)^2 \\ &\quad + \frac{1}{2}g^2(\sigma + v)^2 B_\mu B^\mu\end{aligned} \tag{6.36}$$

Separating now in the last expression its quadratic part and the interaction terms, one has

$$\begin{aligned}\mathscr{L}_{\text{Higgs}}^{(U)} &= \frac{1}{2}\partial_\mu\sigma\partial^\mu\sigma - \lambda v^2\sigma^2 - \frac{1}{4}G_{\mu\nu}G^{\mu\nu} + \frac{1}{2}g^2v^2 B_\mu B^\mu \\ &\quad + g^2 v\sigma B_\mu B^\mu + \frac{1}{2}g^2\sigma^2 B_\mu B^\mu - \lambda v\sigma^3 - \frac{1}{4}\lambda\sigma^4\end{aligned} \tag{6.37}$$

As expected, there is a mass term of the field σ that coincides with (6.25), but the truly remarkable feature of the expression (6.37) is the presence of a **mass term for the vector field B_μ**. Although there was no such thing in the original form (6.29), eventually it has shown up as a consequence of the scalar field shift (6.34). This, in fact, is the essence of the famous "Higgs mechanism" or "Higgs trick", demonstrated here (at the classical level) within the simplest Abelian theory. **When the spontaneously broken**

symmetry of a scalar-field model is gauged, the original Goldstone boson disappears from physical spectrum and the gauge field acquires a mass. (In a common physical "folklore", this situation is sometimes characterized by saying that the would-be Goldstone boson is eaten by the gauge boson, which becomes heavy.) At the same time, a massive scalar field survives in the physical spectrum (we shall call σ the Higgs field). Obviously, the whole mechanism is triggered by the wrong-sign scalar mass term in the original symmetric Lagrangian (6.29). B_μ mass can be easily read off from (6.37); this is

$$m_B = gv \tag{6.38}$$

The reader should note the natural and easy-to-remember structure of the last formula: the induced vector-field mass is a product of the generic mass scale v (the scalar field vacuum value characteristic for spontaneous symmetry breaking) and the gauge interaction strength g. As for the interaction part of (6.37), it basically exhibits a pattern that will be recovered later on within the electroweak standard model. In particular, using (6.38), one can see that the strength of the trilinear coupling σBB is proportional to the B_μ mass, namely

$$g_{\sigma BB} = g m_B \tag{6.39}$$

For the quadrilinear coupling $\sigma\sigma BB$, one has $g_{\sigma\sigma BB} = \frac{1}{2}g^2$ and the coupling constants for the cubic and quartic self-interactions of the Higgs field σ can be easily expressed, e.g., in terms of the m_σ, m_B, and g (needless to say, values of the σ self-couplings are the same as in the Goldstone model).

The U-gauge Higgs model can be quantized in a straightforward way and B_μ propagator then has the canonical form

$$D_{\mu\nu}^{(U)}(k) = \frac{-g_{\mu\nu} + m_B^{-2} k_\mu k_\nu}{k^2 - m_B^2} \tag{6.40}$$

This, in combination with the well-known behaviour of the longitudinal polarization vector for a massive spin-1 particle (cf.(3.29)), may lead to power-like growth of some tree-level Feynman diagrams in the high-energy limit. However, while such divergences indeed occur for individual diagrams, they get cancelled when all relevant contributions to a given physical process are summed. In fact, what one observes here is a simplified variant of the mechanism suggested in Section 6.1. As an instructive exercise, the reader is recommended to verify such a divergence cancellation, e.g., for the process $BB \to \sigma\sigma$. (*Hint*: both σBB and $\sigma\sigma BB$ couplings enter the game in this case.) As we know, the soft high-energy behaviour at the tree level ("tree unitarity") suggests that the theory might be renormalizable. In this case, it is indeed so; quite generally, renormalizability of a gauge theory with the Higgs mechanism[5] was proved first by t Hooft and Veltman [48], and nowadays, this topic is covered by most of the modern textbooks on quantum field theory. For a general proof of renormalizability of a spontaneously broken gauge theory, a different quantization procedure is used, namely the so-called R-gauge formulation (invented originally by 't Hooft [49]). In contrast to the U-gauge, the R-gauge propagator of a massive vector boson behaves for $k^2 \to \infty$ as $1/k^2$ (i.e. in the same way as in the massless case). This in turn means that the convergence properties of higher-order (closed-loop) Feynman diagrams become much better than in the U-gauge and the usual power-counting analysis indicates immediately a renormalizable behaviour (see, e.g., [ChL] or Appendix G in [Hor]). The price to be paid for that is the presence of the unphysical would-be Goldstone boson, which is not eliminated explicitly by means of the gauge choice — instead, it is preserved as an auxiliary field variable. Within such a quantization scheme, a proof of the S-matrix unitarity is consequently more complicated as one has to prove that the additional contributions of unphysical particles are irrelevant. Moreover, since there is in fact a whole class of the R-gauges, one has to demonstrate the gauge independence of the physical S-matrix (in particular, one has to prove an equivalence of the U-gauge with any of the R-gauges). All this has by now become "common wisdom" in modern field theory and the internal consistency of different formulations of a gauge theory with the Higgs mechanism has been firmly established. We are not going into further details here; some technicalities concerning the R-gauges will be described in later sections within the framework of the full standard electroweak model. In any case, one should bear in mind that the basic idea behind the question of renormalizability of a gauge theory with Higgs mechanism is in fact extremely simple: in such a theory, one starts with a massless gauge field, which *a priori* cannot produce any non-renormalizable behaviour of Feynman graphs. The physically interpretable Lagrangian is then obtained by means of a mere redefinition of the relevant dynamical variables (along with an appropriate gauge fixing) and one thus expects, intuitively, that the convergence properties of the S-matrix remain basically the same.

[5]The alternative term "spontaneously broken gauge theory" is also frequently used in this context.

However, it is also fair to stress the following point. Within the Abelian model considered here, it is actually not necessary to invoke the Higgs mechanism for obtaining a renormalizable theory with a massive vector boson — one could as well introduce the corresponding mass term into the scalar QED Lagrangian simply by hand without spoiling renormalizability (this is analogous to the case of spinor QED with massive photon; an essential point is that in both cases, the Abelian gauge field is coupled to a conserved current). The example of the Abelian Higgs model can serve as a prototype for more complicated situations (involving non-Abelian gauge symmetry) encountered within electroweak theory, where the vector field mass term cannot be put in by hand with impunity and an appropriate variant of the Higgs trick becomes necessary.

One more remark is perhaps in order at this place. The basic feature of the Higgs model, namely the appearance of an "induced" vector-boson mass term and simultaneous disappearance of a Goldstone boson, may intuitively be understood as a transformation of the would-be Goldstone boson into the zero-helicity state of the vector boson (i.e. the state corresponding to longitudinal polarization) — such a state is of course absent in the massless case. In this spirit, one can say, somewhat loosely, that the total number of "degrees of freedom" is preserved throughout the Higgs mechanism: at the beginning, there are two real scalar fields and two (transverse) polarizations of a massless gauge field, and we end up with one physical scalar and three polarization states of a massive vector boson. It is interesting that such a vague connection between the unphysical Goldstone boson and physical longitudinal vector boson can be given a more precise meaning within the R-gauge formulation of the Higgs model. This is described by the famous "equivalence theorem" [50], stating roughly that in the high-energy limit, an S-matrix element for longitudinal vector bosons is asymptotically equal (possibly up to a constant factor) to its unphysical counterpart involving the corresponding would-be Goldstone scalars (the asymptotic region here corresponds to energies much larger than the vector boson mass). We will discuss this remarkable statement in more detail in the context of the electroweak standard model.

In closing this section, let us add a brief historical commentary. In fact, the first hint of the Higgs mechanism appeared in the context of non-relativistic condensed-matter physics [51]. Then it was discussed independently by several authors [52–54] within the framework of relativistic quantum field theory without resorting to explicit models of the type described above (a good review of the non-perturbative QFT aspects of spontaneous symmetry breaking and the Higgs phenomenon can be found, e.g., in [Brn]). The explicit model [47] was originally conceived as a mere illustration of the rather abstract field-theory concepts involved, but as we know today, it had an immense heuristic value. In particular, its straightforward non-Abelian generalization [55] was subsequently utilized by Weinberg [40] and Salam [41] for building the first potentially renormalizable unified theory of weak and electromagnetic interactions. In the early 1970s the principles of gauge symmetry and the Higgs mechanism were widely accepted by particle theorists and this led to an explosion of "model building" following the Weinberg–Salam paradigm (cf., e.g., [AbL]). Nevertheless, despite the technical attractiveness of the Higgs mechanism, many theorists were reluctant to accept the real existence of a physical elementary scalar boson as an ingredient of the electroweak gauge theory. In particular, the main controversial point consisted in distinguishing a more general "Higgs mechanism" for generating gauge boson masses[6] and the "Higgs boson" that emerges within a specific model like [47], which involves an elementary scalar field (for a detailed discussion of the subtle issue in question, see, e.g., the nice instructive essay [1]). To put it briefly, a standard statement reads that the Higgs mechanism and the Higgs boson are two different things. This long-standing dilemma was apparently resolved in 2012, when the observation of a Higgs-like boson was announced by two independent experimental collaborations (ATLAS and CMS) working at the Large Hadron Collider (LHC) at CERN (see [63] and [64] for the original discovery papers and [6] for a comprehensive review of current data). So, we may continue with confidence towards constructing the full edifice of the electroweak SM.

6.4 Higgs mechanism for $SU(2) \times U(1)$ gauge theory

Before proceeding to the formulation of the Higgs mechanism that operates within the standard electroweak theory, we will describe some characteristic general features of non-Abelian extensions of the field-theory models considered in preceding two sections. The general statements we are going to specify below will not be proved here as the corresponding proofs can be

[6] For obvious reasons, the extended label Brout–Englert–Higgs (BEH) mechanism is quite frequently used in such a context. The point is that Brout and Englert in their celebrated original paper apparently did not care about the possible existence of a physical scalar boson that may (but need not) occur as a "by-product" of the mass-generation mechanism for gauge fields.

found in many other books (see, e.g., [ChL], [Hua], [Rai]); rather, we will utilize the available general knowledge for motivating the choice of the SM Higgs sector.

Let us start with the Goldstone-type models. In the example discussed in Section 6.2, a particular ground state ("vacuum") belonging to the set (6.17) does not share the one-parametric $U(1)$ symmetry of the Lagrangian (6.12) and, as a result, one Goldstone boson appears. It turns out (see [55] for an original paper) that such a pattern can be generalized as follows. One may consider a model involving a multiplet of scalar fields, with dynamics described by means of a Lagrangian of the type (6.12) possessing a continuous n-parametric internal symmetry group G and with a ground state (defined as a minimum of the corresponding "potential" $V(\varphi)$) that is less symmetric than the Lagrangian. In particular, let us assume that the vacuum state remains invariant under an r-parametric ($r < n$) subgroup $H \subset G$. Then there are $n-r$ massless Goldstone bosons; in other words, the number of Goldstone bosons is equal to the number of broken symmetry generators. Needless to say, some massive scalar bosons always appear as well — their number depends on the dimension of the original multiplet.

Next, let us see what happens when (a part of) the global symmetry of a general Goldstone-type model with the symmetry-breaking pattern indicated above is made local, that is, when a subgroup of G is gauged by introducing a set of Yang–Mills fields associated with the corresponding generators. Let the total number of gauge fields be m ($m \leq n$) and suppose that k of them are coupled to broken symmetry generators, i.e. to those connected with Goldstone bosons (of course, $k \leq n - r$). Then it turns out [55] that upon shifting scalar fields by the relevant vacuum values, one gets k massive vector fields and k Goldstone bosons become unphysical — they can be eliminated by an appropriate choice of gauge (the U-gauge, analogous to that discussed earlier in the Abelian case). The other $m - k$ gauge fields (coupled to unbroken generators) remain massless. Note that the existence of the physical U-gauge in a general case was proved in [56].

Thus, a general scheme of the non-Abelian generalization of the Higgs mechanism that emerges from the preceding discussion is quite elegant and easy to remember: within a gauged Goldstone-type model, the Yang–Mills fields coupled to "spontaneously broken" symmetry generators give rise to massive vector bosons and the associated scalar Goldstone bosons disappear from physical spectrum. Such a result actually implies an important rule for building models of electroweak interactions: **for each vector boson mass, which is to be generated via Higgs mechanism, one needs**

a (would-be) Goldstone boson in the scalar sector. Needless to say, this also represents a certain constraint on the contents of scalar multiplets involved in the theory.

One may employ the above general observations to make a right guess for the Higgs–Goldstone sector of the standard electroweak theory. Since we want to get three massive vector bosons, we must have three Goldstone bosons in the underlying scalar field model. Further, it is also known that at least one physical scalar boson (the Higgs boson) survives the Higgs mechanism. Thus, it is clear that one has to start with at least four real scalar fields. In order to get the desired spectrum of vector boson masses, the scalars must be coupled in a non-trivial way to the $SU(2)$ gauge fields; it means that the two complex scalars should constitute a doublet representation of $SU(2)$. The upshot of these considerations is that the minimal Higgs–Goldstone scalar sector for the $SU(2) \times U(1)$ electroweak theory consists of one complex (weak isospin) $SU(2)$ doublet; this must also be endowed with some specific transformation properties under the $U(1)$ (weak hypercharge) subgroup, as we shall discuss in the sequel. The weak isodoublet can be written as

$$\Phi = \begin{pmatrix} \varphi^+ \\ \varphi^0 \end{pmatrix} \tag{6.41}$$

where the two complex components φ^+ and φ^0 are of course equivalent to four real fields, e.g., through a straightforward parametrization

$$\Phi = \begin{pmatrix} \varphi_1 + i\varphi_2 \\ \varphi_3 + i\varphi_4 \end{pmatrix} \tag{6.42}$$

The superscripts of the components of (6.41) indicate that φ^+ and φ^0 should represent fields carrying charges $+1$ and 0, respectively (this becomes clear when one specifies the interaction terms involving the scalar doublet and other fields with definite charge assignments).

Now, we are in a position to discuss the Higgs mechanism within the $SU(2) \times U(1)$ gauge theory in explicit terms. The starting point of our discussion will be the underlying Goldstone-type model. Using (6.41) as a basic building block, it is easy to construct the corresponding Lagrangian possessing the necessary symmetry. In analogy with (6.12), this can be written as

$$\mathscr{L}_{\text{Goldstone}} = (\partial_\mu \Phi^\dagger)(\partial^\mu \Phi) - V(\Phi) \tag{6.43}$$

with the potential V given by

$$V(\Phi) = -\mu^2 \Phi^\dagger \Phi + \lambda (\Phi^\dagger \Phi)^2 \tag{6.44}$$

It is interesting to note that such a Lagrangian has, in fact, an "accidental" symmetry larger than the originally required $SU(2) \times U(1)$. Indeed, using the parametrization (6.42), one sees that

$$\Phi^\dagger \Phi = \varphi_1^2 + \varphi_2^2 + \varphi_3^2 + \varphi_4^2 \tag{6.45}$$

which means that the full symmetry of V is $O(4)$ (of course, the same is true for the kinetic term in (6.43)). One may observe immediately that this accidental symmetry is due precisely to the doublet character of the basic Higgs–Goldstone field — when starting from (6.42), one must necessarily employ the form (6.45) in order to construct an $SU(2)$ invariant Lagrangian. We shall discuss these deeper symmetry aspects of the standard electroweak theory later in this chapter (see Section 6.8).

It is not difficult to see that the Lagrangian (6.43) describes three massless Goldstone bosons and one massive scalar. Indeed, it can be recast as

$$\mathscr{L}_{\text{Goldstone}} = \text{derivative terms} + \mu^2\rho^2 - \lambda\rho^4 \tag{6.46}$$

where we have denoted

$$\rho^2 = \Phi^\dagger \Phi \tag{6.47}$$

Similar to the Abelian case, one may argue that the minimum of energy density occurs for space–time constant field configurations Φ_0 such that

$$\Phi_0^\dagger \Phi_0 = \frac{v^2}{2} \tag{6.48}$$

where

$$v = \frac{\mu}{\sqrt{\lambda}} \tag{6.49}$$

(cf. (6.18)). Subtracting then from the field variable ρ its "vacuum value" mentioned above, one gets rid of the wrong-sign mass term in (6.46) and the shifted field acquires an ordinary mass in the by now familiar way. The other three real fields that parametrize our complex doublet remain massless as they enter only the derivative terms in (6.46). For an explicit description of the Goldstone bosons and the massive (Higgs) scalar, it is again useful to introduce an exponential parametrization of (6.41) analogous to the relation (6.19) employed in the Abelian case. Now, we can write

$$\Phi(x) = \exp\left(\frac{i}{v}\pi^a(x)\tau^a\right)\begin{pmatrix} 0 \\ \frac{1}{\sqrt{2}}(v + H(x)) \end{pmatrix} \tag{6.50}$$

where we have already marked explicitly the shift of the "radial" variable ρ, defining thus the Higgs field H. The "angular" fields π^a, $a = 1, 2, 3$ represent the Goldstone bosons (τ^a denote, as usual, the Pauli matrices). Using (6.50) in (6.43), it is then elementary to find the Higgs field mass; this is

$$m_H^2 = 2\lambda v^2 \tag{6.51}$$

(which becomes $m_H = \mu\sqrt{2}$ when one takes into account (6.49)). We should note that a particular vacuum field configuration Φ_0 is, for example,

$$\Phi_0^{(0)} = \frac{1}{\sqrt{2}}\begin{pmatrix} 0 \\ v \end{pmatrix} \tag{6.52}$$

Of course, any Φ_0 obtained from (6.52) by means of a global $SU(2)$ transformation can represent the ground state as well, since the potential minimum is determined by the $\Phi_0^\dagger \Phi_0$ value only. In other words, there is a three-parametric degenerate set of vacua associated with the potential (6.44). Thus, in the considered classical field-theory model, we can indeed recognize characteristic features of spontaneous symmetry breakdown: the Lagrangian (6.43) is invariant under $SU(2)$ while the ground state (represented, e.g., by (6.52)) is not. As a result, three Goldstone bosons appear, corresponding to the three generators of $SU(2)$ broken by the vacuum state.

The passage from (6.43) to a Higgs-type Lagrangian with local $SU(2) \times U(1)$ symmetry is accomplished in a similar manner as in the Abelian model discussed in preceding section. From a purely technical point of view, the covariant derivative acting on the scalar doublet Φ can be written in a straightforward analogy with the case of lepton sector described in Chapter 5. Φ, apart from being a doublet under the $SU(2)$, also carries a weak hypercharge Y_Φ associated with the $U(1)$ subgroup (we shall denote it simply as Y in what follows). The gauge invariant Lagrangian can then be written as

$$\begin{aligned}\mathscr{L}_{\text{Higgs}} = \Phi^\dagger \left(\overleftarrow{\partial}_\mu + igA_\mu^a \frac{\tau^a}{2} + ig'YB_\mu\right)\left(\overrightarrow{\partial}^\mu - igA^{b\mu}\frac{\tau^b}{2} - ig'YB^\mu\right)\Phi \\ -\lambda\left(\Phi^\dagger\Phi - \frac{v^2}{2}\right)^2\end{aligned} \tag{6.53}$$

where A_μ^a, $a = 1, 2, 3$ and B_μ are Yang–Mills fields corresponding to $SU(2)$ and $U(1)$, respectively, and g, g' are the associated coupling constants. For the sake of brevity, we have not included here the kinetic terms of gauge fields and their pure self-interactions that have been discussed in detail earlier; these can be retrieved from Chapter 5 whenever necessary. As usual (for later convenience), we have also shifted the bottom of the scalar-field potential (6.44) to zero by adding an otherwise inessential constant. The exponential parametrization (6.50) can be used for fixing the physical U-gauge in the same manner as in the Abelian case discussed in the preceding section. Such a gauge fixing is formally equivalent to a local $SU(2)$ transformation that removes the angular fields from Φ (this indicates the unphysical nature of the would-be Goldstone bosons in the present context). When this is done, one is left with

$$\Phi_U(x) = \begin{pmatrix} 0 \\ \frac{1}{\sqrt{2}}(v + H(x)) \end{pmatrix} \tag{6.54}$$

The Lagrangian (6.53) in the U-gauge can then be written in terms of (6.54) and correspondingly transformed gauge fields without changing its original form. It reads

$$\begin{aligned}\mathscr{L}^{(U)}_{\text{Higgs}} &= \Phi_U^\dagger \left(\overleftarrow{\partial}_\mu + igA^a_\mu \frac{\tau^a}{2} + ig'YB_\mu\right)\left(\vec{\partial}^\mu - igA^{b\mu}\frac{\tau^b}{2} - ig'YB^\mu\right)\Phi_U \\ &\quad - \lambda\left(\Phi_U^\dagger\Phi_U - \frac{v^2}{2}\right)^2\end{aligned} \tag{6.55}$$

where we have used, for notational simplicity, the same symbols for the transformed gauge fields as for the old ones. Writing (6.54) as

$$\Phi_U(x) = \frac{1}{\sqrt{2}}(v + H(x))\xi, \quad \xi = \begin{pmatrix} 0 \\ 1 \end{pmatrix} \tag{6.56}$$

the U-gauge Lagrangian (6.55) can be easily worked out as

$$\begin{aligned}\mathscr{L}^{(U)}_{\text{Higgs}} &= \frac{1}{2}\partial_\mu H \partial^\mu H - \frac{1}{4}\lambda[(v+H)^2 - v^2]^2 \\ &\quad + \frac{1}{2}(v+H)^2 \xi^\dagger \left(gA^a_\mu\frac{\tau^a}{2} + g'YB_\mu\right)\left(gA^{b\mu}\frac{\tau^b}{2} + g'YB^\mu\right)\xi\end{aligned} \tag{6.57}$$

The last expression can be further simplified by means of the relations

$$\begin{aligned}\tau^a\tau^b + \tau^b\tau^a &= 2\delta^{ab}\mathbb{1} \\ \xi^\dagger\tau^a\xi &= -\delta^{3a} \\ \xi^\dagger\xi &= 1\end{aligned} \tag{6.58}$$

and one thus finally gets

$$\begin{aligned}\mathscr{L}^{(U)}_{\text{Higgs}} &= \frac{1}{2}\partial_\mu H\partial^\mu H - \lambda v^2 H^2 - \lambda v H^3 - \frac{1}{4}\lambda H^4 \\ &\quad + \frac{1}{8}(v+H)^2\left(g^2 A^a_\mu A^{a\mu} - 4Ygg'A^3_\mu B^\mu + 4Y^2 g'^2 B_\mu B^\mu\right)\end{aligned} \tag{6.59}$$

Obviously, the Higgs field mass is the same as before (cf. (6.51)). The part of the Lagrangian (6.59) quadratic in gauge fields is diagonalized immediately and mass terms of intermediate vector bosons can thus be identified easily. For the relevant quadratic form, one obtains from (6.59)

$$\begin{aligned}\mathscr{L}_{\text{mass}}^{(\text{IVB})} &= \frac{1}{8}v^2\left[g^2((A_\mu^1)^2+(A_\mu^2)^2)+(gA_\mu^3-2g'YB_\mu)^2\right]\\ &= \frac{1}{8}(g^2+4Y^2g'^2)v^2\left(\frac{g}{\sqrt{g^2+4Y^2g'^2}}A_\mu^3-\frac{2Yg'}{\sqrt{g^2+4Y^2g'^2}}B_\mu\right)^2\\ &\quad+\frac{1}{4}g^2v^2W_\mu^-W^{+\mu}\end{aligned} \tag{6.60}$$

where $W_\mu^\pm$ stand for combinations $\frac{1}{\sqrt{2}}(A_\mu^1\mp iA_\mu^2)$ familiar from our previous analysis of charged current interaction (cf. Section 5.2) and we have also introduced a "normalized" linear combination of A_μ^3 and B_μ, which should presumably be identical with the Z boson field discussed in Section 5.3. Let us recall that for the Z field coupled to weak neutral currents, we had

$$Z_\mu = \cos\theta_W A_\mu^3 - \sin\theta_W B_\mu \tag{6.61}$$

(cf. (5.22)), where the mixing angle θ_W is in general given by $\tan\theta_W = -2Y_Lg'/g$ (see (5.32)) with Y_L being the weak hypercharge of the left-handed leptonic doublet. Comparing this result with (6.60), it is clear that we must set (returning to the notation $Y_\Phi = Y$ for a moment)

$$Y_\Phi = -Y_L \tag{6.62}$$

if Z coming from the mass matrix diagonalization is to be the same as that in (6.61), i.e., if we want the two ends of electroweak theory to be mutually consistent.

The reader may remember that we have eventually set $Y_L = -1/2$ (cf. (5.38)) for the sake of simplicity of the resulting formulae. This implies the conventional choice

$$Y_\Phi = +\frac{1}{2} \tag{6.63}$$

which means that one can always use the rule

$$Q = T_3 + Y \tag{6.64}$$

(cf. Eq. (5.40)). Note that for the scalar doublet (6.41), this reflects the fact that the upper component (with $T_3 = +\frac{1}{2}$) has $Q = +1$. Actually, in most textbooks, the convention (6.64) is usually adopted automatically when the Higgs sector of the standard electroweak theory is described. We have discussed here the general case at some length for completeness; a pragmatically minded reader might omit the analysis involving an arbitrary Y value and use immediately the law (6.64) from the very start.

Thus, we have seen that there is just one massive combination of A^3_μ and B_μ; upon setting $Y = 1/2$ in (6.60), this becomes

$$Z_\mu = \frac{1}{\sqrt{g^2 + g'^2}}(gA^3_\mu - g'B_\mu) \tag{6.65}$$

and coincides with (6.61). Since the total number of gauge fields is four, we will introduce also a combination "orthogonal" to (6.65), namely

$$A_\mu = \frac{1}{\sqrt{g^2 + g'^2}}(g'A^3_\mu + gB_\mu) \tag{6.66}$$

which is obviously massless (simply because there is no such mass term in (6.60)) and coincides with the electromagnetic field appearing in (5.22). Let us recall that the orthogonality is imposed so as to preserve the diagonal structure of the kinetic term for vector fields (cf. (5.25)). Taking into account the normalization of the kinetic term, the non-zero masses can now be read off directly from (6.60). For $Y = +1/2$, one has

$$\mathscr{L}^{(\mathrm{IVB})}_{\mathrm{mass}} = \frac{1}{4}g^2v^2W^-_\mu W^{+\mu} + \frac{1}{8}(g^2 + g'^2)v^2 Z_\mu Z^\mu \tag{6.67}$$

which yields

$$\begin{aligned} m_W^2 &= \frac{1}{4}g^2v^2 \\ \frac{1}{2}m_Z^2 &= \frac{1}{8}(g^2 + g'^2)v^2 \end{aligned} \tag{6.68}$$

Thus, as a result of the Higgs mechanism described above, we have the mass formulae

$$\begin{aligned} m_W &= \frac{1}{2}gv \\ m_Z &= \frac{1}{2}(g^2 + g'^2)^{1/2}v \end{aligned} \tag{6.69}$$

derived first by Weinberg in his celebrated paper [40]. The relations (6.69) imply, in particular,

$$\frac{m_W}{m_Z} = \cos\theta_W \tag{6.70}$$

or, in other words,

$$\frac{m_W^2}{m_Z^2} = 1 - \frac{e^2}{g^2} \tag{6.71}$$

if one uses the relation $e = g \sin\theta_W$ for the electromagnetic coupling constant (see (5.34)). It is easy to realize that the relation (6.70) (or (6.71), respectively) holds for a general value of the weak hypercharge of the scalar isodoublet Φ.

The formula (6.69) for m_W has a rather remarkable consequence that should be emphasized here. When the expression $m_W = \frac{1}{2}gv$ is inserted into the familiar relation for the Fermi constant $G_F/\sqrt{2} = g^2/(8m_W^2)$ (see (3.19)), one immediately gets

$$v = \left(G_\mathrm{F}\sqrt{2}\right)^{-1/2} \doteq 246\,\mathrm{GeV} \tag{6.72}$$

Thus, the vacuum value of the Higgs scalar field turns out to be directly related to the Fermi constant — the parameter of the old weak interaction physics. This may be somewhat surprising at first sight, since the v has originally been expressed (see (6.49)) in terms of the μ and λ, the totally unknown parameters of the "Goldstone potential" $V(\Phi)$. On the other hand, v is obviously the only relevant mass scale that enters the Higgs mechanism and G_F can be considered as the only dimensionful parameter describing weak interactions. Thus, from this point of view, the relation (6.72) appears to be quite natural.

One should note that our specific example of the Higgs mechanism confirms indeed the general statements formulated earlier in this section. We have started with a model that exhibits three Goldstone bosons associated with a spontaneously broken global symmetry $SU(2)$. Within a corresponding Higgs-type model invariant under local $SU(2) \times U(1)$, the erstwhile Goldstone bosons become unphysical (they completely disappear in the U-gauge) and one gets three massive vector bosons W^+, W^- and Z. As an additional bonus, one obtains an interesting relation (6.70) which also means that the parameter $\rho = m_W^2/(m_Z^2 \cos^2\theta_W)$ (cf. (5.63)) is equal to unity at the classical level — we have mentioned this remarkable fact already in Section 5.6. As noted there, the relation $\rho = 1$ (which receives a small correction at the quantum level) is indeed phenomenologically successful, i.e. it is experimentally confirmed with good accuracy. Thus, one can say that long before the experimental discovery of the Higgs boson, there was a clear indirect argument in favour of the assumption that masses of W and Z are generated through the Higgs mechanism implemented by means of a complex scalar doublet.

In closing this section, let us summarize, for reader's convenience, formulae for the W and Z masses written in terms of α, G_{F} and θ_W. We have already found such a formula for m_W in Section 5.4 (see (5.42)), and now, we are able to add the corresponding expression for the m_Z, by making use of (6.70). Thus, we have

$$\begin{aligned} m_W &= \left(\frac{\pi\alpha}{G_F\sqrt{2}}\right)^{1/2} \frac{1}{\sin\theta_W} \\ m_Z &= \left(\frac{\pi\alpha}{G_F\sqrt{2}}\right)^{1/2} \frac{1}{\sin\theta_W \cos\theta_W} \end{aligned} \tag{6.73}$$

Since $\left(\pi\alpha/G_F\sqrt{2}\right)^{1/2} \doteq 37$ GeV and $\sin\theta_W\cos\theta_W = \frac{1}{2}\sin 2\theta_W$, from (6.73), it is obvious that in addition to the lower bound $m_W \gtrsim 37\,\text{GeV}$ derived earlier (cf. (5.44)), one also has $m_Z \gtrsim 74\,\text{GeV}$.

6.5 Higgs boson interactions

Having identified physical scalar and vector fields resulting from the Higgs mechanism within the $SU(2) \times U(1)$ gauge theory, we are now ready to describe their interactions. When the U-gauge Lagrangian (6.59) is recast in terms of $W_\mu^\pm$ and Z_μ (cf. the discussion around (6.60) and the relation (6.67)), we have

$$\begin{aligned} \mathscr{L}_{\text{Higgs}}^{(U)} &= \frac{1}{2}\partial_\mu H \partial^\mu H - \lambda v^2 H^2 - \lambda v H^3 - \frac{1}{4}\lambda H^4 \\ &\quad + \frac{1}{8}(v+H)^2[2g^2 W_\mu^- W^{+\mu} + (g^2 + g'^2) Z_\mu Z^\mu] \end{aligned} \tag{6.74}$$

so that the interaction Lagrangian reads

$$\begin{aligned} \mathscr{L}_{\text{Higgs}}^{(\text{int})} &= \frac{1}{8}(2vH + H^2)[2g^2 W_\mu^- W^{+\mu} + (g^2 + g'^2) Z_\mu Z^\mu] \\ &\quad - \lambda v H^3 - \frac{1}{4}\lambda H^4 \end{aligned} \tag{6.75}$$

Let us now focus on the interactions of W and Z with the Higgs boson H. Similarly, as in the Abelian model of Section 6.3, in (6.75), one can recognize essentially two types of couplings, namely the trilinear and quadrilinear ones. These are

$$\begin{aligned}\mathscr{L}_{WWH} &= g m_W W_\mu^- W^{+\mu} H \\ \mathscr{L}_{ZZH} &= \frac{g m_Z}{2\cos\theta_W} Z_\mu Z^\mu H\end{aligned} \tag{6.76}$$

and

$$\begin{aligned}\mathscr{L}_{WWHH} &= \frac{1}{4} g^2 W_\mu^- W^{+\mu} H^2 \\ \mathscr{L}_{ZZHH} &= \frac{1}{8}\frac{g^2}{\cos^2\theta_W} Z_\mu Z^\mu H^2\end{aligned} \tag{6.77}$$

Note that in writing (6.76) and (6.77), we have eventually used the mass relations (6.69) as well as the familiar expression $\cos\theta_W = g/\sqrt{g^2 + g'^2}$ for the Weinberg mixing angle.

It is quite remarkable that the form of the WWH interaction resulting from the Higgs mechanism coincides with the $WW\sigma$ coupling obtained in Section 6.1 through the analysis of residual high-energy divergences of tree-level Feynman diagrams for $W_L W_L \to W_L W_L$ (cf. (6.2) and (6.6)). This indicates that the Higgs mechanism within a gauge theory is essentially the only means of saving the good asymptotic behaviour of scattering amplitudes involving massive vector bosons and hence is of vital importance for perturbative renormalizability. It is also not difficult to see that the interactions (6.76) and (6.77) lead to the right high-energy behaviour of the tree-level amplitudes for processes $WW \to HH$ and $ZZ \to HH$. In particular, one may observe that the contribution of the direct $WWHH$ interaction (6.77) compensates the high-energy (quadratic) divergences produced by the second-order graph involving the W exchange and two WWH vertices; an analogous mechanism operates in the $ZZ \to HH$ channel as well. The corresponding calculation is left to the reader as an instructive exercise. Note that converse is also true: the set of couplings (6.76) and (6.77) is fixed uniquely by the requirement of tree-level unitarity for the relevant scattering amplitudes (for details, the reader is referred to [Hor]).

Finally, it should be noted that the interaction Lagrangian (6.75) also contains cubic and quartic self-couplings of the Higgs boson. Denoting the corresponding coupling constants as g_{HHH} and g_{HHHH} respectively, one has

$$\begin{aligned}g_{HHH} &= -\lambda v \\ g_{HHHH} &= -\frac{1}{4}\lambda\end{aligned} \tag{6.78}$$

Using now the relations $m_H^2 = 2\lambda v^2$ (see (6.51)) and $v = 2m_W/g = (G_F\sqrt{2})^{-\frac{1}{2}}$ (see (6.72)), one can recast (6.78) as

$$\begin{aligned} g_{HHH} &= -\frac{1}{4}g\frac{m_H^2}{m_W} = -\left(\frac{G_F}{2\sqrt{2}}\right)^{1/2} m_H^2 \\ g_{HHHH} &= -\frac{1}{32}g^2\frac{m_H^2}{m_W^2} = -\frac{G_F m_H^2}{4\sqrt{2}} \end{aligned} \tag{6.79}$$

With current experimental data at hand (see [6]), one may estimate the numerical value of the coupling constant λ: using (6.51), (6.72) and $m_H \doteq 125\,\text{GeV}$, one gets $\lambda \doteq 0.125$. Thus, one may conclude that here one may also rely on the implementation of perturbation theory as in the other parts of the electroweak SM (note, however, that some relevant details of the Higgs boson interactions still require a thorough experimental study). Anyway, it may also be instructive to return briefly to the old times before the Higgs boson discovery, namely to some theoretical (technical) semi-quantitative constraints on the possible value of m_H. A basic hint is based on the simple-minded perturbativity argument: if the relevant Higgs–Goldstone Lagrangian in (6.74) is to be used perturbatively, then any dimensionless coupling should not, roughly speaking, exceed unity (otherwise, the corresponding power expansion would be doubtful *a priori*). In the considered case, the order-of-magnitude estimate $|g_{HHHH}| \lesssim 1$ yields a simple upper bound for the Higgs mass, namely

$$m_H \lesssim 2\sqrt{2}\left(G_F\sqrt{2}\right)^{-1/2} = 2\sqrt{2}\,v \doteq 700\,\text{GeV} \tag{6.80}$$

These considerations can be given more precise quantitative meaning if, e.g., the unitarity condition for partial waves is invoked for an appropriate process at the tree level. Additional numerical factors then modify slightly the straightforward bound (6.80), but the overall scale of the m_H estimate remains the same. Moreover, such an analysis can be further refined if one-loop diagrams are taken into account. A more detailed discussion of these issues would go beyond the scope of the present text and the interested reader is therefore referred to the original literature (see, e.g., [57, 58]).

Coming back to the present-day situation, the experimental value $m_H = 125.25 \pm 0.17\,\text{GeV}$ shown in [6] certainly satisfies the perturbativity criterion, but the technical arguments outlined above are still useful in theoretical considerations concerning possible extensions SM, in particular when contemplating electroweak models involving several Higgs-like doublets (see, e.g., [Gun], and [59–62].

One may also wonder whether the Higgs boson self-interactions specified in (6.78) or (6.79), respectively, play any role in the high-energy divergence cancellations for some specific physical processes. The answer is yes: it turns out that they are necessary to ensure the tree-level unitarity for some $2 \to 3$ reactions, such as, e.g., $WW \to WWH$, $WW \to HHH$, etc. (note that the tree unitarity for five-point amplitudes means that they decrease as $1/E$ in the high-energy limit). For more details, see, e.g., [Hor] and the references therein.

6.6 Yukawa couplings and lepton masses

We will now show that lepton masses can also be generated through appropriate interactions involving the Higgs doublet Φ (the quark sector will be discussed in the next chapter).[7] To this end, we are going to employ a Yukawa-type coupling (that is, an interaction bilinear in lepton fields and linear in Φ). It is not difficult to realize that such a (non-derivative) interaction term is essentially the only renormalizable coupling that can still be added to the Lagrangian considered so far. Following our symmetry principle, we should construct it to be $SU(2) \times U(1)$ invariant. For the moment, let us consider, e.g., only leptons of the electron type. Of course, as the basic building blocks, we have to employ the left-handed doublet

$$L = \begin{pmatrix} \nu_L \\ e_L \end{pmatrix} \tag{6.81}$$

and the right-handed singlet e_R (cf. Section 5.1). From the doublets L and Φ, an $SU(2)$ singlet can be immediately formed as $\bar{L}\Phi$. Multiplying this by e_R, one obtains an $SU(2)$ invariant Yukawa interaction term

$$\mathscr{L}_{\text{Yukawa}} = -h_e \bar{L} \Phi e_R + \text{h.c.} \tag{6.82}$$

where h_e is a (dimensionless) coupling constant and the minus sign has been chosen for later convenience. Note that we suppress the lepton labels whenever it does not lead to confusion. It is easy to see that the Lagrangian (6.82) is automatically invariant under the weak hypercharge $U(1)$ as well. Indeed, one has $Y_L = -1/2$, which in turn means that the Dirac conjugate

[7]Let us recall that, e.g., an electron mass term cannot be added to our $SU(2) \times U(1)$ invariant Lagrangian simply by hand since this would violate the required symmetry. Indeed, $m_e \bar{e} e = m_e(\bar{e}_L e_R + \bar{e}_R e_L)$ and the chiral components e_L and e_R transform differently under the weak isospin $SU(2)$: e_L belongs to an $SU(2)$ doublet while e_R is a singlet.

$\bar{L}$ carries $Y_{\bar{L}} = +1/2$. Further, $Y_R^{(e)} = -1$ and $Y_\Phi = +1/2$. One thus gets $Y_{\bar{L}} + Y_\Phi + Y_R^{(e)} = 0$, which proves our statement. Now, fixing the unitary gauge, (6.82) becomes

$$\begin{aligned}
\mathscr{L}^{(U)}_{\text{Yukawa}} &= -h_e \left(\bar{\nu}_L,\ \bar{e}_L\right) \begin{pmatrix} 0 \\ \frac{1}{\sqrt{2}}(v+H) \end{pmatrix} e_R + \text{h.c.} \\
&= -\frac{1}{\sqrt{2}} h_e (v+H) \bar{e}_L e_R + \text{h.c.} \\
&= -\frac{1}{\sqrt{2}} h_e (v+H) \left(\bar{e}_L e_R + \bar{e}_R e_L\right) \\
&= -\frac{1}{\sqrt{2}} h_e v \bar{e} e - \frac{1}{\sqrt{2}} h_e \bar{e} e H
\end{aligned} \tag{6.83}$$

The last expression contains an electron mass term with

$$m_e = \frac{1}{\sqrt{2}} h_e v \tag{6.84}$$

and a scalar Yukawa coupling

$$\mathscr{L}_{eeH} = g_{eeH} \bar{e} e H \tag{6.85}$$

with $g_{eeH} = -\frac{1}{\sqrt{2}} h_e$. Taking into account (6.84), one then has

$$g_{eeH} = -\frac{m_e}{v} \tag{6.86}$$

that can be recast (by employing the familiar relation $v = 2m_W/g$) as

$$g_{eeH} = -\frac{g}{2} \frac{m_e}{m_W} \tag{6.87}$$

This is seen to coincide with the $g_{ee\sigma}$ coupling obtained in Section 6.1 from the analysis of Feynman diagrams (cf. (6.11)). Thus, similar to that in the previous section, one has another indication that the mass generation through Higgs mechanism is actually necessary for tree-level unitarity (and thereby for perturbative renormalizability) of the electroweak theory. An instructive exercise offered to the interested reader is to check explicitly how the Higgs boson couplings derived here and in the preceding section yield well-behaved tree-level amplitudes, e.g., for the processes $e^+e^- \to Z_L Z_L$ or $e^+e^- \to Z_L H$. Needless to say, mass terms for muon or tau lepton can be produced in a completely analogous way — one only needs different Yukawa coupling constants to account for different lepton masses. Thus, one has in general

$$g_{\ell\ell H} = -\frac{g}{2}\frac{m_\ell}{m_W} \tag{6.88}$$

for $\ell = e, \mu, \tau$, which is characteristic for the standard model Higgs boson. Obviously, the dependence of the interaction strengths on the lepton type embodied in (6.88) is rather dramatic: it means that a reasonably accurate estimate for the leptonic two-body rates would be

$$\Gamma\left(H \to e^+e^-\right) : \Gamma(H \to \mu^+\mu^-) : \Gamma(H \to \tau^+\tau^-) = m_e^2 : m_\mu^2 : m_\tau^2 \tag{6.89}$$

(when writing (6.89), we have taken into account that $m_\ell^2 \ll m_H^2$ according to the current experimental bounds; the lepton-mass dependence of the phase space volume, etc. can then be essentially neglected).

Let us now consider the possibility of giving mass to a neutrino. To begin with, we shall restrict ourselves to a single lepton species (say, the electron type). As we have already noted in Chapter 5, the right-handed component of neutrino field can be introduced without violating any natural requirement of the electroweak theory. Apart from being a weak isospin singlet, it must then carry zero weak hypercharge (see (5.9)). Due to the (mandatory) hypercharge assignments, one cannot construct an $SU(2) \times U(1)$ invariant out of the doublets L and Φ and the singlet ν_R (needless to say, any violation of the $U(1)_Y$ invariance would lead to the non-conservation of the electric charge). However, it is possible to employ the following trick. It can be shown that the quantity $\widetilde{\Phi}$ defined in terms of the original Higgs doublet Φ as

$$\widetilde{\Phi} = i\tau_2\Phi^* \tag{6.90}$$

with τ_2 being the Pauli matrix

$$\tau_2 = \begin{pmatrix} 0 & -i \\ i & 0 \end{pmatrix} \tag{6.91}$$

transforms under $SU(2)$ in the same way as Φ, i.e. $\widetilde{\Phi}$ is another scalar doublet. (Note that this observation is also crucial for giving masses to all types of quarks and we will utilize it in the next chapter as well.) We defer a formal proof of the transformation properties of $\widetilde{\Phi}$ to the end of this section, and now, let us proceed to see how it can be exploited for our purpose. Since the definition (6.90) involves complex conjugation, it is clear that $\widetilde{\Phi}$ carries weak hypercharge $Y_{\widetilde{\Phi}} = -Y_\Phi = -\frac{1}{2}$ and one is then able to construct a desired invariant form containing ν_R. Indeed, one can write

$$\widetilde{\mathscr{L}}_{\text{Yukawa}} = -h_\nu \bar{L} \widetilde{\Phi} \nu_R + \text{h.c.} \tag{6.92}$$

which is clearly $SU(2) \times U(1)$ invariant as $Y_{\bar{L}} + Y_{\widetilde{\Phi}} + Y_R^{(\nu)} = -Y_L - Y_\Phi + Y_R^{(\nu)} = \frac{1}{2} - \frac{1}{2} + 0 = 0$. In the unitary gauge, (6.92) becomes

$$\begin{aligned}
\widetilde{\mathscr{L}}^{(U)}_{\text{Yukawa}} &= -h_\nu \left(\bar{\nu}_L, \, \bar{e}_L\right) \begin{pmatrix} \frac{1}{\sqrt{2}}(v+H) \\ 0 \end{pmatrix} \nu_R + \text{h.c.} \\
&= -\frac{1}{\sqrt{2}} h_\nu (\bar{\nu}_L \nu_R + \bar{\nu}_R \nu_L)(v + H) \\
&= -\frac{1}{\sqrt{2}} h_\nu v \bar{\nu}\nu - \frac{1}{\sqrt{2}} h_\nu \bar{\nu}\nu H
\end{aligned} \tag{6.93}$$

where we may identify immediately the neutrino mass term with $m_\nu = \frac{1}{\sqrt{2}} h_\nu v$ and a scalar Yukawa interaction

$$\begin{aligned}
\mathscr{L}_{\nu\nu H} &= -\frac{m_\nu}{v} \bar{\nu}\nu H \\
&= -\frac{g}{2} \frac{m_\nu}{m_W} \bar{\nu}\nu H
\end{aligned} \tag{6.94}$$

Now, it is natural to ask, among other things, what is the impact of such an additional neutrino interaction on the divergence cancellations demonstrated earlier for various processes (see Section 5.8). In particular, we can reconsider the process $\bar{\nu}\nu \to W_L^- W_L^+$. For massive neutrinos in the initial state, one finds easily that the sum of the two diagrams shown in Fig. 5.4 (the exchange of electron and Z) still contains a residual $O(m_\nu E/m_W^2)$ divergence for $E \to \infty$. Once the coupling (6.94) is present, there is an additional graph involving s-channel Higgs boson exchange and its contribution cancels exactly the linear divergence in close resemblance with the case of the $e^+e^- \to W^+W^-$ process discussed earlier in this chapter. The reader is recommended to verify this by means of an explicit calculation; another instructive exercise would be to check that an analogous mechanism also works for the process $\nu\bar{\nu} \to Z_L Z_L$. In this context, it should be emphasized that independent of its mass, the neutrino neutral current remains purely left-handed — ν_R remains uncoupled to the Z boson.

It is obvious that the simple mechanism for generating neutrino masses described above can be used for any lepton type. In fact, it can be generalized in a more substantial way by also producing possible mixings between different lepton species. We shall come back to this issue later on, in connection with the discussion of the quark sector of standard electroweak theory. To close this section, let us now prove formally that $\widetilde{\Phi}$ defined in (6.90) is indeed an $SU(2)$ doublet. In particular, we are going to prove that if

$$\Phi' = e^{i\omega_a \tau_a}\Phi \tag{6.95}$$

(where ω_a denote three arbitrary transformation parameters), then

$$\widetilde{\Phi}' = e^{i\omega_a \tau_a}\widetilde{\Phi} \tag{6.96}$$

where $\widetilde{\Phi}'$ is of course defined as $i\tau_2\Phi'^*$. The crucial technical ingredient of the proof is a simple identity for complex conjugation of the Pauli matrices, namely

$$\tau_a^* = -\tau_2\tau_a\tau_2 \tag{6.97}$$

The verification of (6.97) is straightforward and we leave it to the reader. Now, according to our definitions, the left-hand side of (6.96) can be written as

$$i\tau_2\left(e^{i\omega_a \tau_a}\Phi\right)^* \tag{6.98}$$

Expanding the exponential in (6.98) in power series and employing (6.97), $\widetilde{\Phi}'$ is worked out as

$$\begin{aligned}
\widetilde{\Phi}' &= i\tau_2\left(\mathbb{1} + \frac{i}{1!}\omega_a\tau_a + \frac{i^2}{2!}\left(\omega_a\tau_a\right)^2 + \cdots\right)^*\Phi^* \\
&= i\tau_2\left(\mathbb{1} + \frac{(-i)}{1!}\omega_a\tau_a^* + \frac{(-i)^2}{2!}\left(\omega_a\tau_a^*\right)^2 + \cdots\right)\Phi^* \\
&= i\tau_2\cdot\tau_2\left(\mathbb{1} + \frac{i}{1!}\omega_a\tau_a + \frac{i^2}{2!}\left(\omega_a\tau_a\right)^2 + \cdots\right)\tau_2\Phi^* \\
&= ie^{i\omega_a\tau_a}\tau_2\Phi^* = e^{i\omega_a\tau_a}\widetilde{\Phi}
\end{aligned} \tag{6.99}$$

and (6.96) is thus proved.

6.7 Higgs–Yukawa mechanism and parity violation

The mechanism employed for generating masses within the standard GWS model has another interesting aspect that deserves attention. In particular, it turns out that the "Higgs–Yukawa scheme" adopted here leads quite naturally to the familiar parity-violating weak interactions as well as to the parity-conserving electromagnetic current (the properties of the weak neutral currents then follow automatically in the usual way). This statement, that may seem somewhat surprising at first sight, will be explained below (the argument is essentially due to Veltman [65]).

For the sake of simplicity, let us restrict ourselves to the electron-type leptons ν_e, e — in fact, adding further fermion species does not bring anything new in the present context. As before, we shall assume that the Higgs mechanism is realized via one complex doublet Φ (let us recall that this is the minimum option giving the right values of vector boson masses). Now, the Yukawa interaction that is supposed to produce the electron mass must involve both e_L and e_R, and it is also clear that the two chiral components of the electron field must have different transformation properties under the weak isospin $SU(2)$. Indeed, if they were e.g., both singlets, then by coupling them to the doublet Φ one could not get an $SU(2)$ singlet interaction term; a similar problem would occur if both e_L and e_R belonged to doublets (note that one is certainly not able to make a singlet out of three doublets — mathematically, this would be tantamount to adding three spins-1/2 to a resulting zero value). Thus, if one considers only the lowest-dimensional representations of the $SU(2)$, e_L should belong to a doublet and e_R to a singlet or the other way round. Conventionally, we choose the first possibility, i.e. we place e_L into the usual doublet L (cf. (6.81)) and e_R is taken to be singlet under $SU(2)$; for simplicity, we shall ignore here ν_R. Remembering now how the gauge interactions are constructed (see Chapter 5, in particular the formulae (5.14) and (5.17)), it becomes clear that weak interactions necessarily exhibit maximum parity violation (the charged currents are left-handed owing to our option). Of course, had we chosen the other possibility (namely a doublet consisting of ν_R and e_R, with e_L being an $SU(2)$ singlet), the charged weak currents would be purely right-handed — but this would mean a maximum parity violation anyway. Thus, these simple considerations show that the familiar pattern of parity violation in weak interactions emerges quite naturally: if one insists on generating the lepton mass through a Yukawa coupling involving the Higgs doublet, different $SU(2)$ transformation properties of the left- and right-handed lepton fields are inevitable.

Next, let us examine the consequences of the Higgs–Yukawa mechanism for the parity properties of the electromagnetic interaction. We shall denote the weak hypercharges of the L, e_R, and Φ as Y_L, Y_R, and Y. Since the Yukawa interaction has the form $\bar{L}\Phi e_R$, the invariance under the hypercharge $U(1)$ gauge subgroup requires that

$$-Y_L + Y + Y_R = 0 \tag{6.100}$$

We already know (cf. the discussion around the formula (6.60)) that the diagonalization of the mass matrix for neutral vector bosons leads to the massive field Z_μ and a massless A_μ, expressed in terms of the original gauge fields A^3_μ and B_μ as

$$\begin{aligned} Z_\mu &= cA^3_\mu - sB_\mu \\ A_\mu &= sA^3_\mu + cB_\mu \end{aligned} \tag{6.101}$$

where c and s are shorthand notation for $\cos\theta_W$ and $\sin\theta_W$, respectively; one has

$$c = \frac{g}{\sqrt{g^2 + 4Y^2 g'^2}}, \qquad s = \frac{2Yg'}{\sqrt{g^2 + 4Y^2 g'^2}} \tag{6.102}$$

To identify the interactions of A_μ with leptons, one can employ the formulae derived previously in Chapter 5 (see, in particular, (5.21), where we shall ignore the term involving ν_R). Expressing A^3_μ and B_μ in terms of Z_μ and A_μ, one obtains (cf. (5.27))

$$\begin{aligned} \mathscr{L}^{(A)}_{\text{int}} &= \left(\frac{1}{2}gs + Y_L g'c\right)\bar{\nu}_L\gamma^\mu\nu_L A_\mu \\ &\quad + \left(-\frac{1}{2}gs + Y_L g'c\right)\bar{e}_L\gamma^\mu e_L A_\mu + Y_R g'c\bar{e}_R\gamma^\mu e_R A_\mu \end{aligned} \tag{6.103}$$

The condition of vanishing neutrino charge reads

$$\frac{1}{2}gs + Y_L g'c = 0 \tag{6.104}$$

and substituting into (6.104) the expressions (6.102) for c and s, one readily gets

$$Y_L = -Y \tag{6.105}$$

This, in combination with (6.100), yields

$$Y_R = 2Y_L \tag{6.106}$$

Using all the relations shown above, the interaction of A_μ with fermions can be worked out as

$$\begin{aligned} \mathscr{L}^{(A)}_{\text{int}} &= \left(-\frac{1}{2}gs - Yg'c\right)\bar{e}_L\gamma^\mu e_L A_\mu - 2Yg'c\bar{e}_R\gamma^\mu e_R A_\mu \\ &= -\frac{2Ygg'}{\sqrt{g^2 + 4Y^2 g'^2}}\left(\bar{e}_L\gamma^\mu e_L + \bar{e}_R\gamma^\mu e_R\right)A_\mu \end{aligned} \tag{6.107}$$

and the last line of (6.107) exhibits clearly the envisaged parity-conserving nature of the A_μ (electromagnetic) interaction.

Thus, the preceding considerations can be summarized briefly as follows:

(i) The maximum parity violation in charged-current weak interactions emerges naturally within the GWS standard model as a consequence of the Higgs–Yukawa mechanism for the generation of fermion masses. The essential point is that the doublet character of the Higgs field enforces different $SU(2)$ transformation properties upon the left- and right-handed chiral components of fermion fields.[8] In this sense, the parity violation in weak currents is intimately connected with properties of the Higgs sector of the standard GWS model; the connection is straightforward, though it may appear somewhat surprising at first sight.

(ii) The Higgs–Yukawa mechanism leads automatically to the parity-conserving electromagnetic interaction. More precisely, one gets a vector-like interaction of the massless physical gauge field emerging from the standard Higgs mechanism if one assumes that the Yukawa interaction responsible for the lepton mass generation is $SU(2) \times U(1)$ invariant and the neutrino charge is fixed to be zero.

6.8 Custodial symmetry

Let us now turn to a discussion of some deeper symmetry aspects of the standard "minimal" Higgs system. As we have already noted earlier (see the remarks around the relation (6.45)), the symmetry of the Higgs–Goldstone potential $V(\Phi)$ is in fact $O(4)$, i.e. it is larger than the mandatory $SU(2) \times U(1)$: the reason is simply that a V with required properties must inevitably depend on $\Phi^\dagger\Phi = \varphi_1^2 + \varphi_2^2 + \varphi_3^2 + \varphi_4^2$, where $\varphi_1, \ldots, \varphi_4$ are the four real scalar fields that parametrize the complex doublet Φ according to (6.42). For our present purpose, we shall employ such a four-dimensional real parametrization explicitly, describing the Higgs multiplet as

$$\Phi = \begin{pmatrix} \varphi_1 \\ \varphi_2 \\ \varphi_3 \\ \varphi_4 \end{pmatrix} \tag{6.108}$$

[8] Of course, we always assume tacitly that only the lowest-dimensional $SU(2)$ representations of fermion fields — namely the singlets and doublets — are relevant.

A "vacuum configuration" Φ_0 corresponding to the Lagrangian (6.43) is in general given by

$$\Phi_0 = \begin{pmatrix} v_1 \\ v_2 \\ v_3 \\ v_4 \end{pmatrix} \tag{6.109}$$

with $v_1^2 + v_2^2 + v_3^2 + v_4^2 = v^2/2$ (cf. (6.48)). For convenience, we will often use a particular representant of the manifold (6.109), namely

$$\Phi_0^{(4)} = \begin{pmatrix} 0 \\ 0 \\ 0 \\ v/\sqrt{2} \end{pmatrix} \tag{6.110}$$

(strictly speaking, this differs slightly from our previous conventional choice (6.52), but it does not matter).

The familiar Goldstone-type symmetry breakdown occurring in the considered model can now be described in a concise way (in fact, the algebraic clarity is the main virtue of the real-field formalism in the present context). Let us start with the specification of the $O(4)$ symmetry generators. A "canonical" set is represented by the six real antisymmetric matrices

$$M_1 = \begin{pmatrix} 0 & 0 & 0 & 0 \\ 0 & 0 & -1 & 0 \\ 0 & 1 & 0 & 0 \\ 0 & 0 & 0 & 0 \end{pmatrix}, \quad N_1 = \begin{pmatrix} 0 & 0 & 0 & -1 \\ 0 & 0 & 0 & 0 \\ 0 & 0 & 0 & 0 \\ 1 & 0 & 0 & 0 \end{pmatrix}$$

$$M_2 = \begin{pmatrix} 0 & 0 & 1 & 0 \\ 0 & 0 & 0 & 0 \\ -1 & 0 & 0 & 0 \\ 0 & 0 & 0 & 0 \end{pmatrix}, \quad N_2 = \begin{pmatrix} 0 & 0 & 0 & 0 \\ 0 & 0 & 0 & -1 \\ 0 & 0 & 0 & 0 \\ 0 & 1 & 0 & 0 \end{pmatrix}$$

$$M_3 = \begin{pmatrix} 0 & -1 & 0 & 0 \\ 1 & 0 & 0 & 0 \\ 0 & 0 & 0 & 0 \\ 0 & 0 & 0 & 0 \end{pmatrix}, \quad N_3 = \begin{pmatrix} 0 & 0 & 0 & 0 \\ 0 & 0 & 0 & 0 \\ 0 & 0 & 0 & -1 \\ 0 & 0 & 1 & 0 \end{pmatrix} \tag{6.111}$$

satisfying the commutation relations

$$
\begin{aligned}
[M_j, M_k] &= \epsilon_{jkl} M_l \\
[M_j, N_k] &= \epsilon_{jkl} N_l \\
[N_j, N_k] &= \epsilon_{jkl} M_l
\end{aligned}
\tag{6.112}
$$

Occasionally, we will also use the shorthand notation $\vec{M}$, $\vec{N}$ for the two triplets of M_j and N_j.[9] To examine the action of the generators (6.111) on the vacuum state, one may consider first, e.g., the particular choice $\Phi_0^{(4)}$ shown in (6.110). It is immediately seen that the three four-component vectors $\vec{N}\Phi_0^{(4)}$ are non-zero and linearly independent, while all $\vec{M}\Phi_0^{(4)}$ vanish. Thus, as expected, $\Phi_0^{(4)}$ breaks the considered symmetry in three directions, i.e. with respect to the three generators $\vec{N}$. In fact, such a result can be easily generalized — for an arbitrary representant Φ_0 of the vacuum manifold (as given by (6.109)), one can show that the space made of the linear combinations of the six real four-component vectors $\vec{M}\Phi_0$, $\vec{N}\Phi_0$ is *three-dimensional* (the proof is left to the reader as an exercise in linear algebra). Note that this is actually the precise contents of the statement about the number of broken symmetry generators (associated with Goldstone bosons), i.e. about the symmetry-breaking pattern occurring within our model.

Next, one would like to choose an appropriate electroweak $SU(2)\times U(1)$ subgroup of our $O(4)$; in other words, one has to identify the generators corresponding to the weak isospin and hypercharge (that are to be gauged subsequently). To this end, it is convenient to pass from $\vec{M}$, $\vec{N}$ to another set (basis) of generators defined by

$$
\begin{aligned}
\vec{L} &= \frac{1}{2}\left(\vec{M} + \vec{N}\right) \\
\vec{R} &= \frac{1}{2}\left(\vec{M} - \vec{N}\right)
\end{aligned}
\tag{6.113}
$$

Using (6.112), one then easily gets

$$
\begin{aligned}
[L_j, L_k] &= \epsilon_{jkl} L_l \\
[R_j, R_k] &= \epsilon_{jkl} R_l \\
[L_j, R_k] &= 0
\end{aligned}
\tag{6.114}
$$

[9]Note that it is most natural to use real antisymmetric matrices as the generators of real orthogonal transformations. Equivalently, one could work with the hermitian generators $i\vec{M}$, $i\vec{N}$; in fact, these would fit better into the gauge theory formalism that we have developed so far. We will return to the hermitian representation of the symmetry generators later in this section when we reconsider the Higgs mechanism.

which means that the sets $\vec{L}$ and $\vec{R}$ generate two independent (commuting) $O(3)$ subalgebras of the original $O(4)$. Mathematically, this remarkable fact corresponds to the known statement that the group $O(4)$ is locally (i.e. at the level of Lie algebras) isomorphic to the direct product $O(3) \times O(3)$; in the common symbolic notation, $O(4) \simeq O(3) \times O(3)$. In the present context, it is important to recall that the $O(3)$ algebra is isomorphic to that of $SU(2)$ (indeed, the hermitian matrices $i\vec{L}$ and $i\vec{R}$ obviously satisfy the familiar $SU(2)$ commutation relations) and the considered decomposition is therefore usually also written as $O(4) \simeq SU(2) \times SU(2)$. Thus, in view of (6.114), one can take, e.g., $\vec{L}$ to be the isospin generators and the hypercharge (that has to commute with isospin) is then selected among the matrices $\vec{R}$ (conventionally, R_3 is chosen). For practical purposes, we shall employ the hermitian generators $\vec{T}$ and Y defined as

$$\vec{T} = i\vec{L}, \quad Y = iR_3 \tag{6.115}$$

Note that their explicit matrix representation can then be written (using (6.111) and (6.113)) as

$$T_1 = \frac{i}{2}\begin{pmatrix} 0 & 0 & 0 & -1 \\ 0 & 0 & -1 & 0 \\ 0 & 1 & 0 & 0 \\ 1 & 0 & 0 & 0 \end{pmatrix}, \quad T_2 = \frac{i}{2}\begin{pmatrix} 0 & 0 & 1 & 0 \\ 0 & 0 & 0 & -1 \\ -1 & 0 & 0 & 0 \\ 0 & 1 & 0 & 0 \end{pmatrix}$$

$$T_3 = \frac{i}{2}\begin{pmatrix} 0 & -1 & 0 & 0 \\ 1 & 0 & 0 & 0 \\ 0 & 0 & 0 & -1 \\ 0 & 0 & 1 & 0 \end{pmatrix}, \quad Y = \frac{i}{2}\begin{pmatrix} 0 & -1 & 0 & 0 \\ 1 & 0 & 0 & 0 \\ 0 & 0 & 0 & 1 \\ 0 & 0 & -1 & 0 \end{pmatrix} \tag{6.116}$$

Now, considering again $\Phi_0^{(4)}$ as a vacuum state, one sees immediately that both $\vec{L}\Phi_0^{(4)}$ and $\vec{R}\Phi_0^{(4)}$ are different from zero. However, as noted earlier, the sum $\vec{L} + \vec{R} = \vec{M}$ does annihilate $\Phi_0^{(4)}$ (let us recall that there can be just three independent broken symmetry generators). In particular, $L_3 + R_3 = M_3$ yields the electric charge in correspondence with the relation $T_3 + Y = Q$. From now on, we shall use the hermitian generators only, so in addition to (6.115), let us also introduce the notation

$$\vec{K} = i\vec{M} \tag{6.117}$$

For $\vec{K}$ and $\vec{T}$, we then have a set of commutation relations

$$\begin{aligned} [K_j, K_k] &= i\epsilon_{jkl} K_l \\ [T_j, T_k] &= i\epsilon_{jkl} T_l \\ [T_j, K_k] &= i\epsilon_{jkl} T_l \end{aligned} \tag{6.118}$$

that follow immediately from (6.114) and from the above definitions. These relations mean, among other things, that the weak isospin generators $\vec{T}$ form a three-component vector under rotations generated by $\vec{K}$ (note that the commutation relation between $\vec{T}$ and $\vec{K}$ are formally the same as, e.g., those between momentum and angular momentum in ordinary quantum mechanics). Passing from the algebra of commutators (i.e. from infinitesimal transformations) to finite rotations, one can write

$$U^\dagger(\vec{\omega})\, T_j U(\vec{\omega}) = \mathscr{D}_{jk}(\vec{\omega})\, T_k \tag{6.119}$$

where $U(\vec{\omega}) = \exp(-i\vec{\omega}\cdot\vec{K})$ is unitary 4×4 matrix and $\mathscr{D}_{jk}(\vec{\omega})$ represent a (real orthogonal) matrix of three-dimensional rotation described by the parameters $\vec{\omega}$.[10] The relation (6.119) is of crucial importance for our further considerations and we will return to it shortly. At this point, let us summarize briefly the essential features of the symmetry pattern discussed so far:

(i) The conventionally chosen vacuum $\Phi_0^{(4)}$ (denoted in what follows simply as Φ_0) is invariant under $SU(2)$ transformations generated by the matrices $\vec{K}$, i.e.

$$U(\vec{\omega})\,\Phi_0 = \Phi_0 \tag{6.120}$$

for arbitrary $\vec{\omega}$ (of course, this is equivalent to $\vec{K}\Phi_0 = 0$).

(ii) The weak isospin generators $\vec{T}$ (corresponding to the broken symmetry) constitute a triplet with respect to the vacuum symmetry subgroup — in other words, they behave as a three-component vector under rotations generated by $\vec{K}$.

We are now in a position to reconsider the Higgs mechanism for $SU(2)\times U(1)$ gauge bosons. The relevant Lagrangian can be written in analogy with

[10]Let us recall that the generators $\vec{K}$ are hermitian and purely imaginary by construction, so $U(\vec{\omega})$ is in fact a real orthogonal matrix as well.

the formula (6.53), replacing there $\vec{\tau}/2$ by $\vec{T}$ and substituting the matrix

$$Y = Q - T_3 = K_3 - T_3 \tag{6.121}$$

for the weak hypercharge. From the discussion, following (6.53), it is then clear that the mass term for the vector bosons A^a_μ and B_μ is given by

$$\mathscr{L}^{(\mathrm{IVB})}_{\mathrm{mass}} = \Phi_0^\dagger \left(gA^a_\mu T_a + g' B_\mu Y\right)\left(gA^{b\mu}T_b + g'B^\mu Y\right)\Phi_0 \tag{6.122}$$

Taking into account (6.121) and using the identities $Q\Phi_0 = 0$, $\Phi_0^\dagger Q = 0$, the expression (6.122) can be worked out as

$$\begin{aligned}\mathscr{L}^{(\mathrm{IVB})}_{\mathrm{mass}} &= g^2\Phi_0^\dagger T_a T_b \Phi_0 A^a_\mu A^{b\mu} - gg'\Phi_0^\dagger T_a T_3 \Phi_0 A^a_\mu B^\mu \\ &\quad - gg'\Phi_0^\dagger T_3 T_b \Phi_0 A^b_\mu B^\mu + g'^2\Phi_0^\dagger (T_3)^2 \Phi_0 B_\mu B^\mu\end{aligned} \tag{6.123}$$

All coefficients in (6.123) are of the form $\Phi_0^\dagger T_a T_b \Phi_0$, so let us now examine the properties of such an algebraic expression. Invoking the invariance of Φ_0 under the unitary transformations $U(\vec{\omega})$ (see (6.120)) and making use of the fundamental symmetry relation (6.119), one thus obtains

$$\begin{aligned}\Phi_0^\dagger T_a T_b \Phi_0 &= \Phi_0^\dagger U^\dagger(\vec{\omega})\, T_a T_b U(\vec{\omega})\,\Phi_0 \\ &= \Phi_0^\dagger U^\dagger(\vec{\omega})\, T_a U(\vec{\omega})\, U^\dagger(\vec{\omega})\, T_b U(\vec{\omega})\,\Phi_0 \\ &= \mathscr{D}_{aj}(\vec{\omega})\,\mathscr{D}_{bk}(\vec{\omega})\,\Phi_0^\dagger T_j T_k \Phi_0\end{aligned} \tag{6.124}$$

which means that the numerical coefficients $\Phi_0^\dagger T_a T_b \Phi_0$ behave as components of a second rank tensor under three-dimensional rotations. On the other hand, these coefficients are pure numbers (i.e. they are obviously independent of the "reference frame" characterized by the transformation parameters $\vec{\omega}$); in other words, the expression $\Phi_0^\dagger T_a T_b \Phi_0$ represents an "isotropic tensor" (i.e. such that its components are the same in all reference frames). This, of course, is a rather severe restriction and it is not surprising that the second rank tensor endowed with this property is essentially unique: it is the Kronecker delta (up to a constant multiplicative factor). Thus, solely on the basis of our symmetry arguments, we can write

$$\Phi_0^\dagger T_a T_b \Phi_0 = N\delta_{ab} \tag{6.125}$$

where N is a numerical (normalization) constant. The value of N can be fixed by taking into account the explicit representation of the generators

$\vec{T}$ shown in (6.116); one thus finds easily that $N = v^2/8$. Substituting now the result (6.125) into (6.123), one gets

$$\mathscr{L}_{\text{mass}}^{(\text{IVB})} = g^2 N A_\mu^a A^{a\mu} - 2gg' N A_\mu^3 B^\mu + g'^2 N B_\mu B^\mu \tag{6.126}$$

and the last expression is immediately diagonalized as

$$\mathscr{L}_{\text{mass}}^{(\text{IVB})} = N \left[g^2 \left(A_\mu^1 A^{1\mu} + A_\mu^2 A^{2\mu} \right) + \left(g A_\mu^3 - g' B_\mu \right)^2 \right] \tag{6.127}$$

Identifying the physical vector fields $W_\mu^\pm, Z_\mu$ and their masses in an analogous manner as in Section 6.4, one has finally

$$\mathscr{L}_{\text{mass}}^{(\text{IVB})} = m_W^2 W_\mu^- W^{+\mu} + \frac{1}{2} m_Z^2 Z_\mu Z^\mu \tag{6.128}$$

with

$$m_W^2 = 2g^2 N, \quad m_Z^2 = 2\left(g^2 + g'^2\right) N \tag{6.129}$$

Note that for the relevant value $N = v^2/8$ specified above, one reproduces, as expected, the standard formulae (6.69). Even without using an explicit value of N, the result (6.129) obviously yields the famous Weinberg relation

$$m_W^2 / m_Z^2 = g^2 / \left(g^2 + g'^2\right) \tag{6.130}$$

In other words (taking into account that $g/(g^2 + g'^2)^{1/2} = \cos\theta_W$), one recovers the value $\rho = 1$ for the parameter $\rho = m_W^2/(m_Z^2 \cos^2\theta_W)$ introduced earlier (cf. (5.63) and the end of Section 6.4).

It is important to realize that Eq. (6.130) has been derived here on the symmetry grounds only[11] and one thus gains a deeper insight into the origin of the observed pattern of vector boson masses. **In particular, our argument relied substantially on the fact that the vacuum symmetry is a global $SU(2)$, under which the (gauged) weak isospin generators behave as a triplet.** In this sense, the vacuum symmetry $SU(2)$ controls, or "protects", the value $\rho = 1$, and therefore, it is usually called the **custodial symmetry**. It should be stressed that the existence of such a symmetry is not tied with a particular choice of the vacuum: for the sake of technical simplicity, we have chosen here $\Phi_0^{(4)}$ shown in (6.110), but in fact, any Φ_0 belonging to the set (6.109) is invariant under an $SU(2)$ generated by appropriately "rotated" matrices $\vec{K}$ (the gauged generators must then also be modified accordingly). Note finally that the

[11]Remember that in the elementary treatment of Section 6.4, we had to invoke some specific algebraic properties of the Pauli matrices, etc. to arrive at the same result.

concept of custodial $SU(2)$ symmetry has appeared for the first time in [66], and because of its rather general nature, one can also utilize it in some schemes of electroweak symmetry breaking that go beyond the standard model, e.g. when one considers a generic model of "dynamical symmetry breaking" described by an effective Lagrangian not involving elementary physical Higgs fields (see, e.g., [67]).

6.9 Non-standard Higgs multiplets

Up to now, we have focused our attention on the Higgs mechanism operating within the standard electroweak theory. Despite the current success of SM involving the solitary Higgs boson, it is not excluded that future experiments will reveal the existence of some siblings of H, i.e. some extra scalar bosons belonging to a broader Higgs-like family (in fact, many theorists and experimentalists do hope so). Thus, it may be instructive to discuss briefly extended Higgs-like scalar systems that go beyond SM, but still could basically fit into the overall picture of present-day phenomenology. In particular, it is useful to know how the mass relation for the vector bosons W and Z is modified in the presence of higher scalar $SU(2)$ multiplets — in other words, how such a relation depends on the values of weak isospin T and hypercharge Y labelling the Higgs multiplet in question.

To begin with, let us recall that the standard doublet carries $T = 1/2$ and $Y = 1/2$ and its electrically neutral component (which acquires a non-zero vacuum value) has the third component of isospin $T_3 = -Y = -1/2$, in accordance with the relation $Q = T_3 + Y$. For a general value of T (integer or half-integer), we have a multiplet consisting of $2T + 1$ complex components[12] (corresponding to $T_3 = -T, \cdots, T$) that can be written as

$$\Phi = \begin{pmatrix} \varphi_{T,\,T} \\ \varphi_{T,\,T-1} \\ \vdots \\ \varphi_{T,\,-T} \end{pmatrix} \tag{6.131}$$

Let the weak hypercharge of this multiplet be Y. The neutral component (to be shifted away from the vacuum value) then has the weak isospin

[12]We return here to the complex Higgs fields since such a notation is the most compact and very convenient for our present purpose.

projection $T_3 = T_3^{(0)} = -Y$ and the vacuum value of the multiplet (6.131) (which minimizes the Higgs–Goldstone potential) is, in analogy with the standard case,

$$\Phi_0 = \frac{v}{\sqrt{2}} \begin{pmatrix} 0 \\ \vdots \\ 1 \\ \vdots \\ 0 \end{pmatrix} \tag{6.132}$$

i.e. the Φ_0 is an eigenvector of the $SU(2)$ generator T_3 corresponding to the eigenvalue $T_3^{(0)}(= -Y)$. Let us now see what are the vector boson masses resulting from the Higgs mechanism based on the scalar multiplet (6.131). As before, from the Higgs–Goldstone Lagrangian of the type (6.53), one gets the quadratic mass term

$$\mathscr{L}_{\text{mass}}^{\text{IVB}} = \Phi_0^\dagger \left(gA_\mu^a T_a + g'YB_\mu\right) \left(gA^{b\mu}T_b + g'YB^\mu\right) \Phi_0 \tag{6.133}$$

where T_a, $a = 1,2,3$ are $(2T+1)\times(2T+1)$ matrices representing the $SU(2)$ generators, with T_3 taken to be diagonal and Y is a multiple of unit matrix. Introducing the isospin raising and lowering operators $T_\pm = T_1 \pm \mathrm{i}T_2$, as well as the $W^\pm$ fields (cf. the discussion around Eq. (5.17)), the expression (6.133) is recast as

$$\mathscr{L}_{\text{mass}}^{\text{IVB}} = \Phi_0^\dagger \left\{ g\left[\frac{1}{\sqrt{2}}\left(T_-W_\mu^- + T_+W_\mu^+\right) + T_3A_\mu^3\right] + g'YB_\mu \right\}$$
$$\times \left\{ g\left[\frac{1}{\sqrt{2}}\left(T_-W^{-\mu} + T_+W^{+\mu}\right) + T_3A^{3\mu}\right] + g'YB^\mu \right\} \Phi_0$$

and this becomes, after simple algebraic manipulations,

$$\mathscr{L}_{\text{mass}}^{\text{IVB}} = \Phi_0^\dagger \left[\frac{1}{2}g^2\left(T_-T_+ + T_+T_-\right)W_\mu^- W^{+\mu} + Y^2\left(gA_\mu^3 - g'B_\mu\right)^2\right] \Phi_0 \tag{6.134}$$

Note that in arriving at (6.134), we have utilized the familiar properties of the ladder operators $T_\pm$ and the fact that Φ_0 is an eigenvector of T_3 (with the eigenvalue $-Y$); one thus has, in particular, $\Phi_0^\dagger(T_\pm)^2\Phi_0 = 0$ and $\Phi_0^\dagger T_\pm T_3 \Phi_0 = \Phi_0^\dagger T_3 T_\pm \Phi_0 = 0$. Finally, using the identities $T_-T_+ + T_+T_- = 2\left(\vec{T}^2 - T_3^2\right)$ and $\vec{T}^2\Phi_0 = T(T+1)\Phi_0$, we get from (6.134)

$$\mathscr{L}_{\text{mass}}^{\text{IVB}} = \frac{1}{2}g^2v^2[T(T+1) - Y^2]W_\mu^- W^{+\mu} + \frac{1}{2}\left(g^2 + g'^2\right)v^2Y^2 Z_\mu Z^\mu \tag{6.135}$$

where $Z_\mu = \cos\theta_W A_\mu^3 - \sin\theta_W B_\mu$, with $\cos\theta_W = g/\left(g^2 + g'^2\right)^{1/2}$.

Thus, the corresponding masses can be identified as follows:

$$\begin{aligned} m_W^2 &= \frac{1}{2} g^2 v^2 [T(T+1) - Y^2] \\ m_Z^2 &= \left(g^2 + g'^2\right) v^2 Y^2 \end{aligned} \tag{6.136}$$

These formulae represent our desired goal. Let us now discuss the contents of (6.136) in more detail. Obviously, for the standard model values $T = 1/2, Y = 1/2$, one recovers our previous result (cf. (6.68)). An interesting feature of the formulae (6.136) is the proportionality of the m_Z to Y. This means that, e.g., for a Higgs triplet ($T = 1$) with $Y = 0$ (i.e. with neutral middle component), one gets $m_Z = 0$. In other words, a real scalar triplet

$$\Phi = \begin{pmatrix} \varphi^+ \\ \varphi^0 \\ \varphi^- \end{pmatrix}, \quad \varphi^- = \left(\varphi^+\right)^*, \quad \varphi^0 \text{ real} \tag{6.137}$$

can only give mass to $W^\pm$ but not to Z. Such an observation is in fact quite instructive: it demonstrates explicitly that three real scalar fields involved in a Goldstone-type potential are not enough for generating realistic masses of the three vector bosons W^-, W^+, and Z[13] — as we have already noted in Section 6.4, one needs at least four real scalars (e.g. those contained within the standard complex doublet).

Assuming generally that $Y \neq 0$, one obtains from (6.136) a simple formula for the ratio of the vector boson masses

$$\frac{m_W^2}{m_Z^2} = \frac{g^2}{g^2 + g'^2} \frac{T(T+1) - Y^2}{2Y^2} \tag{6.138}$$

In terms of the parameter $\rho = m_W^2 / \left(m_Z^2 \cos^2 \theta_W\right)$, it means that

$$\rho = \frac{T(T+1) - Y^2}{2Y^2} \tag{6.139}$$

[13] Of course, this is equivalent to the fact that only two of the three real scalar fields contained in (6.137) can be identified as Goldstone bosons when the corresponding potential $V(\Phi)$ is worked out. Thus, in accordance with the general theorems, the Higgs mechanism results in two massive vector bosons corresponding to the (unphysical) Goldstone bosons and one real scalar acquires a mass and becomes physical. We shall return to the example of the real Higgs triplet at the end of this section.

As we have already noted, the value of ρ is very close to unity in the real world, so it is desirable to have $\rho = 1$ at the classical level. From (6.139), it is clear that such a relation is valid whenever the Higgs multiplet is chosen so that T and Y satisfy the equation

$$T(T+1) - 3Y^2 = 0 \tag{6.140}$$

(obviously, T and Y must be either both integer or both half-integer). The first few solutions of Eq. (6.140) are

$$(T, Y) = \left(\frac{1}{2}, \frac{1}{2}\right), \ (3, 2), \ \left(\frac{25}{2}, \frac{15}{2}\right), \ldots \tag{6.141}$$

It is amusing to observe (see [68]) that there are 11 solutions of Eq. (6.140) less than 10^6, the biggest one being $T = 489060\frac{1}{2}, Y = 282359\frac{1}{2}$.

The above result (6.139) can be slightly generalized as follows. If there are several Higgs scalar multiplets having in general different vacuum values, one gets, instead of (6.139),

$$\rho = \frac{\sum_{T,Y} |v_{T,Y}|^2 \left[T(T+1) - Y^2\right]}{2\sum_{T,Y} |v_{T,Y}|^2 Y^2} \tag{6.142}$$

From the last expression, it is particularly clear that for a Higgs sector consisting of doublets only, one always has $\rho = 1$, independent of the values of $v_{T,Y}$ (note that for $T = \frac{1}{2}$, the only possible values of Y are $\pm\frac{1}{2}$).

In closing this section, let us remark that we have not discussed the problem of defining a U-gauge (i.e. that in which the would-be Goldstone bosons are eliminated) for an extended Higgs sector considered here. As we have noted before, there is a general proof that such a U-gauge always exists (this can be found in Ref. [56]; see also [AbL] and [Hua]). Let us give at least an example of a non-standard Higgs multiplet for which a U-gauge can be defined explicitly in a straightforward manner similar to the case of the standard doublet (cf. (6.50)). The example to be considered here is the real triplet (6.137) (note that such a Higgs sector was relevant, e.g., in the old Georgi–Glashow $SU(2)$ (or $O(3)$) electroweak model [69] that avoided a neutral vector boson Z in favour of heavy leptons, see also [Hor]). When working with (6.137), one has to choose a corresponding basis of the $SU(2)$ generators carefully so that Φ be transformed into the same form. It is easy to find out that a suitable $SU(2)$ basis is

$$T_1 = \frac{1}{\sqrt{2}} \begin{pmatrix} 0 & -1 & 0 \\ -1 & 0 & 1 \\ 0 & 1 & 0 \end{pmatrix}, \quad T_2 = \frac{1}{\sqrt{2}} \begin{pmatrix} 0 & i & 0 \\ -i & 0 & -i \\ 0 & i & 0 \end{pmatrix},$$

$$T_3 = \begin{pmatrix} 1 & 0 & 0 \\ 0 & 0 & 0 \\ 0 & 0 & -1 \end{pmatrix} \tag{6.143}$$

Then it is not difficult to show that Φ defined by (6.137) can be written as

$$\Phi = \exp[i(\xi_- T_+ + \xi_+ T_-)] \begin{pmatrix} 0 \\ \eta \\ 0 \end{pmatrix} \tag{6.144}$$

where $T_\pm = T_1 \pm \mathrm{i}T_2$, $\xi_- = \xi_+^*$ and η is real. Of course, $\xi_\pm$ then represent the would-be Goldstone bosons if a scalar potential $V^\pm(\Phi)$ of the usual type is considered and η (when shifted appropriately) becomes a physical Higgs scalar boson.

Problems

6.1 Show that the tree-level matrix element for the process $W_L^+ W_L^- \to HH$ behaves well in the high-energy limit (i.e. for $s \gg m_W^2, m_H^2$).

6.2 Prove that an analogous statement also holds for the process $e^+e^- \to Z_L H$. Keep $m_e \neq 0$ in order to appreciate the mechanism of cancellations of high-energy divergences arising from the individual diagrams.

6.3 Calculate the cross-section $\sigma(e^+e^- \to ZH)$ as a function of the c.m. energy and of the Higgs boson mass. For simplicity, set $m_e = 0$ throughout the calculation (obviously, such an approximation is absolutely safe because of the high threshold energy for the considered process).

6.4 Show that the tree-level matrix element for the process $Z_L Z_L \to Z_L Z_L$ is free of high-energy divergences. Examine also the dependence of the scattering amplitude in question on the Higgs boson mass.

6.5 For the SM Higgs boson with mass $m_H = 125\,\text{GeV}$, make an order-of-magnitude estimate of the cross-section for $e^+e^- \to HH$ at the energy

$E_{\text{c.m.}} = 500\,\text{GeV}$ (at the tree level). Compare your estimate with the cross-section for $e^+e^- \to ZH$ and also with the QED cross-section for $\sigma(e^+e^- \to \mu^+\mu^-)$.

6.6 Write down the most general Higgs–Goldstone potential in the SM extension involving two complex scalar doublets (this is currently popular under the label THDM, an acronym for "two-Higgs doublet model").

Hint: Consult [Gun].

Chapter 7

Standard Model of Electroweak Interactions

7.1 Leptonic world — brief recapitulation

In previous chapters, we have discussed in some detail the basic principles upon which the GWS electroweak theory is built. As regards the spectrum of elementary fermions, we have restricted ourselves — for simplicity of the exposition — to its leptonic part. In the following sections, we will complete the edifice of the standard electroweak model by incorporating its quark sector. Before doing it, let us summarize here very briefly (mostly for reference purposes and for reader's convenience) the relevant results that we have achieved so far in our description of the "leptonic world".

As we know, in building the GWS electroweak theory, one relies on two basic principles:

(1) gauge symmetry $SU(2) \times U(1)$;
(2) Higgs mechanism realized via a complex scalar doublet.

For convenience, one may fix the physical U-gauge, which means that the Higgs doublet $\Phi = \binom{\varphi^+}{\varphi^0}$ becomes

$$\Phi = \Phi_U = \begin{pmatrix} 0 \\ \frac{1}{\sqrt{2}}(v + H) \end{pmatrix} \tag{7.1}$$

(cf. (6.54)). Let us also recall that fixing the U-gauge is formally equivalent to an $SU(2)$ gauge transformation and the U-gauge GWS Lagrangian can thus be obtained from its gauge invariant form simply by replacing Φ with Φ_U.

We have seen that the GWS Lagrangian of the leptonic world can be written, schematically, as

$$\mathscr{L}^{\text{GWS}} = \mathscr{L}_{\text{gauge}} + \mathscr{L}_{\text{lepton}} + \mathscr{L}_{\text{Higgs}} + \mathscr{L}_{\text{Yukawa}} \tag{7.2}$$

and the individual terms appearing in (7.2) were described in great detail in preceding chapters. Now, we are going to focus our attention on the term $\mathscr{L}_{\text{lepton}}$ that describes the interactions of leptons with vector bosons (it will serve as a starting point for our preliminary discussion of quark sector in the next section). The $\mathscr{L}_{\text{lepton}}$ for a particular lepton species ℓ ($\ell = e, \mu$ or τ) can be written as

$$\mathscr{L}^{(\ell)}_{\text{lepton}} = i\bar{L}^{(\ell)}\gamma^{\mu}\left(\partial_{\mu} - igA^{a}_{\mu}\frac{\tau^{a}}{2} - ig'Y^{(\ell)}_{L}B_{\mu}\right)L^{(\ell)} + i\bar{R}^{(\ell)}\gamma^{\mu}(\partial_{\mu} - ig'Y^{(\ell)}_{R}B_{\mu})R^{(\ell)} \tag{7.3}$$

where

$$L^{(\ell)} = \begin{pmatrix} \nu_{\ell L} \\ \ell_L \end{pmatrix}, \quad R^{(\ell)} = \ell_R \tag{7.4}$$

(we ignore here momentarily the right-handed neutrino field $\nu_{\ell R}$). The weak hypercharges $Y^{(\ell)}_{L}$ and $Y^{(\ell)}_{R}$ are fixed by the rule

$$Q = T_3 + Y \tag{7.5}$$

(cf. (5.40)); in this way, one gets $Y^{(\ell)}_{L} = -\frac{1}{2}$ and $Y^{(\ell)}_{R} = -1$. (Needless to say, in a full Lagrangian incorporating all lepton species, one has to take the sum of the expressions (7.3) over $\ell = e, \mu, \tau$.) Working out (7.3), one easily recovers the interactions of familiar charged $V - A$ weak currents with vector bosons $W^{\pm}$, namely

$$\begin{aligned}\mathscr{L}^{(\ell)}_{\text{CC}} &= \frac{g}{\sqrt{2}}\bar{\nu}_{\ell L}\gamma^{\mu}\ell_L W^{+}_{\mu} + \text{h.c.} \\ &= \frac{g}{2\sqrt{2}}\bar{\nu}_{\ell}\gamma^{\mu}(1-\gamma_5)\ell W^{+}_{\mu} + \text{h.c.}\end{aligned} \tag{7.6}$$

where

$$W^{\pm}_{\mu} = \frac{1}{\sqrt{2}}(A^{1}_{\mu} \mp iA^{2}_{\mu}) \tag{7.7}$$

Let us recall that (7.6) descends from the part of the covariant derivative in (7.3) involving non-diagonal matrices τ^1 and τ^2 (cf. (5.15)–(5.18)). Further, in the neutral sector of (7.3) (containing diagonal matrices τ^3 and $\mathbb{1}$), neither A^{3}_{μ} nor B_{μ} can be interpreted as the electromagnetic field. Therefore, an orthogonal transformation

$$\begin{aligned} A^{3}_{\mu} &= \cos\theta_W Z_{\mu} + \sin\theta_W A_{\mu} \\ B_{\mu} &= -\sin\theta_W Z_{\mu} + \cos\theta_W A_{\mu} \end{aligned} \tag{7.8}$$

introducing physical fields A_μ and Z_μ must be performed and if the θ_W is chosen so that

$$\cos\theta_W = \frac{g}{\sqrt{g^2 + g'^2}}, \quad \sin\theta_W = \frac{g'}{\sqrt{g^2 + g'^2}} \tag{7.9}$$

A_μ is coupled to the ordinary electromagnetic current. Z_μ interacts with the weak neutral current; one gets

$$\mathscr{L}_{\mathrm{NC}}^{(\ell)} = \frac{g}{\cos\theta_W} \sum_{f=\ell_L,\ell_R,\nu_{\ell L}} \varepsilon_f \bar{f}\gamma_\mu f Z^\mu \tag{7.10}$$

where

$$\varepsilon_f = T_{3f} - Q_f \sin^2\theta_W \tag{7.11}$$

It is important to realize that (owing to the simplicity of algebraic manipulations leading from (7.3) to (7.6)) there is a straightforward connection between the contents of the leptonic doublet $L^{(\ell)}$ in (7.4) and the structure of the current: the charged current is simply composed of the upper and lower components of $L^{(\ell)}$. In a similar way, the neutral current is made of upper and lower components of $L^{(\ell)}$ separately and it also gets a contribution from right-handed singlets.

Now, one would like to generalize the GWS gauge theory construction of lepton currents so as to reproduce the phenomenologically successful Cabibbo form of the quark weak current. We already know that recovering a desired form of the charged current is just a matter of proper choice of the basic fermion building blocks ($SU(2)$ doublets and singlets), but, once such a choice is made, a definite structure of the weak neutral current already follows as a pure theoretical prediction. On the other hand, experimental data put severe constraints on the phenomenology of hadronic neutral current interactions. Thus, it is clear *a priori* that in any attempt at extending the $SU(2) \times U(1)$ electroweak gauge theory to the quark sector, one has to deal seriously with the issue of neutral currents.

7.2 Difficulties with three quarks

In Chapter 2, we have written the hadronic part of the charged weak current in terms of the quark fields u, d, s as

$$\bar{u}\gamma_\mu(1 - \gamma_5)(d\cos\theta_C + s\sin\theta_C) \tag{7.12}$$

(see (2.59)) where θ_C is the Cabibbo angle. It means that the interaction of quarks with charged vector bosons can be described by the Lagrangian

$$\begin{aligned}\mathscr{L}_{\text{CC}}^{(u,d,s)} &= \frac{g}{2\sqrt{2}}\bar{u}\gamma^\mu(1-\gamma_5)(d\cos\theta_C + s\sin\theta_{\text{C}})W_\mu^+ + \text{h.c.} \\ &= \frac{g}{\sqrt{2}}\bar{u}_L\gamma^\mu(d_L\cos\theta_{\text{C}} + s_L\sin\theta_{\text{C}})W_\mu^+ + \text{h.c.}\end{aligned} \tag{7.13}$$

where the coupling constant g is related to the Fermi constant through $G_{\text{F}}/\sqrt{2} = g^2/(8m_W^2)$. Comparing the last line of (7.13) with the leptonic Lagrangian (7.6), it is obvious that the result (7.13) is reproduced automatically within the $SU(2)\times U(1)$ electroweak gauge theory if one chooses as one of the basic building blocks of the quark sector an $SU(2)$ doublet of left-handed fields

$$U_L = \begin{pmatrix} u_L \\ d_L\cos\theta_{\text{C}} + s_L\sin\theta_{\text{C}}\end{pmatrix} \tag{7.14}$$

Of course, right-handed components of quark fields

$$u_R, \quad d_R, \quad s_R \tag{7.15}$$

are taken to be $SU(2)$ singlets. Weak hypercharges specifying the $U(1)$ transformation properties of (7.14) and (7.15) are given by the relation (7.5). Taking into account the charge assignments

$$Q_u = \frac{2}{3}, \quad Q_d = Q_s = -\frac{1}{3} \tag{7.16}$$

one thus gets

$$Y_{U_L} = \frac{1}{6}, \quad Y_{u_R} = \frac{2}{3}, \quad Y_{d_R} = Y_{s_R} = -\frac{1}{3} \tag{7.17}$$

The $SU(2)\times U(1)$ invariant Lagrangian for quarks can be written down in analogy with (7.3) (using U_L instead of $L^{(\ell)}$ etc.) and physical vector fields are then introduced according to (7.7) and (7.8). However, when the interaction Lagrangian is worked out in detail, one finds out that the result has a serious flaw: while the weak charged current comes out right, the current coupled to the electromagnetic field does not have the correct form; apart from the desired term

$$\frac{2}{3}\bar{u}\gamma_\mu u - \frac{1}{3}\bar{d}\gamma_\mu d - \frac{1}{3}\bar{s}\gamma_\mu s \tag{7.18}$$

it contains "flavour non-diagonal" pieces of the type $\bar{d}_L\gamma_\mu s_L$ (a verification of this statement is left to the reader as an exercise). Of course, these non-diagonal contributions are a direct consequence of the Cabibbo mixing embodied in the doublet (7.14). Such terms do conserve electric charge, but they are in flagrant contradiction with the empirical fact that electromagnetic interactions conserve strangeness. To make things worse, a contribution like $\bar{d}_L \gamma_\mu s_L$ would produce parity violation in the electromagnetic current.

In fact, these defects can be cured quite easily. It turns out that if the naive model of electroweak quark interactions outlined above is supplemented with an additional $SU(2)$ singlet

$$s'_L = -d_L \sin\theta_C + s_L \cos\theta_C \tag{7.19}$$

(carrying the hypercharge $Y_{s'_L} = -\frac{1}{3}$), the unwanted pieces of the electromagnetic current are cancelled and one ends up with (7.18) as it should be (again, an independent verification of this result is left to the reader as a rewarding exercise). However, a problem still persists. Working out the interaction of Z_μ with weak neutral current, one gets, after somewhat lengthy but elementary calculations,

$$\mathscr{L}_{\text{NC}}^{(u,d,s)} = \frac{g}{\cos\theta_W}\left(\bar{U}_L \gamma_\mu \frac{\tau^3}{2} U_L - \sin^2\theta_W J_\mu^{(\text{em})}\right) Z^\mu \tag{7.20}$$

where

$$J_\mu^{(\text{em})} = \frac{2}{3}\bar{u}\gamma_\mu u - \frac{1}{3}\bar{d}\gamma_\mu d - \frac{1}{3}\bar{s}\gamma_\mu s \tag{7.21}$$

Let us now evaluate the first term of the neutral current in (7.20). This becomes

$$\begin{aligned}
\bar{U}_L \gamma_\mu \frac{\tau^3}{2} U_L &= \frac{1}{2}\bar{u}_L \gamma_\mu u_L - \frac{1}{2}(\bar{d}_L \cos\theta_C + \bar{s}_L \sin\theta_C)\gamma_\mu (d_L \cos\theta_C + s_L \sin\theta_C) \\
&= \frac{1}{2}\bar{u}_L \gamma_\mu u_L - \frac{1}{2}\cos^2\theta_C \bar{d}_L \gamma_\mu d_L - \frac{1}{2}\sin\theta_C \cos\theta_C \bar{d}_L \gamma_\mu s_L \\
&\quad - \frac{1}{2}\sin\theta_C \cos\theta_C \bar{s}_L \gamma_\mu d_L - \frac{1}{2}\sin^2\theta_C \bar{s}_L \gamma_\mu s_L
\end{aligned} \tag{7.22}$$

It means that such a provisional model with three quarks u, d, s leads inevitably to neutral-current interactions of the type

$$\mathscr{L}_{dsZ} = g_{dsZ} \bar{d}_L \gamma_\mu s_L Z^\mu \tag{7.23}$$

where the coupling strength is of the order

$$g_{dsZ} \simeq \frac{g}{\cos\theta_W} \sin\theta_C \cos\theta_C \tag{7.24}$$

The weak current appearing in (7.20) is an example of the so-called "strangeness-changing neutral current" (more generally, "flavour-changing neutral current", usually referred to by the acronym FCNC). The presence of a term like (7.23) would be a phenomenological disaster (remember the empirical selection rule $\Delta S = \Delta Q$ for semileptonic weak decays!). For an instructive example, let us recall the case of kaon decays $K^+ \to \pi^0 e^+ \nu_e$ and $K^+ \to \pi^+ e^+ e^-$ mentioned earlier (see Section 2.5). The former process, where $\Delta S = \Delta Q = -1$, can be viewed at the quark level as

$$\bar{s} \to \bar{u} + e^+ + \nu_e \tag{7.25}$$

(since K^+ has the quark composition $u\bar{s}$ while π^0 is made of $u\bar{u}$ and $d\bar{d}$) and the latter, for which $\Delta S = -1$ and $\Delta Q = 0$, may be represented as

$$\bar{s} \to \bar{d} + e^+ + e^- \tag{7.26}$$

(since the quark contents of π^+ is $u\bar{d}$). Now, (7.25) proceeds at the tree level via W^+ exchange and an overall coupling factor associated with such a diagram is of the order

$$g^2 \sin\theta_C \tag{7.27}$$

(cf. (7.6) and (7.13)). If the FCNC interaction (7.23) were present, the process (7.26) would proceed at the tree level through the Z^0 exchange and the overall coupling factor associated with the corresponding diagram would be, in view of (7.10) and (7.24), of the order

$$\left(\frac{g}{\cos\theta_W}\right)^2 \sin\theta_C \cos\theta_C \tag{7.28}$$

which is numerically rather close to (7.27). In other words, the reactions (7.25) and (7.26) would occur with roughly equal probability and this in turn means that one would then expect comparable branching ratios for the two kaon decays. However, as we already noted in Section 2.5, $K^+ \to \pi^+ e^+ e^-$ is in fact much less probable than $K^+ \to \pi^0 e^+ \nu_e$, by about five orders of magnitude! Thus, the coupling (7.23) is clearly unacceptable from the phenomenological point of view.

One final remark is in order here. We have seen that a strangeness-changing weak neutral current necessarily appears within the $SU(2) \times U(1)$

gauge theory of electroweak interactions incorporating three quarks u, d, s. In fact, there is another instructive argument showing that such an effect is essentially unavoidable within an electroweak theory involving $W^\pm, Z^0$ and three quarks with Cabibbo mixing. Requiring the "good high-energy behaviour" for all tree-level scattering amplitudes (in the sense elucidated in previous chapters), one may consider, in particular, the process $d\bar{s} \to W^+W^-$. This certainly gets a contribution from a u-quark exchange diagram (descending from charged-current interactions), which produces a quadratic high-energy divergence if both W^+ and W^- are longitudinally polarized; in order to compensate this divergence, one has to introduce a neutral-current dsZ coupling with the above-mentioned strength. More about this line of argument can be found in [Hor].

Thus, the moral of this story is as follows. **Cabibbo mixing in a world built upon just three quarks is not compatible with the $SU(2)\times U(1)$ gauge symmetry of electroweak interactions because of the appearance of phenomenologically unacceptable strangeness-changing neutral currents.** In the early days of the GWS model, this pathological feature was indeed a mortal danger for the whole concept of gauge theories of fundamental interactions. Fortunately, a simple and elegant solution of the problem emerged in the early 1970s and this in fact played a substantial role in the subsequent establishment of the GWS theory as a true "standard model" of electroweak interactions.

7.3 Fourth quark and GIM construction

The idea of how to get rid of the strangeness-changing neutral currents within the GWS theory originated from the work of Glashow, Iliopoulos and Maiani [70]. They postulated a fourth quark, carrying the charge $+2/3$ and labelled as c, which stands for "charm" as the new flavour was named. Such a scheme has an obvious aesthetic appeal because of a nice lepton–quark symmetry (four quarks u, d, c, s as counterparts of the four leptons ν_e, e, ν_μ, μ known then), but, what is more important, it enables one to introduce a new doublet into the $SU(2)\times U(1)$ gauge theory of electroweak interactions. Thus, within the model due to Glashow, Iliopoulos and Maiani (GIM), one may consider a set of basic building blocks for the quark sector consisting of two left-handed $SU(2)$ doublets

$$U_L = \begin{pmatrix} u_L \\ d_L \cos\theta_C + s_L \sin\theta_C \end{pmatrix}, \quad C_L = \begin{pmatrix} c_L \\ -d_L \sin\theta_C + s_L \cos\theta_C \end{pmatrix} \tag{7.29}$$

and four right-handed singlets

$$u_R, \quad c_R, \quad d_R, \quad s_R \tag{7.30}$$

The choice of the "orthogonal" combinations of d_L and s_L in the two doublets in (7.29) is motivated by the presumed cancellation of the unwanted strangeness-changing terms in the electromagnetic and weak neutral currents. We will show now that such a cancellation is indeed achieved.

To this end, let us start with the $SU(2) \times U(1)$ gauge invariant Lagrangian made of the quark fields (7.29) and (7.30). This can be written as

$$\begin{aligned}\mathscr{L}^{(GIM)} &= i\bar{U}_L\gamma^\mu\left(\partial_\mu - igA^a_\mu\frac{\tau^a}{2} - ig'Y_{U_L}B_\mu\right)U_L \\ &\quad + i\bar{C}_L\gamma^\mu\left(\partial_\mu - igA^a_\mu\frac{\tau^a}{2} - ig'Y_{C_L}B_\mu\right)C_L \\ &\quad + \sum_{f=u,c,d,s} i\bar{f}_R\gamma^\mu(\partial_\mu - ig'Y_{f_R}B_\mu)f_R\end{aligned} \tag{7.31}$$

with the weak hypercharges

$$Y_{U_L} = Y_{C_L} = \frac{1}{6}, \quad Y_{u_R} = Y_{c_R} = \frac{2}{3}, \quad Y_{d_R} = Y_{s_R} = -\frac{1}{3} \tag{7.32}$$

that follow from (7.5) and from the quark charge assignments. The evaluation of the relevant interaction Lagrangian repeats essentially the steps that were already necessary in the preceding section, but here we will be more explicit (for the reader's convenience) as the envisaged result is rather important. The interactions in the neutral current sector (i.e. those involving A^3_μ and B_μ) descending from (7.31) are then

$$\begin{aligned}\mathscr{L}^{(N)}_{\text{int}} &= \bar{U}_L\gamma^\mu\left(\frac{1}{2}g\tau^3A^3_\mu + \frac{1}{6}g'B_\mu\right)U_L + \bar{C}_L\gamma^\mu\left(\frac{1}{2}g\tau^3A^3_\mu + \frac{1}{6}g'B_\mu\right)C_L \\ &\quad + \frac{2}{3}g'\bar{u}_R\gamma^\mu u_RB_\mu + \frac{2}{3}g'\bar{c}_R\gamma^\mu c_RB_\mu \\ &\quad - \frac{1}{3}g'\bar{d}_R\gamma^\mu d_RB_\mu - \frac{1}{3}g'\bar{s}_R\gamma^\mu s_RB_\mu\end{aligned} \tag{7.33}$$

Introducing now A_μ and Z_μ according to (7.8), the interaction of quarks with A_μ becomes

$$\mathscr{L}^{(\text{em})}_{\text{int}} = eJ^{(\text{em})}_\mu A^\mu \tag{7.34}$$

where $e = g \sin\theta_W$ and

$$J_\mu^{(\text{em})} = \frac{1}{2}\bar{U}_L\gamma_\mu\tau^3 U_L + \frac{1}{2}\bar{C}_L\gamma_\mu\tau^3 C_L + \frac{1}{6}\bar{U}_L\gamma_\mu U_L + \frac{1}{6}\bar{C}_L\gamma_\mu C_L + \frac{2}{3}\bar{u}_R\gamma_\mu u_R + \frac{2}{3}\bar{c}_R\gamma_\mu c_R - \frac{1}{3}\bar{d}_R\gamma_\mu d_R - \frac{1}{3}\bar{s}_R\gamma_\mu s_R \tag{7.35}$$

The last expression is worked out as

$$\begin{aligned} J_\mu^{(\text{em})} &= \frac{1}{2}\bar{u}_L\gamma_\mu u_L - \frac{1}{2}(\bar{d}_L\cos\theta_C + \bar{s}_L\sin\theta_C)\gamma_\mu(d_L\cos\theta_C + s_L\sin\theta_C) \\ &+ \frac{1}{2}\bar{c}_L\gamma_\mu c_L - \frac{1}{2}(-\bar{d}_L\sin\theta_C + \bar{s}_L\cos\theta_C)\gamma_\mu(-d_L\sin\theta_C + s_L\cos\theta_C) \\ &+ \frac{1}{6}\bar{u}_L\gamma_\mu u_L + \frac{1}{6}(\bar{d}_L\cos\theta_C + \bar{s}_L\sin\theta_C)\gamma_\mu(d_L\cos\theta_C + s_L\sin\theta_C) \\ &+ \frac{1}{6}\bar{c}_L\gamma_\mu c_L + \frac{1}{6}(-\bar{d}_L\sin\theta_C + \bar{s}_L\cos\theta_C)\gamma_\mu(-d_L\sin\theta_C + s_L\cos\theta_C) \\ &+ \frac{2}{3}\bar{u}_R\gamma_\mu u_R + \frac{2}{3}\bar{c}_R\gamma_\mu c_R - \frac{1}{3}\bar{d}_R\gamma_\mu d_R - \frac{1}{3}\bar{s}_R\gamma_\mu s_R \end{aligned} \tag{7.36}$$

which is readily simplified to

$$J_\mu^{(\text{em})} = \frac{2}{3}\bar{u}\gamma_\mu u + \frac{2}{3}\bar{c}\gamma_\mu c - \frac{1}{3}\bar{d}\gamma_\mu d - \frac{1}{3}\bar{s}\gamma_\mu s \tag{7.37}$$

Thus, the electromagnetic current has indeed the desired flavour-diagonal form. For the Z_μ interaction, one gets, after some algebraic manipulations,

$$\mathscr{L}_{\text{NC}}^{(\text{GIM})} = \frac{g}{\cos\theta_W} J_\mu^{(\text{GIM})} Z^\mu \tag{7.38}$$

where the current

$$J_\mu^{(\text{GIM})} = \frac{1}{2}\bar{U}_L\gamma_\mu\tau^3 U_L + \frac{1}{2}\bar{C}_L\gamma_\mu\tau^3 C_L - \sin^2\theta_W J_\mu^{(\text{em})} \tag{7.39}$$

has a form analogous to (7.20), with the additional contribution of the second ("charmed") doublet shown in (7.29). Of course, it is an expected result and one can now show easily (in full analogy with what we have done for electromagnetic current) that the $d - s$ mixing terms cancel as needed. Indeed, one has

$$\begin{aligned} &\frac{1}{2}\bar{U}_L\gamma_\mu\tau^3 U_L + \frac{1}{2}\bar{C}_L\gamma_\mu\tau^3 C_L \\ &= \frac{1}{2}\bar{u}_L\gamma_\mu u_L - \frac{1}{2}(\bar{d}_L\cos\theta_C + \bar{s}_L\sin\theta_C)\gamma_\mu(d_L\cos\theta_C + s_L\sin\theta_C) \end{aligned}$$

$$+\frac{1}{2}\bar{c}_L\gamma_\mu c_L - \frac{1}{2}(-\bar{d}_L\sin\theta_C + \bar{s}_L\cos\theta_C)\gamma_\mu(-d_L\sin\theta_C + s_L\cos\theta_C)$$

$$= \frac{1}{2}\bar{u}_L\gamma_\mu u_L + \frac{1}{2}\bar{c}_L\gamma_\mu c_L - \frac{1}{2}\bar{d}_L\gamma_\mu d_L - \frac{1}{2}\bar{s}_L\gamma_\mu s_L \tag{7.40}$$

Thus, the form (7.40) is diagonal in quark flavours and its coefficients obviously coincide with the weak isospin values for u_L, c_L, d_L, s_L. The GIM neutral current (7.39) can therefore be written as

$$J_\mu^{(\mathrm{GIM})} = \sum_{f=u,c,d,s}\left(\varepsilon_L^{(f)}\bar{f}_L\gamma_\mu f_L + \varepsilon_R^{(f)}\bar{f}_R\gamma_\mu f_R\right) \tag{7.41}$$

with

$$\varepsilon^{(f)} = T_3^{(f)} - Q^{(f)}\sin^2\theta_W \tag{7.42}$$

In other words, weak neutral currents for quarks and leptons now have the same structure (cf. (7.10)).

Let us also add that using (7.29) and our previous experience, the interactions of charged currents can be written down almost immediately; obviously, one gets

$$\mathscr{L}_{\mathrm{CC}}^{(\mathrm{GIM})} = \frac{g}{\sqrt{2}}[\bar{u}_L\gamma^\mu(d_L\cos\theta_C + s_L\sin\theta_C)$$
$$+\bar{c}_L\gamma^\mu(-d_L\sin\theta_C + s_L\cos\theta_C)]W_\mu^+ + \mathrm{h.c.} \tag{7.43}$$

Note that (7.43) represents a specific prediction for flavour-changing processes mediated by $W^\pm$: the $c \to d$ transitions are Cabibbo-suppressed similar to $u \to s$ while $c \to s$ goes unsuppressed, in analogy with $u \to d$.

Coming back to the crucial result (7.41), one may say that the GIM construction solved the problem of strangeness-changing currents in the early 1970s and it saved, at least conceptually, the idea of the GWS gauge electroweak theory. However, a historical remark is in order here. When the fourth "charmed" quark has been postulated as a remedy for the difficulties described above, there was absolutely no experimental sign of a possible existence of such a particle. Fortunately enough, in 1974, a major discovery came. Two experimental teams, led by Richter and Ting, respectively, observed independently [71] a new meson, denoted as J/ψ, which found a natural interpretation as a bound state $c\bar{c}$ (with $m_c \doteq 1.5$ GeV); for this reason, J/ψ is also called charmonium (in analogy with positronium e^+e^-). More precisely, J/ψ represents charmonium ground state and soon after its discovery, the Richter group revealed the existence of further mesons that could be interpreted as the corresponding excited states. This spectacular result (which in fact came not long after the discovery of weak neutral currents in 1973) was a real breakthrough, as it removed a major obstacle to the recognition of the GWS gauge theory as a physically meaningful model of electroweak interactions. In fact, the discovery of hadrons containing the c-quark provided a great support to the whole concept of gauge theories of fundamental interactions as well as to the quark model itself. A nice and rather detailed description of the J/ψ discovery and related matters can be found, e.g., in [CaG].

There is still one point to be mentioned here. Similarly, as in the case of leptons, quarks should acquire masses through Yukawa-type interactions. However, a generalization of the procedure described in Section 6.6 is not entirely straightforward in the quark case owing to the flavour mixing embodied in the basic left-handed doublets. From Yukawa interactions made of (7.29), (7.30) and the Higgs field (7.1), one gets $d - s$ mixing terms in the resulting quark mass matrix and these have to be eliminated by imposing some appropriate relations that should hold for the relevant coupling constants and masses. This can be done, but we shall not proceed in this way. Instead, we are going to put forward a slightly different formulation of the GIM construction, based primarily on the discussion of general quark mass matrices arising from Yukawa couplings *without* assuming Cabibbo mixing *a priori*. A great virtue of such an alternative approach is, among other things, that one thus arrives at a rather natural understanding of the very existence of the Cabibbo angle.

7.4 GIM construction via diagonalization of quark mass matrices

To begin with, let us pinpoint some essential aspects of the GIM construction as formulated in preceding section. It relies on the empirical fact of the existence of Cabibbo angle, which is then taken as an input parameter in the basic $SU(2)$ doublets (7.29). While the structure of U_L reflects the old phenomenology of 1960s (actually, it defines the θ_C), the form of the $d - s$ mixing appearing in C_L is picked by hand so as to ensure the elimination of strangeness-changing neutral currents. Quark fields entering the relevant Lagrangian are supposed to be the physical ones (i.e. corresponding to mass eigenstates). Within such a scheme, the intriguing problem of a deeper origin of the Cabibbo angle remains totally obscure and also the remarkable "orthogonality" of the lower components of

U_L and C_L appears as a rather *ad hoc* choice, enforced upon us by demands of hadron phenomenology — one might wonder rightfully whether it has a more profound explanation.

As we have already noted, at least partial answer to these questions can be found quite naturally if the whole GIM construction is formulated in a slightly different way. So, this is what we are going to do now. The main idea is to start with quark fields that need not, in general, coincide with the physical ones; the latter will only emerge as a result of a mass matrix diagonalization. Thus, let the basic building blocks for the relevant quark Lagrangians be two left-handed $SU(2)$ doublets

$$U_{0L} = \begin{pmatrix} u_{0L} \\ d_{0L} \end{pmatrix}, \quad C_{0L} = \begin{pmatrix} c_{0L} \\ s_{0L} \end{pmatrix} \tag{7.44}$$

(corresponding to two "generations" of quarks) and four right-handed singlets

$$u_{0R}, \quad d_{0R}, \quad s_{0R}, \quad c_{0R} \tag{7.45}$$

where the label '0' indicates the presumed unphysical nature of the fields in question. Again, the corresponding weak hypercharges are determined according to (7.5), so that

$$Y_{U_{0L}} = Y_{C_{0L}} = \frac{1}{6}, \quad Y_{u_{0R}} = Y_{c_{0R}} = \frac{2}{3}, \quad Y_{d_{0R}} = Y_{s_{0R}} = -\frac{1}{3} \tag{7.46}$$

and the interactions of quarks with the $SU(2) \times U(1)$ gauge fields thus become

$$\begin{aligned}
\mathscr{L}_{\text{int}}^{(\text{quark})} &= \bar{U}_{0L}\gamma^\mu \left(\frac{1}{2}gA_\mu^a\tau^a + \frac{1}{6}g'B_\mu\right) U_{0L} \\
&\quad + \bar{C}_{0L}\gamma^\mu \left(\frac{1}{2}gA_\mu^a\tau^a + \frac{1}{6}g'B_\mu\right) C_{0L} \\
&\quad + \frac{2}{3}g'\bar{u}_{0R}\gamma^\mu u_{0R}B_\mu + \frac{2}{3}g'\bar{c}_{0R}\gamma^\mu c_{0R}B_\mu \\
&\quad - \frac{1}{3}g'\bar{d}_{0R}\gamma^\mu d_{0R}B_\mu - \frac{1}{3}g'\bar{s}_{0R}\gamma^\mu s_{0R}B_\mu
\end{aligned} \tag{7.47}$$

Now, the most general $SU(2) \times U(1)$ invariant Yukawa-type interaction involving the quark fields (7.44) and (7.45) and the Higgs doublet Φ has the form

$$\begin{aligned}
\mathscr{L}_{\text{Yukawa}}^{(d,s)} &= -h_{11}\bar{U}_{0L}\Phi d_{0R} - h_{12}\bar{U}_{0L}\Phi s_{0R} \\
&\quad - h_{21}\bar{C}_{0L}\Phi d_{0R} - h_{22}\bar{C}_{0L}\Phi s_{0R} + \text{h.c.}
\end{aligned} \tag{7.48}$$

where $h_{ij},\ i,j = 1,2$ are arbitrary (dimensionless) coupling constants; for simplicity, we may assume that they are real (we shall see later on that such a restriction does not mean any loss of generality). Note that the $SU(2)$ symmetry of the expression (7.48) is obvious and the hypercharge values shown in (7.46) guarantee its invariance under $U(1)$ (remember that $\bar{U}_{0L}$ and $\bar{C}_{0L}$ carry $Y = -\frac{1}{6}$!). In this context, it is also important to realize that an analogous coupling that would involve u_{0R} and c_{0R} is forbidden precisely by the requirement of hypercharge $U(1)$ symmetry. Working out (7.48) in the U-gauge, one gets

$$\begin{aligned}\mathscr{L}^{(d,s)}_{\text{Yukawa}} &= -\frac{1}{\sqrt{2}}(v+H)(h_{11}\bar{d}_{0L}d_{0R} + h_{12}\bar{d}_{0L}s_{0R} \\ &\quad + h_{21}\bar{s}_{0L}d_{0R} + h_{22}\bar{s}_{0L}s_{0R}) + \text{h.c.} \\ &= -\frac{1}{\sqrt{2}}(v+H)\begin{pmatrix}\bar{d}_{0L}, & \bar{s}_{0L}\end{pmatrix}\begin{pmatrix}h_{11} & h_{12} \\ h_{21} & h_{22}\end{pmatrix}\begin{pmatrix}d_{0R} \\ s_{0R}\end{pmatrix} + \text{h.c.}\end{aligned} \tag{7.49}$$

In (7.49), one may identify both interactions and mass terms. The latter can be collected in a compact form

$$\mathscr{L}^{(d,s)}_{\text{mass}} = -\begin{pmatrix}\bar{d}_{0L}, & \bar{s}_{0L}\end{pmatrix} M \begin{pmatrix}d_{0R} \\ s_{0R}\end{pmatrix} + \text{h.c.} \tag{7.50}$$

where

$$M = \frac{v}{\sqrt{2}}\begin{pmatrix}h_{11} & h_{12} \\ h_{21} & h_{22}\end{pmatrix} \tag{7.51}$$

Thus, it becomes clear that only the "down-type" quarks d and s can acquire masses through Yukawa couplings involving the Higgs doublet Φ. We shall explain a bit later how the mass terms for u and c are generated, and now, let us discuss a diagonalization of the mass matrix (7.51) (this, of course, is necessary for the proper identification of physical quark fields).

At first sight, this may seem rather problematic, since (7.51) is not, in general, hermitian and thus it cannot be diagonalized simply by means of a unitary transformation matrix. Fortunately, the job can be done with the help of a *biunitary* transformation.[1] In particular, one may rely on the following theorem:

[1] An erudite reader may note that the technique of biunitary transformations is in fact the construction that mathematicians call the "singular value decomposition" (SVD). Although in mathematics this has been known since the 19th century, particle physicists apparently developed it for their pragmatic needs independently in the early 1970s.

Any non-singular square complex matrix M can be decomposed as

$$M = \mathcal{U}^\dagger \mathfrak{M} \mathcal{V} \tag{7.52}$$

where $\mathcal{U}$, $\mathcal{V}$ are unitary matrices and $\mathfrak{M}$ is diagonal and positive.

When M is real, $\mathcal{U}$ and $\mathcal{V}$ are real orthogonal matrices.

A proof of this theorem is quite simple and we defer it to the end of this section; now, we are going to apply its statement to the above quark mass term (we assume *a priori* that the matrix (7.51) is non-singular in order to get non-zero quark masses). Using (7.52) in (7.50), one has

$$\mathscr{L}_{\text{mass}}^{(d,s)} = -\left(\bar{d}_{0L},\ \bar{s}_{0L}\right)\mathcal{U}^\dagger \mathfrak{M} \mathcal{V}\begin{pmatrix} d_{0R} \\ s_{0R} \end{pmatrix} + \text{h.c.} \tag{7.53}$$

where the diagonal matrix $\mathfrak{M}$ shall be written, for obvious reasons, as

$$\mathfrak{M} = \begin{pmatrix} m_d & 0 \\ 0 & m_s \end{pmatrix} \tag{7.54}$$

Let us now define new quark fields d_L, s_L and d_R, s_R through unitary transformations

$$\begin{pmatrix} d_L \\ s_L \end{pmatrix} = \mathcal{U}\begin{pmatrix} d_{0L} \\ s_{0L} \end{pmatrix}, \quad \begin{pmatrix} d_R \\ s_R \end{pmatrix} = \mathcal{V}\begin{pmatrix} d_{0R} \\ s_{0R} \end{pmatrix} \tag{7.55}$$

Then (7.53) becomes

$$\begin{aligned} \mathscr{L}_{\text{mass}}^{(d,s)} &= -\left(\bar{d}_L,\ \bar{s}_L\right)\mathfrak{M}\begin{pmatrix} d_R \\ s_R \end{pmatrix} + \text{h.c.} \\ &= -m_d\bar{d}_L d_R - m_s\bar{s}_L s_R + \text{h.c.} \\ &= -m_d(\bar{d}_L d_R + \bar{d}_R d_L) - m_s(\bar{s}_L s_R + \bar{s}_R s_L) \\ &= -m_d\bar{d}d - m_s\bar{s}s \end{aligned} \tag{7.56}$$

It means that — as anticipated in (7.54) — d and s correspond to quark mass eigenstates; in other words, they can be identified with physical fields.

It turns out that mathematicians usually do not know that SVD has such an important application within SM, and particle physicists are rarely aware of the mathematical context and history of the currently popular algebraic tool of biunitary transformations.

In this context, it is also important to realize that the kinetic terms remain diagonal, e.g., in terms of the original variables, we had

$$\begin{aligned}\mathscr{L}_{\text{kin}}^{(d,s)} &= i\bar{d}_{0L}\gamma^\mu\partial_\mu d_{0L} + i\bar{s}_{0L}\gamma^\mu\partial_\mu s_{0L} + i\bar{d}_{0R}\gamma^\mu\partial_\mu d_{0R} + i\bar{s}_{0R}\gamma^\mu\partial_\mu s_{0R} \\ &= i\left(\bar{d}_{0L},\ \bar{s}_{0L}\right)\gamma^\mu\partial_\mu\begin{pmatrix} d_{0L} \\ s_{0L}\end{pmatrix} + i\left(\bar{d}_{0R},\ \bar{s}_{0R}\right)\gamma^\mu\partial_\mu\begin{pmatrix} d_{0R} \\ s_{0R}\end{pmatrix}\end{aligned} \tag{7.57}$$

and this becomes

$$\begin{aligned}\mathscr{L}_{\text{kin}}^{(d,s)} &= i\left(\bar{d}_L,\ \bar{s}_L\right)\gamma^\mu\partial_\mu\begin{pmatrix} d_L \\ s_L\end{pmatrix} + i\left(\bar{d}_R,\ \bar{s}_R\right)\gamma^\mu\partial_\mu\begin{pmatrix} d_R \\ s_R\end{pmatrix} \\ &= i\bar{d}\gamma^\mu\partial_\mu d + i\bar{s}\gamma^\mu\partial_\mu s\end{aligned} \tag{7.58}$$

when the transformation (7.55) is implemented (simply because $\mathcal{U}\mathcal{U}^\dagger = \mathcal{V}\mathcal{V}^\dagger = \mathbb{1}$).

Since the interaction terms descending from (7.49) have an algebraic structure completely analogous to that of the mass terms, it is obvious that the Lagrangian describing the interactions of quarks d and s with the physical Higgs boson H has the form

$$\mathscr{L}_{\text{int}}^{(d,s)} = -\frac{m_d}{v}\bar{d}dH - \frac{m_s}{v}\bar{s}sH \tag{7.59}$$

Thus, the relevant coupling constants, denoted in a self-explanatory way as g_{ddH} and g_{ssH} respectively, are given by

$$g_{ddH} = -\frac{g}{2}\frac{m_d}{m_W}, \quad g_{ssH} = -\frac{g}{2}\frac{m_s}{m_W} \tag{7.60}$$

in full analogy with the result valid for leptons (cf. (6.88)).

As the next step, one should generate mass terms for the "up-type" quarks u and c. This is done by means of the trick that we have already used for neutrinos in Chapter 6 (see the discussion following after the formula (6.89)). Thus, let us consider the conjugate doublet $\widetilde{\Phi}$ defined in (6.90). It carries weak hypercharge $-\frac{1}{2}$ and this in turn means that one can employ $\widetilde{\Phi}$ for constructing an $SU(2) \times U(1)$ invariant Yukawa interaction

$$\begin{aligned}\mathscr{L}_{\text{Yukawa}}^{(u,c)} &= -\tilde{h}_{11}\bar{U}_{0L}\widetilde{\Phi}u_{0R} - \tilde{h}_{12}\bar{U}_{0L}\widetilde{\Phi}c_{0R} \\ &\quad - \tilde{h}_{21}\bar{C}_{0L}\widetilde{\Phi}u_{0R} - \tilde{h}_{22}\bar{C}_{0L}\widetilde{\Phi}c_{0R} + \text{h.c.}\end{aligned} \tag{7.61}$$

needed for our purpose (one may check readily that the relevant weak hypercharge values fit precisely the invariance requirement for (7.61)).

Of course, the coupling constants $\tilde{h}_{ij}$ are completely independent of h_{ij} appearing in (7.48). For simplicity, we are again assuming that all of $\tilde{h}_{ij}$ are real. Now, in U-gauge, $\widetilde{\Phi}$ becomes

$$\widetilde{\Phi}_U = \begin{pmatrix} \frac{1}{\sqrt{2}}(v+H) \\ 0 \end{pmatrix} \tag{7.62}$$

(cf. (6.93)) and substituting (7.62) into (7.61), one gets

$$\mathscr{L}^{(u,c)}_{\text{Yukawa}} = -\frac{1}{\sqrt{2}}(v+H)\left(\bar{u}_{0L},\ \bar{c}_{0L}\right)\begin{pmatrix} \tilde{h}_{11} & \tilde{h}_{12} \\ \tilde{h}_{21} & \tilde{h}_{22} \end{pmatrix}\begin{pmatrix} u_{0R} \\ c_{0R} \end{pmatrix} + \text{h.c.} \tag{7.63}$$

In particular, (7.63) contains the expected mass term

$$\mathscr{L}^{(u,c)}_{\text{mass}} = -\left(\bar{u}_{0L},\ \bar{c}_{0L}\right)\widetilde{M}\begin{pmatrix} u_{0R} \\ c_{0R} \end{pmatrix} + \text{h.c.} \tag{7.64}$$

where

$$\widetilde{M} = \frac{v}{\sqrt{2}}\begin{pmatrix} \tilde{h}_{11} & \tilde{h}_{12} \\ \tilde{h}_{21} & \tilde{h}_{22} \end{pmatrix} \tag{7.65}$$

Again, (7.65) can be diagonalized by means of a biunitary (here, in fact, real biorthogonal) transformation, i.e. one can write

$$\widetilde{M} = \widetilde{\mathcal{U}}^{\dagger}\widetilde{\mathfrak{M}}\widetilde{\mathcal{V}} \tag{7.66}$$

with

$$\widetilde{\mathfrak{M}} = \begin{pmatrix} m_u & 0 \\ 0 & m_c \end{pmatrix} \tag{7.67}$$

and define new fields $u_{L,R}$, $c_{L,R}$ by rotating the original ones according to

$$\begin{pmatrix} u_L \\ c_L \end{pmatrix} = \widetilde{\mathcal{U}}\begin{pmatrix} u_{0L} \\ c_{0L} \end{pmatrix}, \quad \begin{pmatrix} u_R \\ c_R \end{pmatrix} = \widetilde{\mathcal{V}}\begin{pmatrix} u_{0R} \\ c_{0R} \end{pmatrix} \tag{7.68}$$

Then the mass term (7.64) is recast as

$$\mathscr{L}^{(u,c)}_{\text{mass}} = -\left(\bar{u}_L,\ \bar{c}_L\right)\widetilde{\mathfrak{M}}\begin{pmatrix} u_R \\ c_R \end{pmatrix} + \text{h.c.} = -m_u\bar{u}u - m_c\bar{c}c \tag{7.69}$$

so that the u, c may be identified with physical fields in complete analogy with the preceding discussion of quarks d and s. Of course, the corresponding interaction with the Higgs field becomes

$$\mathscr{L}_{\text{int}}^{(u,c)} = -\frac{m_u}{v}\bar{u}uH - \frac{m_c}{v}\bar{c}cH = -\frac{g}{2}\left(\frac{m_u}{m_W}\bar{u}u + \frac{m_c}{m_W}\bar{c}c\right)H \tag{7.70}$$

similar to (7.59) and (7.60).

Now, it remains to be seen how the weak interactions of charged currents are expressed in terms of physical quark fields and what happens in the sector of neutral currents. First, from (7.47), one gets readily the charged-current interaction written in terms of the original fields (7.44):

$$\begin{aligned}\mathscr{L}_{CC}^{(\text{quark})} &= \frac{g}{\sqrt{2}}(\bar{u}_{0L}\gamma^\mu d_{0L} + \bar{c}_{0L}\gamma^\mu s_{0L})W_\mu^+ + \text{h.c.}\\ &= \frac{g}{\sqrt{2}}\left(\bar{u}_{0L},\ \bar{c}_{0L}\right)\gamma^\mu\begin{pmatrix} d_{0L}\\ s_{0L}\end{pmatrix}W_\mu^+ + \text{h.c.}\end{aligned} \tag{7.71}$$

When the physical quark fields are introduced through the rotations

$$\begin{pmatrix} d_{0L}\\ s_{0L}\end{pmatrix} = \mathcal{U}^\dagger\begin{pmatrix} d_L\\ s_L\end{pmatrix},\quad \begin{pmatrix} u_{0L}\\ c_{0L}\end{pmatrix} = \widetilde{\mathcal{U}}^\dagger\begin{pmatrix} u_L\\ c_L\end{pmatrix} \tag{7.72}$$

(see (7.55) and (7.68)), the expression (7.71) is recast as

$$\mathscr{L}_{CC}^{(\text{quark})} = \frac{g}{\sqrt{2}}\left(\bar{u}_L,\ \bar{c}_L\right)\gamma^\mu\widetilde{\mathcal{U}}\mathcal{U}^\dagger\begin{pmatrix} d_L\\ s_L\end{pmatrix}W_\mu^+ + \text{h.c.} \tag{7.73}$$

According to our conventions, $\mathcal{U}$ and $\widetilde{\mathcal{U}}$ are real orthogonal matrices; thus, each of them is characterized by a rotation angle. Denoting these angles as θ_1 and θ_2, respectively, the product $\widetilde{\mathcal{U}}\mathcal{U}^\dagger$ is consequently parametrized by the difference $\theta_2 - \theta_1$, which can justly be called θ_C. Indeed, it is seen immediately that $\theta_2 - \theta_1$ plays the role of Cabibbo angle because in terms of $\theta_C = \theta_2 - \theta_1$, (7.73) is written as

$$\begin{aligned}\mathscr{L}_{CC}^{(\text{quark})} &= \frac{g}{\sqrt{2}}\left(\bar{u}_L,\ \bar{c}_L\right)\gamma^\mu\begin{pmatrix}\cos\theta_C & \sin\theta_C\\ -\sin\theta_C & \cos\theta_C\end{pmatrix}\begin{pmatrix} d_L\\ s_L\end{pmatrix}W_\mu^+ + \text{h.c.}\\ &= \frac{g}{\sqrt{2}}[\bar{u}_L\gamma^\mu(d_L\cos\theta_C + s_L\sin\theta_C)\\ &\quad + \bar{c}_L\gamma^\mu(-d_L\sin\theta_C + s_L\cos\theta_C)]W_\mu^+ + \text{h.c.}\end{aligned} \tag{7.74}$$

In other words, we have reproduced the GIM construction of the charged-current interactions (cf. (7.43)). The crucial aspect of our analysis is a natural appearance of the Cabibbo angle, which originates in quark field rotations necessary for diagonalization of their quark matrices; more precisely, it is due to a mismatch between the rotations performed on (left-handed) up-type and down-type quarks. At the same time, the reason for "orthogonality" of the d_L and s_L combinations coupled to u_L and c_L in the original form (7.43) becomes manifest: the pattern of $d - s$ mixing is determined by the orthogonal matrix $\widetilde{\mathcal{U}}\mathcal{U}^\dagger$ shown explicitly in the first line of (7.74); for the purpose of later references, we will denote it as U_{GIM}, i.e.

$$U_{\mathrm{GIM}} = \begin{pmatrix} \cos\theta_C & \sin\theta_C \\ -\sin\theta_C & \cos\theta_C \end{pmatrix} = \begin{pmatrix} U_{ud} & U_{us} \\ U_{cd} & U_{cs} \end{pmatrix} \tag{7.75}$$

It should be stressed that within such an approach, separate $d - s$ and $u - c$ mixings (characterized, e.g., by the above-mentioned angles θ_1 and θ_2) would not make physical sense: only the difference $\theta_2 - \theta_1$ is physically relevant and this is conventionally taken as the $d - s$ mixing angle.

Next, let us see what the results for neutral currents are. Working out (7.47), interaction terms involving the gauge fields A^3_μ, B_μ become

$$\begin{aligned}\mathscr{L}_{\mathrm{NC}} &= \frac{1}{2}g(\bar{u}_{0L}\gamma^\mu u_{0L} - \bar{d}_{0L}\gamma^\mu d_{0L} + \bar{c}_{0L}\gamma^\mu c_{0L} - \bar{s}_{0L}\gamma^\mu s_{0L})A^3_\mu \\ &+ \frac{1}{6}g'(\bar{u}_{0L}\gamma^\mu u_{0L} + \bar{d}_{0L}\gamma^\mu d_{0L} + \bar{c}_{0L}\gamma^\mu c_{0L} + \bar{s}_{0L}\gamma^\mu s_{0L})B_\mu \\ &+ \frac{2}{3}g'(\bar{u}_{0R}\gamma^\mu u_{0R} + \bar{c}_{0R}\gamma^\mu c_{0R})B_\mu - \frac{1}{3}g'(\bar{d}_{0R}\gamma^\mu d_{0R} + \bar{s}_{0R}\gamma^\mu s_{0R})B_\mu\end{aligned} \tag{7.76}$$

and this can be conveniently rewritten as

$$\begin{aligned}\mathscr{L}_{\mathrm{NC}} &= \frac{1}{2}g\left[(\bar{u}_{0L},\ \bar{c}_{0L})\gamma^\mu\begin{pmatrix} u_{0L} \\ c_{0L}\end{pmatrix} - (\bar{d}_{0L},\ \bar{s}_{0L})\gamma^\mu\begin{pmatrix} d_{0L} \\ s_{0L}\end{pmatrix}\right]A^3_\mu \\ &+ \frac{1}{6}g'\left[(\bar{u}_{0L},\ \bar{c}_{0L})\gamma^\mu\begin{pmatrix} u_{0L} \\ c_{0L}\end{pmatrix} + (\bar{d}_{0L},\ \bar{s}_{0L})\gamma^\mu\begin{pmatrix} d_{0L} \\ s_{0L}\end{pmatrix}\right]B_\mu \\ &+ \frac{2}{3}g'\,(\bar{u}_{0R},\ \bar{c}_{0R})\gamma^\mu\begin{pmatrix} u_{0R} \\ c_{0R}\end{pmatrix}B_\mu - \frac{1}{3}g'\,(\bar{d}_{0R},\ \bar{s}_{0R})\gamma^\mu\begin{pmatrix} d_{0R} \\ s_{0R}\end{pmatrix}B_\mu\end{aligned} \tag{7.77}$$

When one passes to the physical quark fields via (7.55) and (7.68), it is clear that only the products $\mathcal{U}\mathcal{U}^\dagger$, $\mathcal{V}\mathcal{V}^\dagger$, $\widetilde{\mathcal{U}}\widetilde{\mathcal{U}}^\dagger$ and $\widetilde{\mathcal{V}}\widetilde{\mathcal{V}}^\dagger$ can appear in the expression (7.77). However, any such product reduces to the unit matrix.

Thus, (7.77) becomes immediately

$$\begin{aligned}\mathscr{L}_{\mathrm{NC}} &= \frac{1}{2}g(\bar{u}_L\gamma^\mu u_L + \bar{c}_L\gamma^\mu c_L - \bar{d}_L\gamma^\mu d_L - \bar{s}_L\gamma^\mu s_L)A^3_\mu \\ &\quad + \frac{1}{6}g'(\bar{u}_L\gamma^\mu u_L + \bar{c}_L\gamma^\mu c_L + \bar{d}_L\gamma^\mu d_L + \bar{s}_L\gamma^\mu s_L)B_\mu \\ &\quad + \frac{2}{3}g'(\bar{u}_R\gamma^\mu u_R + \bar{c}_R\gamma^\mu c_R)B_\mu - \frac{1}{3}g'(\bar{d}_R\gamma^\mu d_R + \bar{s}_R\gamma^\mu s_R)B_\mu\end{aligned} \tag{7.78}$$

It is clear that by introducing A_μ, Z_μ instead of A^3_μ, B_μ (through (7.8)), one cannot spoil the flavour-diagonal structure of (7.78); in fact, when this is done, the desired expression for the electromagnetic current is recovered along with the GIM result (7.39) for the weak neutral current.

Thus, we have arrived at the most transparent formulation of the GIM mechanism that can be succinctly summarized as follows. Neutral currents are manifestly (by construction) flavour-diagonal in the basis of unphysical quark fields u_0, d_0, c_0, s_0 and quarks of an equal charge are grouped in pairs that can be subsequently transformed into the corresponding physical fields. The currents remain diagonal under such transformations because these are implemented by means of unitary matrices and each transformation matrix eventually gets multiplied by its inverse inside the current (remember that in the case of charged currents, we had products like $\widetilde{\mathcal{U}}\mathcal{U}^\dagger$, involving two *different* matrices!).

In closing this section, we are going to prove the mathematical theorem on biunitary transformations (expressed by the relation (7.52)), which played a central role in our considerations. Let M be an arbitrary non-singular complex square $n \times n$ matrix. Then $MM^\dagger$ is hermitian and positive, and consequently, it may be diagonalized by means of a unitary transformation, i.e.

$$MM^\dagger = \mathcal{U}^\dagger \mathfrak{M}^2 \mathcal{U} \tag{7.79}$$

where $\mathcal{U}\mathcal{U}^+ = \mathcal{U}^+\mathcal{U} = \mathbb{1}$ and $\mathfrak{M}^2$ can be written as

$$\mathfrak{M}^2 = \mathrm{diag}(m_1^2, \ldots, m_n^2) \tag{7.80}$$

with all m_j^2, $j = 1, \ldots, n$ being positive numbers. Obviously, for M real, $\mathcal{U}$ can be taken as a real orthogonal matrix. Let us also define $\mathfrak{M}$ as

$$\mathfrak{M} = \sqrt{\mathfrak{M}^2} = \mathrm{diag}(|m_1|, \ldots, |m_n|) \tag{7.81}$$

Now, with the relation (7.52) in mind, one can define

$$\mathcal{V} = \mathfrak{M}^{-1}\mathcal{U}M \tag{7.82}$$

It is easy to show that $\mathcal{V}$ is unitary. Indeed,

$$\mathcal{V}\mathcal{V}^\dagger = \mathfrak{M}^{-1}\mathcal{U}MM^\dagger\mathcal{U}^\dagger\mathfrak{M}^{-1} = \mathfrak{M}^{-1}\cdot\mathfrak{M}^2\cdot\mathfrak{M}^{-1} = \mathbb{1} \tag{7.83}$$

and $\mathcal{V}^\dagger\mathcal{V} = \mathbb{1}$ then follows automatically, since we are dealing with finite-dimensional matrices (reality of $\mathcal{V}$ for a real M is also obvious from (7.82)). Thus, if for a given M, one chooses the matrices $\mathcal{U}$ and $\mathcal{V}$ as defined above, (7.52) is satisfied and our theorem is thereby proved.

7.5 Kobayashi–Maskawa matrix

So far, we have considered electroweak interactions within a model involving four quark flavours. However, as we know, the existence of six quarks is now firmly established by experiments; more precisely, the present-day picture of standard model of particle physics incorporates six leptons and six quarks, i.e. three "generations" of elementary fermions. Therefore, we should generalize our discussion so as to include two more quark types in addition to the four considered previously.

In fact, this can be done quite easily if one adopts the strategy described in the preceding section. It means that in building the quark sector of the $SU(2)\times U(1)$ gauge theory of electroweak forces, one starts with three left-handed $SU(2)$ doublets

$$U_{0L} = \begin{pmatrix} u_{0L} \\ d_{0L} \end{pmatrix}, \quad C_{0L} = \begin{pmatrix} c_{0L} \\ s_{0L} \end{pmatrix}, \quad T_{0L} = \begin{pmatrix} t_{0L} \\ b_{0L} \end{pmatrix} \tag{7.84}$$

and six right-handed singlets

$$u_{0R}, \quad d_{0R}, \quad c_{0R}, \quad s_{0R}, \quad t_{0R}, \quad b_{0R} \tag{7.85}$$

where b stands for "bottom" (or "beauty"), t for "top" and the other symbols have the by now familiar meaning. Quarks b and t carry electric charges $-\frac{1}{3}$ and $+\frac{2}{3}$, respectively, and the weak hypercharges of the fields (7.84) and (7.85) are fixed by (7.5) as usual. Now, we can proceed in full analogy with the two-generation model. The relevant Yukawa interactions are written as

$$\begin{aligned}\mathscr{L}^{(d,s,b)}_{\text{Yukawa}} = &- h_{11}\bar{U}_{0L}\Phi d_{0R} - h_{12}\bar{U}_{0L}\Phi s_{0R} - h_{13}\bar{U}_{0L}\Phi b_{0R} \\ &- h_{21}\bar{C}_{0L}\Phi d_{0R} - h_{22}\bar{C}_{0L}\Phi s_{0R} - h_{23}\bar{C}_{0L}\Phi b_{0R} \\ &- h_{31}\bar{T}_{0L}\Phi d_{0R} - h_{32}\bar{T}_{0L}\Phi s_{0R} - h_{33}\bar{T}_{0L}\Phi b_{0R} + \text{h.c.}\end{aligned} \tag{7.86}$$

and

$$\begin{aligned}\mathscr{L}^{(u,c,t)}_{\text{Yukawa}} = & -\tilde{h}_{11}\bar{U}_{0L}\widetilde{\Phi}u_{0R} - \tilde{h}_{12}\bar{U}_{0L}\widetilde{\Phi}c_{0R} - \tilde{h}_{13}\bar{U}_{0L}\widetilde{\Phi}t_{0R} \\ & -\tilde{h}_{21}\bar{C}_{0L}\widetilde{\Phi}u_{0R} - \tilde{h}_{22}\bar{C}_{0L}\widetilde{\Phi}c_{0R} - \tilde{h}_{23}\bar{C}_{0L}\widetilde{\Phi}t_{0R} \\ & -\tilde{h}_{31}\bar{T}_{0L}\widetilde{\Phi}u_{0R} - \tilde{h}_{32}\bar{T}_{0L}\widetilde{\Phi}c_{0R} - \tilde{h}_{33}\bar{T}_{0L}\widetilde{\Phi}t_{0R} + \text{h.c.} \qquad (7.87)\end{aligned}$$

where the coupling constants h_{ij} and $\tilde{h}_{ij}$ are, in general, complex numbers. Substituting into (7.86) and (7.87) the U-gauge values for the Φ and $\widetilde{\Phi}$, one gets

$$\mathscr{L}^{(d,s,b)}_{\text{Yukawa}} = -\frac{1}{\sqrt{2}}(v+H)\left(\bar{d}_{0L},\ \bar{s}_{0L},\ \bar{b}_{0L}\right)\begin{pmatrix} h_{11} & h_{12} & h_{13} \\ h_{21} & h_{22} & h_{23} \\ h_{31} & h_{32} & h_{33}\end{pmatrix}\begin{pmatrix} d_{0R} \\ s_{0R} \\ b_{0R}\end{pmatrix} + \text{h.c.} \tag{7.88}$$

and

$$\mathscr{L}^{(u,c,t)}_{\text{Yukawa}} = -\frac{1}{\sqrt{2}}(v+H)\left(\bar{u}_{0L},\ \bar{c}_{0L},\ \bar{t}_{0L}\right)\begin{pmatrix} \tilde{h}_{11} & \tilde{h}_{12} & \tilde{h}_{13} \\ \tilde{h}_{21} & \tilde{h}_{22} & \tilde{h}_{23} \\ \tilde{h}_{31} & \tilde{h}_{32} & \tilde{h}_{33}\end{pmatrix}\begin{pmatrix} u_{0R} \\ c_{0R} \\ t_{0R}\end{pmatrix} + \text{h.c.} \tag{7.89}$$

The mass terms for down- and up-type quarks contained in (7.88) and (7.89) are diagonalized by means of appropriate biunitary transformations. Denoting the relevant 3×3 transformation matrices as $\mathcal{U}, \mathcal{V}$ (for down-type quarks) and $\widetilde{\mathcal{U}}, \widetilde{\mathcal{V}}$ (for up-type quarks), one is thus led to redefine the quark fields as

$$\begin{pmatrix} d_L \\ s_L \\ b_L\end{pmatrix} = \mathcal{U}\begin{pmatrix} d_{0L} \\ s_{0L} \\ b_{0L}\end{pmatrix}, \qquad \begin{pmatrix} d_R \\ s_R \\ b_R\end{pmatrix} = \mathcal{V}\begin{pmatrix} d_{0R} \\ s_{0R} \\ b_{0R}\end{pmatrix} \tag{7.90}$$

and

$$\begin{pmatrix} u_L \\ c_L \\ t_L\end{pmatrix} = \widetilde{\mathcal{U}}\begin{pmatrix} u_{0L} \\ c_{0L} \\ t_{0L}\end{pmatrix}, \qquad \begin{pmatrix} u_R \\ c_R \\ t_R\end{pmatrix} = \widetilde{\mathcal{V}}\begin{pmatrix} u_{0R} \\ c_{0R} \\ t_{0R}\end{pmatrix} \tag{7.91}$$

The d, s, b, u, c, t then represent a set of physical quark fields; along with mass terms, also the H interactions appearing in (7.88) and (7.89) become flavour-diagonal, with coupling constants obeying the law

$$g_{ffH} = -\frac{g}{2}\frac{m_f}{m_W} \tag{7.92}$$

(where f is a generic label for any flavour in question).

The charged-current interaction written in terms of the original (unphysical) quark fields has the form

$$\begin{aligned}\mathscr{L}_{CC}^{(\text{quark})} &= \frac{g}{\sqrt{2}}(\bar{u}_{0L}\gamma^\mu d_{0L} + \bar{c}_{0L}\gamma^\mu s_{0L} + \bar{t}_{0L}\gamma^\mu b_{0L})W_\mu^+ + \text{h.c.} \\ &= \frac{g}{\sqrt{2}}\left(\bar{u}_{0L},\ \bar{c}_{0L},\ \bar{t}_{0L}\right)\gamma^\mu \begin{pmatrix} d_{0L} \\ s_{0L} \\ b_{0L} \end{pmatrix} W_\mu^+ + \text{h.c.}\end{aligned} \tag{7.93}$$

and when one passes to the physical basis according to (7.90) and (7.91), this becomes

$$\mathscr{L}_{CC}^{(\text{quark})} = \frac{g}{\sqrt{2}}\left(\bar{u}_L,\ \bar{c}_L,\ \bar{t}_L\right)\gamma^\mu \widetilde{\mathcal{U}}\mathcal{U}^\dagger \begin{pmatrix} d_L \\ s_L \\ b_L \end{pmatrix} W_\mu^+ + \text{h.c.} \tag{7.94}$$

in full analogy with (7.73). Thus, the flavour mixing occurring in quark interactions with W bosons is now represented by a 3×3 unitary matrix $\widetilde{\mathcal{U}}\mathcal{U}^\dagger$ that has replaced U_{GIM} discussed previously.

U_{GIM} was eventually described with the help of a single real parameter — the Cabibbo angle — and one may wonder what is a physically relevant parametrization of the 3×3 matrix $\widetilde{\mathcal{U}}\mathcal{U}^\dagger$ in (7.94). Let us start our counting with a general unitary matrix 3×3. This has nine complex (i.e. 18 real) elements, which are constrained by three real and three complex conditions (normalization of columns to unit length and their mutual orthogonality). A complex condition is equivalent to two real ones, so one can also say that the elements of a matrix in question are subject to $3 + 2 \times 3 = 9$ real constraints. Thus, a unitary 3×3 matrix is parametrized by means of $18 - 9 = 9$ real numbers. However, when one has in mind the matrix $\widetilde{\mathcal{U}}\mathcal{U}^\dagger$ entering the Lagrangian (7.94), the number of its independent parameters can be further reduced: some phase factors become physically irrelevant, as they can be absorbed into appropriate redefinitions of the quark fields. We are now going to show explicitly how this is done. Denoting

$$\widetilde{\mathcal{U}}\mathcal{U}^\dagger = V = \begin{pmatrix} V_{11} & V_{12} & V_{13} \\ V_{21} & V_{22} & V_{23} \\ V_{31} & V_{32} & V_{33} \end{pmatrix} \tag{7.95}$$

the essential part of the expression (7.94) reads

$$\left(\bar{u},\ \bar{c},\ \bar{t}\right) \begin{pmatrix} V_{11} & V_{12} & V_{13} \\ V_{21} & V_{22} & V_{23} \\ V_{31} & V_{32} & V_{33} \end{pmatrix} \begin{pmatrix} d \\ s \\ b \end{pmatrix} \tag{7.96}$$

(for the moment, we may ignore Dirac gamma matrices within the weak current, since in the present context, these play the role of an overall numerical factor). One may now factor out possible complex phase factors from the first column of (7.95) and, having in mind a later redefinition of the quark fields u, c, t, recast the V identically as

$$V = \begin{pmatrix} e^{i\delta_{11}} & 0 & 0 \\ 0 & e^{i\delta_{21}} & 0 \\ 0 & 0 & e^{i\delta_{31}} \end{pmatrix} \begin{pmatrix} e^{-i\delta_{11}} & 0 & 0 \\ 0 & e^{-i\delta_{21}} & 0 \\ 0 & 0 & e^{-i\delta_{31}} \end{pmatrix} \times \begin{pmatrix} R_{11}e^{i\delta_{11}} & V_{12} & V_{13} \\ R_{21}e^{i\delta_{21}} & V_{22} & V_{23} \\ R_{31}e^{i\delta_{31}} & V_{32} & V_{33} \end{pmatrix} \tag{7.97}$$

where R_{11}, R_{21} and R_{31} are real numbers (as well as the $\delta_{11}, \delta_{21}, \delta_{31}$). The matrix product (7.96) then becomes

$$\begin{aligned} &\left(\bar{u}e^{i\delta_{11}},\ \bar{c}e^{i\delta_{21}},\ \bar{t}e^{i\delta_{31}}\right) \begin{pmatrix} R_{11} & V_{12}e^{-i\delta_{11}} & V_{13}e^{-i\delta_{11}} \\ R_{21} & V_{22}e^{-i\delta_{21}} & V_{23}e^{-i\delta_{21}} \\ R_{31} & V_{32}e^{-i\delta_{31}} & V_{33}e^{-i\delta_{31}} \end{pmatrix} \begin{pmatrix} d \\ s \\ b \end{pmatrix} \\ &= \left(\bar{u}',\ \bar{c}',\ \bar{t}'\right) \begin{pmatrix} R_{11} & V'_{12} & V'_{13} \\ R_{21} & V'_{22} & V'_{23} \\ R_{31} & V'_{32} & V'_{33} \end{pmatrix} \begin{pmatrix} d \\ s \\ b \end{pmatrix} \end{aligned} \tag{7.98}$$

where

$$u' = e^{-i\delta_{11}}u, \quad c' = e^{-i\delta_{21}}c, \quad t' = e^{-i\delta_{31}}t \tag{7.99}$$

and we have introduced an obvious shorthand notation for the matrix elements in the last expression of (7.98). Next, we make a second step in this direction and write

$$\begin{aligned} &\begin{pmatrix} R_{11} & V'_{12} & V'_{13} \\ R_{21} & V'_{22} & V'_{23} \\ R_{31} & V'_{32} & V'_{33} \end{pmatrix} \\ &= \begin{pmatrix} R_{11} & R_{12}e^{i\delta'_{12}} & R_{13}e^{i\delta'_{13}} \\ R_{21} & V'_{22} & V'_{23} \\ R_{31} & V'_{32} & V'_{33} \end{pmatrix} \begin{pmatrix} 1 & 0 & 0 \\ 0 & e^{-i\delta'_{12}} & 0 \\ 0 & 0 & e^{-i\delta'_{13}} \end{pmatrix} \begin{pmatrix} 1 & 0 & 0 \\ 0 & e^{i\delta'_{12}} & 0 \\ 0 & 0 & e^{i\delta'_{13}} \end{pmatrix} \end{aligned} \tag{7.100}$$

Using this, the expression (7.98) can finally be recast as

$$(\bar{u}',\ \bar{c}',\ \bar{t}')\begin{pmatrix} R_{11} & R_{12} & R_{13} \\ R_{21} & V''_{22} & V''_{23} \\ R_{31} & V''_{32} & V''_{33} \end{pmatrix}\begin{pmatrix} d \\ s' \\ b' \end{pmatrix} \tag{7.101}$$

where

$$s' = \mathrm{e}^{i\delta'_{12}} s, \quad b' = \mathrm{e}^{i\delta'_{13}} b \tag{7.102}$$

and the meaning of the other symbols should be clear.

Needless to say, the quark field redefinitions (7.99) and (7.102) have no physical consequences and we will drop the primes in what follows. What we have achieved is that we got rid of five complex phase factors that could generally occur in the first column and the first row of (7.95). In other words, we have reduced the number of physically relevant real parameters describing our unitary matrix $V = \widetilde{\mathcal{U}}\mathcal{U}^\dagger$ from nine to four. If V were purely real (i.e. real orthogonal), it would be parametrized by just three rotation angles. Thus, the fourth real parameter left over corresponds to the phase of a complex factor $\mathrm{e}^{i\delta}$. Summing up these considerations, we see that the six-flavour mixing matrix entering the interactions of quark charged currents (7.94) can be parametrized by means of three rotation angles $\theta_1, \theta_2, \theta_3$ and one complex phase.

As regards the neutral currents, it is quite clear that within the considered three-generation scheme, they must exhibit essentially the same properties as in the four-quark model discussed previously. Indeed, one has a naturally diagonal structure in terms of the unphysical quark fields, which now become grouped in triplets u_0, c_0, t_0 and d_0, s_0, b_0, respectively. The unitary transformations (7.90) and (7.91) are then implemented in a straightforward manner and, as we know, they preserve automatically the diagonal character of the currents in question — the reason is that one encounters only products like $\mathcal{U}\mathcal{U}^\dagger$, etc. equal to unit matrix. This is gratifying, as all the available experimental data clearly show that processes in which, e.g., a b quark would change into s are strongly suppressed, similarly as in the case of $d - s$ transitions mentioned earlier.

The six-quark version of the $SU(2) \times U(1)$ electroweak theory described above was formulated for the first time by Kobayashi and Maskawa in their celebrated paper [72]. This remarkable work (especially its timing) certainly deserves an additional historical commentary, but we postpone it to the end of this section. Now, let us come back to some important physical aspects of the charged-current interactions

$$\mathscr{L}_{CC}^{(\text{quark})} = \frac{g}{2\sqrt{2}} \left(\bar{u},\ \bar{c},\ \bar{t}\right) \gamma^{\mu}(1-\gamma_5) V \begin{pmatrix} d \\ s \\ b \end{pmatrix} W_{\mu}^{+} + \text{h.c.} \tag{7.103}$$

involving the flavour-mixing matrix

$$V = \begin{pmatrix} V_{ud} & V_{us} & V_{ub} \\ V_{cd} & V_{cs} & V_{cb} \\ V_{td} & V_{ts} & V_{tb} \end{pmatrix} \tag{7.104}$$

parametrized, as indicated above (cf. (7.101)). This is called **Kobayashi–Maskawa** (or **Cabibbo–Kobayashi–Maskawa**) **matrix**[2] and its essential feature is that it has, in general, a non-trivial imaginary part — as we have seen, it contains a complex phase that cannot be removed by redefinitions of quark fields. The original parametrization [72] of V relies on a simple generalization of Euler-type rotations[3]; it is written as

$$V = \begin{pmatrix} 1 & 0 & 0 \\ 0 & c_2 & s_2 \\ 0 & -s_2 & c_2 \end{pmatrix} \begin{pmatrix} c_1 & s_1 & 0 \\ -s_1 & c_1 & 0 \\ 0 & 0 & \mathrm{e}^{i\delta} \end{pmatrix} \begin{pmatrix} 1 & 0 & 0 \\ 0 & c_3 & s_3 \\ 0 & -s_3 & c_3 \end{pmatrix} \tag{7.105}$$

where $c_i = \cos\theta_i$, $s_i = \sin\theta_i$ for $i = 1, 2, 3$, so that the resulting form is

$$V = \begin{pmatrix} c_1 & s_1 c_3 & s_1 s_3 \\ -s_1 c_2 & c_1 c_2 c_3 - s_2 s_3 \mathrm{e}^{i\delta} & c_1 c_2 s_3 + s_2 c_3 \mathrm{e}^{i\delta} \\ s_1 s_2 & -c_1 s_2 c_3 - c_2 s_3 \mathrm{e}^{i\delta} & -c_1 s_2 s_3 + c_2 c_3 \mathrm{e}^{i\delta} \end{pmatrix} \tag{7.106}$$

According to (7.103), the pattern of W boson couplings to quarks is determined, up to an overall real factor, by the elements of the CKM matrix. In this way, at least some coupling constants in the charged-current sector can be imaginary and this in turn has dramatic consequences for symmetry properties of the relevant interaction Lagrangian: in general, it is no longer invariant under $\mathcal{CP}$, the combination of charge conjugation C and space inversion $\mathcal{P}$. In other words, apart from the separate violation of $\mathcal{C}$ and $\mathcal{P}$ (which is due to the $V - A$ nature of charged currents), one can have a $\mathcal{CP}$ violation as well if the phase δ is different from zero (of course, the

[2]In what follows, we will usually employ the acronym "CKM matrix" that has become customary in the current literature.

[3]For other parametrizations that have become more common in current literature, see [6].

interaction Lagrangian is still invariant under $\mathcal{CPT}$, so one can also say that $\delta \neq 0$ is tantamount to a violation of $\mathcal{T}$, the time-reversal invariance). We will now explain in more detail how the fact that a coupling constant is not real implies the $\mathcal{CP}$ violation in considered interactions; to this end, the results of Section 2.9 will be utilized in a substantial way (it is sufficient to stay at the level of classical fields).

Let us consider a part of the interaction Lagrangian (7.103) written as

$$\begin{aligned}\mathscr{L}_{\text{int}}^{(12)} &= g_{12}\bar{\psi}_1\gamma^\mu(1-\gamma_5)\psi_2 W_\mu^+ + \text{h.c.}\\ &= g_{12}\bar{\psi}_1\gamma^\mu(1-\gamma_5)\psi_2 W_\mu^+ + g_{12}^*\bar{\psi}_2\gamma^\mu(1-\gamma_5)\psi_1 W_\mu^- \end{aligned} \tag{7.107}$$

where the fields ψ_1 and ψ_2 represent two different quark flavours and the coupling constant g_{12} is, in general, complex. According to (2.140), a $V-A$ current transforms as

$$\bar{\psi}_1\gamma_\mu(1-\gamma_5)\psi_2(x) \xrightarrow{\mathcal{CP}} \bar{\psi}_2(\tilde{x})\gamma^\mu(1-\gamma_5)\psi_1(\tilde{x}) \tag{7.108}$$

where $\tilde{x} = (x_0, -\vec{x})$ i.e. $\tilde{x}^\mu = x_\mu$. The transformation law for the W boson field should reflect its four-vector character and the requirement that W^+ is changed into W^- under charge conjugation; thus, one has

$$W_\mu^\pm(x) \xrightarrow{\mathcal{CP}} W^{\mp\mu}(\tilde{x}) \tag{7.109}$$

(see, e.g., [Bra] for details). The resulting transformation of the Lagrangian (7.107) can therefore be written as

$$\mathscr{L}_{\text{int}}^{(12)} \xrightarrow{\mathcal{CP}} \mathscr{L}_{\text{int}}^{(12)'} = g_{12}\bar{\psi}_2\gamma_\mu(1-\gamma_5)\psi_1 W^{-\mu} + g_{12}^*\bar{\psi}_1\gamma_\mu(1-\gamma_5)\psi_2 W^{+\mu} \tag{7.110}$$

(with the fields taken at the point $\tilde{x}$). Thus, it is seen that for $g_{12}^* = g_{12}$, the original form of the Lagrangian (7.107) is not changed, but if $g_{12}^* \neq g_{12}$, the $\mathcal{CP}$ invariance is lost. The possible non-invariance of the Lagrangian (7.103) under $\mathcal{CP}$, embodied in the CKM matrix (7.104), is very important from the phenomenological point of view. $\mathcal{CP}$ violation in weak interactions (in particular in the system of neutral kaons) has been an experimental fact for a long time [27], and also, now it is a topic of paramount importance in connection with experimental studies of mesons containing the b quark. We will add more remarks on the history of the problem later in this section, and now, as a last technical point, let us generalize slightly our previous discussion.

The generalization we have in mind is a model involving an arbitrary number (n) generations of quarks, i.e. n left-handed doublets and $2n$

right-handed singlets, as a straightforward extension of the pattern (7.84) and (7.85). Proceeding in analogy with the three-generation model described above, one arrives at charged-current interaction involving an $n \times n$ flavour-mixing matrix, which is unitary by construction. Obviously, one can also repeat the previous considerations concerning the V parametrization. As a unitary $n \times n$ matrix, V is in general described in terms of n^2 independent real parameters, since the $2n^2$ real numbers representing its elements are subject to n real and $\binom{n}{2} = \frac{1}{2}n(n-1)$ complex constraints (normalization and orthogonality of columns). Further, the first column and row can be made real by appropriate redefinitions of quark fields; in such a way, $n + (n-1) = 2n - 1$ parameters become unphysical. Thus, an $n \times n$ generalization of the CKM matrix involves $n^2 - (2n-1) = (n-1)^2$ physically relevant real parameters. Obviously, these comprise $\frac{1}{2}n(n-1)$ rotation angles (that would provide complete description of a purely real V) and the remaining $(n-1)^2 - \frac{1}{2}n(n-1) = \frac{1}{2}(n-1)(n-2)$ parameters correspond to complex phases. Thus, one can conclude that for n generations of quarks (i.e. for $2n$ flavours), one has

$$\#\ \text{CKM phases} = \frac{1}{2}(n-1)(n-2) \tag{7.111}$$

In particular, (7.111) implies immediately that for $n = 2$ (i.e. for the four-quark model considered in the preceding section), there is no physically relevant complex phase; in other words, the mixing matrix for four flavours can always be made purely real and reduced thus to the GIM matrix (7.75). Thereby, it is confirmed that our earlier result, obtained for the GIM model in Section 7.4, is in fact completely general (i.e. it holds even in the case of complex Yukawa couplings), though originally we have restricted ourselves to real mass matrices. An important lesson to be learnt from the above discussion is that a four-quark model cannot accommodate naturally $\mathcal{CP}$ violation within the Lagrangian built according to the principles of GWS theory.[4]

Historically, the need for an incorporation of $\mathcal{CP}$ violation into the $SU(2) \times U(1)$ gauge theory of electroweak interactions was the prime motive that led Kobayashi and Maskawa to consider a six-quark model as early as in 1973 — at a time when only three quarks u, d and s were recognized "officially". Of course, achieving right theoretical description

[4]It should be stressed that another source of $\mathcal{CP}$ violation within a gauge theory of electroweak interactions could be an extended Higgs sector [73], but such a possibility has a highly speculative status at present.

of $\mathcal{CP}$ violation is an important goal. As we have already noted, $\mathcal{CP}$ violating effects in weak interactions have been known since 1964; the crucial discovery was an observation of the decay of the long-lived neutral kaon K_L^0 into two pions [27], a process that would be strictly forbidden if the $\mathcal{CP}$ symmetry were exact. Following the GIM scheme, Kobayashi and Maskawa simply noticed that the four-quark model is $\mathcal{CP}$ conserving, while an extra generation of quarks would solve the problem quite naturally. One should realize that this was a really bold proposal, taking into account that even the fourth quark c was discovered only one year later in 1974! Thus, it is not surprising that the work of Kobayashi and Maskawa (KM) went almost unnoticed when published, but it did gain some popularity after the breakthrough discovery [71] of the charmed quark c (that we have already mentioned at the end of Section 7.3). Soon after that, the "heavy lepton" τ with mass $m_\tau \doteq 1.8$ GeV was observed [74] (quite unexpectedly at that time) and it has also become clear that τ is accompanied by its own neutrino ν_τ.[5] Then, in 1977, Lederman and collaborators [76] found a new resonance called Υ that has been interpreted readily as a bound state of a quark denoted as b ($Q_b = -\frac{1}{3}$, $m_b \doteq 4.5$ GeV) and its antiquark; mesons carrying the b-flavour were subsequently discovered during the 1980s. Thus, the spectrum of elementary fermions known since the late 1970s comprised six leptons $\nu_e, e, \nu_\mu, \mu, \nu_\tau, \tau$ and five quarks u, d, s, c, b. Of course, such a development provided much support for the KM scheme, which has thus become a widely recognized and trusted candidate for a realistic model of the quark sector of electroweak theory. In 1980s, experimental data for the production of $b\bar{b}$ pairs in e^+e^- annihilation indicated clearly (though indirectly) that the b quark must in fact belong to a weak isospin $SU(2)$ doublet; moreover, decays involving flavour-changing neutral currents (i.e. transitions like $b \to s$ or $b \to d$) were conspicuously absent. In a sense, the situation of the early 1970s described in Section 7.2 thus repeated itself: an *odd* number of quarks could not match the demands of phenomenology and, at the same time, the elegant and simple KM theory has already been at hand. The hunting for the sixth quark t, expected eagerly since the late 1970s, was rather long and ended successfully in 1995 when its discovery was confirmed officially [77].[6] A remarkable quark–lepton symmetry thus has been restored (three generations of quarks and leptons). Let us note already

[5]However, it should be noted that ν_τ has been detected directly (see [75]) only in 2000!

[6]Note that the top quark is much heavier than intermediate vector bosons W and Z: with its rest mass of about 175 GeV, it is as heavy as another W, the atom of tungsten!

here that apart from the absence of FCNC and an obvious aesthetic appeal, such a scheme has another rather deep aspect: equal number of quarks and leptons within each generation of elementary fermions guarantees the cancellation of the so-called ABJ anomalies and this in turn is crucial for internal consistency of the considered $SU(2) \times U(1)$ gauge theory. This technical aspect of the electroweak standard model will be discussed in some detail in Section 7.9.

The KM model can serve as another example of a theoretical scheme going far beyond the experimental knowledge of its time, yet confirmed eventually in quite an amazing way. The story of heavy quark flavours described briefly in this chapter exhibits a remarkable and typical feature of the electroweak standard model — an interplay of bold theoretical ideas and ingenious experiments that resulted in a truly realistic and predictive description of phenomena at a deep subnuclear level.

Finally, let us note that further study of $\mathcal{CP}$ violation is a very important area of research; the present and forthcoming experiments should determine in detail the elements of the flavour-mixing matrix and tell us whether the KM mechanism is indeed sufficient for explaining all relevant phenomena. Of course, sufficiently accurate data could also open up a window on a possible new physics beyond the standard model. The literature concerning the phenomenology of $\mathcal{CP}$ violation is enormous and the subject is growing fast. As the present text is concerned primarily with basic principles of the theory, the reader interested in phenomenology (and/or in further technical details, such as the various parametrizations of CKM matrix, etc.) is referred, e.g., to [Bra].

7.6 *R*-gauges

As the last topic, we are going to discuss here and in the following three sections some deeper gauge theory aspects of the GWS standard model. From the technical point of view, our treatment will be far from complete; nevertheless, it could serve, hopefully, as a useful introduction to the concepts and techniques involved. The main bonus to be gained is an additional insight into some particular properties of the electroweak standard model encountered in previous chapters.

In our formulation of the SM, we have employed so far the U-gauge, in which the would-be Goldstone bosons are eliminated explicitly and the interaction Lagrangian is written solely in terms of the fields corresponding to physical particles. However, it turns out that an appropriate gauge can be fixed consistently in a completely different way, such that the unphysical

Goldstone boson fields are kept in the Lagrangian and their effects only disappear at the level of physical scattering amplitudes. As we have already noted in Section 6.3, the main advantage of such an approach (developed originally by 't Hooft [49]) is that the propagators of massive vector bosons then exhibit the same asymptotic behaviour as the photon propagator and this makes the power-counting of the relevant ultraviolet divergences in higher-order Feynman diagrams (related intimately to the problem of perturbative renormalizability) much more transparent. Let us now describe such a gauge-fixing procedure in detail. Since its formulation is essentially inspired by the familiar case of quantum electrodynamics, we shall start with a recapitulation of the covariant gauges in QED.

The Lagrangian density for the free electromagnetic (Maxwell) field has the familiar form

$$\mathscr{L}_M = -\frac{1}{4} F_{\mu\nu} F^{\mu\nu} \tag{7.112}$$

with

$$F_{\mu\nu} = \partial_\mu A_\nu - \partial_\nu A_\mu \tag{7.113}$$

For quantizing it, one has to cope with the problem of redundant (unphysical) degrees of freedom involved in the four-potential A_μ — this, of course, is intimately related to the gauge invariance of the Lagrangian (7.112). In particular, within a straightforward canonical approach based on (7.112), one cannot treat all components A_μ, $\mu = 0, 1, 2, 3$ on an equal footing, since the conjugate momentum corresponding to A_0 vanishes identically. To see this, one should first note that

$$\frac{\delta \mathscr{L}_M}{\delta(\partial_\rho A_\sigma)} = -F^{\rho\sigma} \tag{7.114}$$

which means that the momentum π_μ associated with A_μ is

$$\pi_\mu = \frac{\delta \mathscr{L}_M}{\delta(\partial_0 A_\mu)} = -F^{0\mu} \tag{7.115}$$

and thus, $\pi_0 = -F^{00} = 0$. One well-known way out is to quantize only the physical degrees of freedom, but a manifest Lorentz covariance is then lost. For accomplishing a covariant quantization, one has to keep all components A_μ in the game. This can be consistently implemented by modifying the basic Lagrangian (7.112) (so as to break its gauge invariance) and imposing subsequently an appropriate subsidiary condition on the physical solutions.

In particular, instead of (7.112), one can consider the Lagrangian

$$\widetilde{\mathscr{L}}_M = -\frac{1}{4} F_{\mu\nu} F^{\mu\nu} - \frac{1}{2}(\partial \cdot A)^2 \tag{7.116}$$

where $\partial \cdot A$ is a shorthand notation for $\partial_\mu A^\mu$. The equation of motion following from (7.116) reads

$$\partial_\mu F^{\mu\nu} + \partial^\nu(\partial \cdot A) = 0 \tag{7.117}$$

which reduces to

$$\Box A^\nu = 0 \tag{7.118}$$

when one takes into account (7.113). Thus, one must add the Lorenz condition

$$\partial \cdot A = 0 \tag{7.119}$$

to the d'Alembert equation (7.118) if one wants to recover the original Maxwell equations $\partial_\mu F^{\mu\nu} = 0$. In other words, the modified Lagrangian (7.116) has to be supplemented with the constraint (7.119) if it is to describe eventually the Maxwell field at the classical level. Note also that the second term in (7.116) is usually called the "gauge-fixing term" (and denoted correspondingly $\mathscr{L}_{g.f.}$) as it violates the gauge invariance.[7]

In quantum theory, one can employ the Lagrangian (7.116) and postulate a canonical commutation relation for any A_μ, $\mu = 0, 1, 2, 3$; note that the conjugate momenta are given by

$$\frac{\delta \widetilde{\mathscr{L}}_M}{\delta(\partial_0 A_\mu)} = -F^{0\mu} - g^{0\mu}\partial \cdot A \tag{7.120}$$

The Lorenz condition is then imposed, in an appropriate form, on the physical states and the resulting (manifestly covariant) theory is physically equivalent to a non-covariant formulation, in which the unphysical degrees of freedom are eliminated from the very beginning. Such a covariant procedure is originally due to Gupta and Bleuler (see, e.g., [Ryd]); note also that for its formulation, one has to introduce the state-vector space with indefinite metric. The propagator of covariantly quantized field A_μ can be calculated in a straightforward way and the result (in the momentum space) reads

$$D_{\mu\nu}(q) = \frac{-g_{\mu\nu}}{q^2 + i\varepsilon} \tag{7.121}$$

[7]Maxwell equations $\partial_\mu F^{\mu\nu} = 0$ written in terms of A_ρ read $\Box A^\nu - \partial^\nu(\partial \cdot A) = 0$ and when the gauge freedom is constrained by $\partial \cdot A = 0$, one is led to the d'Alembert equation that follows directly from (7.116). While this observation may provide some additional justification of the term "gauge fixing" in the present context, one should keep in mind that fixing a definite gauge for the electromagnetic four-potential is in general a rather subtle matter.

The above procedure can be generalized in such a way that the gauge-fixing term in (7.116) is taken with an arbitrary coefficient. Thus, the Lagrangian

$$\mathscr{L}_M^{(\alpha)} = -\frac{1}{4}F_{\mu\nu}F^{\mu\nu} - \frac{1}{2\alpha}(\partial \cdot A)^2 \tag{7.122}$$

is considered with α being a real "gauge-fixing parameter"; in order to make its physical contents equivalent to the Maxwell field, an appropriate subsidiary condition has to be added. Note that the Lagrangian (7.116) considered previously represents a particular case of (7.122) with $\alpha = 1$. For a general α, the canonical operator quantization based on (7.122) is more complicated than for $\alpha = 1$ because the equation of motion corresponding to (7.122) becomes

$$\partial_\lambda F^{\lambda\mu} + \frac{1}{\alpha}\partial^\mu(\partial \cdot A) = 0 \tag{7.123}$$

or, equivalently,

$$\Box A^\mu + \Big(\frac{1}{\alpha} - 1\Big)\partial^\mu(\partial \cdot A) = 0 \tag{7.124}$$

and this obviously does not coincide with the simple d'Alembert equation for $\alpha \neq 1$. An appropriate generalization of the Gupta–Bleuler method is the so-called Nakanishi–Lautrup formalism (see, e.g., [Nak]), but we will not need such a detailed treatment for our purpose.[8] The object of our primary interest is the propagator, which can be obtained as a Green's function of Eq. (7.124) (in analogy with what we do, e.g., for the massive vector field, cf. Appendix D): to find it, one simply has to solve the equation

$$\Box\mathcal{D}^\mu_\nu(x) + \left(\frac{1}{\alpha} - 1\right)\partial^\mu(\partial_\rho \mathcal{D}^\rho_\nu(x)) = \delta^\mu_\nu \delta^4(x) \tag{7.125}$$

with $\mathcal{D}^\mu_\nu$ denoting the propagator in the coordinate representation. Passing to the momentum space via Fourier transformation, (7.125) yields an algebraic (matrix) equation

$$L^\mu_\rho(q) D^\rho_\nu(q) = \delta^\mu_\nu \tag{7.126}$$

where D^ρ_ν stands for the Fourier transform of $\mathcal{D}^\rho_\nu$ and the coefficient matrix L is given by

$$L^\mu_\rho(q) = -q^2 g^\mu_\rho + \left(1 - \frac{1}{\alpha}\right) q^\mu q_\rho \tag{7.127}$$

[8]Let us remark that the modified Maxwell Lagrangian (7.122) is also a convenient starting point for a covariant quantization by means of the path-integral method.

where we have taken into account that $\delta^\mu_\rho = g^\mu_\rho$. Equation (7.126) is solved by inverting the matrix L by means of the method explained in Appendix D and one thus gets

$$D^{(\alpha)}_{\mu\nu} = \frac{1}{q^2 + i\varepsilon}\left[-g_{\mu\nu} + (1-\alpha)\frac{q_\mu q_\nu}{q^2}\right] \tag{7.128}$$

where we have also introduced the usual $+i\varepsilon$ prescription for the Feynman propagator.[9] In this way, we get a one-parametric set of photon propagators that can be used in Feynman-diagram calculations. Different choices of the parameter α are — for reasons indicated above — referred to as different "gauges". In particular, the value $\alpha = 1$ (for which one recovers the previous result (7.121)) corresponds to the **Feynman gauge**, as it is called in common parlance. It is also interesting to note that for $\alpha = 0$, one gets a purely transverse propagator

$$D^{(\alpha=0)}_{\mu\nu}(q) = \frac{1}{q^2 + i\varepsilon}\left(-g_{\mu\nu} + \frac{q_\mu q_\nu}{q^2}\right) \tag{7.129}$$

that corresponds to the so-called **Landau gauge**. Obviously, such a value of α cannot be accommodated directly in the Lagrangian (7.122); the result (7.129) should be understood as a limiting case of Eq. (7.128) within our straightforward approach.

In the calculations of physical scattering amplitudes, all gauges should be equivalent, i.e. the results should be independent of α. For tree-level diagrams, it is elementary to verify such a statement explicitly; as an instructive example, one can consider, e.g., the process $e^+e^- \to \mu^+\mu^-$ in the lowest order of spinor QED and show that the $q_\mu q_\nu$ part of the photon propagator (7.128) does not contribute within the relevant Feynman diagram (the reader is recommended to prove it directly by using equations of motion for Dirac spinors in external fermion lines). Of course, the crucial underlying fact is that in a QED interaction vertex, the electromagnetic four-potential is coupled to conserved current.

After this somewhat long introduction, we are now in a position to examine an appropriate class of covariant gauges for the GWS electroweak

[9]It is instructive to note that for the original Maxwell Lagrangian (7.112), which corresponds formally to the limit $\alpha \to \infty$ in (7.122), the coefficient matrix L in (7.126) would become $-q^2 g^\mu_\rho + q^\mu q_\rho$ and this is singular, i.e. has no inverse. Of course, such a "pathological" behaviour is due to the gauge invariance of $\mathscr{L}_M$. By adding $\mathscr{L}_{\text{g.f.}}$ one breaks the original gauge symmetry and the singularity is thus removed. From this point of view, $\mathscr{L}_{\text{g.f.}}$ can be understood as a simple device for fixing a propagator of the electromagnetic field in a consistent way.

theory. The scheme [49,78] we have in mind is conceptually rather similar to that discussed above, but there is also a significant difference in comparison with the QED case: the vector fields W and Z become massive through the Higgs mechanism and thus they are mixed, in a sense, with the would-be Goldstone bosons (that eventually disappear from the physical spectrum). The key idea of the original papers [49, 78] is to retain the unphysical Goldstone bosons as auxiliary fields in the Lagrangian and add subsequently a suitable gauge-fixing term to the original gauge invariant Lagrangian. For implementing this, let us start with the Higgs–Goldstone scalar doublet (6.41)

$$\Phi = \begin{pmatrix} \varphi^+ \\ \varphi^0 \end{pmatrix} = \begin{pmatrix} \varphi_1 + i\varphi_2 \\ \varphi_3 + i\varphi_4 \end{pmatrix} \tag{7.130}$$

and reparametrize it as

$$\Phi = \begin{pmatrix} -iw^+ \\ \frac{1}{\sqrt{2}}(v + H + iz) \end{pmatrix} \tag{7.131}$$

where H stands for the physical Higgs boson and the constant v has the familiar meaning. Of course, the hermitian conjugate $\Phi^\dagger$ is then written as

$$\Phi^\dagger = \left(iw^-, \ \frac{1}{\sqrt{2}}(v + H - iz) \right) \tag{7.132}$$

The auxiliary fields $w^\pm$ and z correspond to the would-be Goldstone bosons. Now, the Higgs part of the GWS Lagrangian (see (6.53)) reads[10]

$$\begin{aligned} \mathscr{L}_{\text{Higgs}} = {} & \Phi^\dagger \left(\overleftarrow{\partial}_\mu + igA^a_\mu \frac{\tau^a}{2} + \frac{1}{2} ig' B_\mu \right) \left(\overrightarrow{\partial}^\mu - igA^{b\mu} \frac{\tau^b}{2} - \frac{1}{2} ig' B^\mu \right) \Phi \\ & - \lambda \left(\Phi^\dagger \Phi - \frac{v^2}{2} \right)^2 \end{aligned} \tag{7.133}$$

When the expressions (7.131) and (7.132) are substituted into (7.133), the vacuum shift v yields mass terms for vector bosons and the gauge fields A^a_μ and B_μ can be replaced by their physical combinations $W^\pm_\mu$, Z_μ and A_μ in full analogy with what has been done in Section 6.4. Of course, the shift of the lower component of the Higgs doublet Φ would lead to mass terms *in*

---

[10]Throughout this discussion, we are ignoring the fermionic sector of the GWS Lagrangian as this has no impact on the problem of gauge fixing. We will retrieve the interactions of fermions later on.

any parametrization, but an advantage of the representation (7.131) over (7.130) is that $w^\pm$ and z are, technically, direct counterparts of the vector fields $W_\mu^\pm$ and Z_μ. This becomes clear when one considers the relevant mixing terms, i.e. terms bilinear in scalar and vector fields that show up in the gauge invariant Lagrangian (7.133) upon substitutions (7.131), (7.132). A straightforward (though somewhat tedious) calculation yields the result

$$(D_\mu\Phi)^\dagger(D^\mu\Phi) = m_W(\partial^\mu w^- W_\mu^+ + \partial^\mu w^+ W_\mu^-) + m_Z \partial^\mu z Z_\mu + \cdots \tag{7.134}$$

where "$\cdots$" denotes all remaining contributions to (7.133), including mass terms and true interactions. Now, one has to add a gauge-fixing term. It is clearly desirable to get rid of the mixing contributions (7.134), since the quadratic part of the Lagrangian should be diagonal in order to define a conventional perturbation theory. An appropriate choice of $\mathscr{L}_{\text{g.f.}}$ can indeed do the job. To see this, let us define

$$\begin{aligned}\mathscr{L}_{\text{g.f.}} = &-\frac{1}{2\xi}|\partial\cdot W^- - \xi m_W w^-|^2 - \frac{1}{2\xi}|\partial\cdot W^+ - \xi m_W w^+|^2 \\ &-\frac{1}{2\eta}(\partial\cdot Z - \eta m_Z z)^2 - \frac{1}{2\alpha}(\partial\cdot A)^2\end{aligned} \tag{7.135}$$

where the ξ, η and α are arbitrary real parameters. The expression (7.135) can immediately be recast as

$$\begin{aligned}\mathscr{L}_{\text{g.f.}} = &-\frac{1}{\xi}(\partial\cdot W^- - \xi m_W w^-)(\partial\cdot W^+ - \xi m_W w^+) \\ &-\frac{1}{2\eta}(\partial\cdot Z - \eta m_Z z)^2 - \frac{1}{2\alpha}(\partial\cdot A)^2\end{aligned} \tag{7.136}$$

which, in turn, is easily worked out as

$$\begin{aligned}\mathscr{L}_{\text{g.f.}} = &-\frac{1}{\xi}(\partial\cdot W^-)(\partial\cdot W^+) + m_W w^+ \partial\cdot W^- + m_W w^- \partial\cdot W^+ \\ &- \xi m_W^2 w^- w^+ - \frac{1}{2\eta}(\partial\cdot Z)^2 + m_Z z\partial\cdot Z - \frac{1}{2}\eta m_Z^2 z^2 \\ &-\frac{1}{2\alpha}(\partial\cdot A)^2\end{aligned} \tag{7.137}$$

When this is combined with (7.134), one gets

$$\begin{aligned}\mathscr{L}_{\text{Higgs}} + \mathscr{L}_{\text{g.f.}} = &\, m_W w^+ \partial\cdot W^- + m_W w^- \partial\cdot W^+ + m_Z z\partial\cdot Z \\ &+ m_W W^{-\mu}\partial_\mu w^+ + m_W W^{+\mu}\partial_\mu w^- + m_Z Z^\mu \partial_\mu z + \cdots\end{aligned} \tag{7.138}$$

where we have singled out explicitly only the total contribution to the scalar — vector boson mixing, suppressing the other terms for the moment. Obviously, the last expression can be recast as

$$\mathscr{L}_{\text{Higgs}} + \mathscr{L}_{\text{g.f.}} = m_W \partial_\mu (w^+ W^{-\mu}) + m_W \partial_\mu (w^- W^{+\mu}) + m_Z \partial_\mu (z Z^\mu) + \cdots \tag{7.139}$$

Thus, the bilinear terms in question are combined into four-divergences and therefore can be discarded from the Lagrangian.[11]

Let us now analyze the remaining quadratic terms in the considered Lagrangian. First, for the scalar fields, one gets

$$\begin{aligned}\mathscr{L}_{\text{Higgs}} + \mathscr{L}_{\text{g.f.}} = \partial_\mu w^- \partial^\mu w^+ &+ \frac{1}{2}\partial_\mu z \partial^\mu z + \frac{1}{2}\partial_\mu H \partial^\mu H \\ &- \xi m_W^2 w^- w^+ - \frac{1}{2}\eta m_Z^2 z^2 - \lambda v^2 H^2 + \cdots\end{aligned} \tag{7.140}$$

Note that the kinetic terms for $w^\pm$, z and H (as well as the H mass term) descend from $\mathscr{L}_{\text{Higgs}}$ (using (7.133) and (7.131), (7.132)), while the mass terms for $w^\pm$ and z originate from the $\mathscr{L}_{\text{g.f.}}$ (see (7.137)). The result (7.140) means that the masses (or, more accurately, "mass parameters") of $w^\pm$ and z can be identified as

$$m_{w^\pm}^2 = \xi m_W^2, \quad m_z^2 = \eta m_Z^2 \tag{7.141}$$

while the H mass is given by $m_H^2 = 2\lambda v^2$ as before (see (6.51)). The dependence of the "masses" (7.141) on the gauge parameters ξ and η reflects clearly the unphysical nature of $w^\pm$ and z. Next, collecting all quadratic terms for vector fields (including also the kinetic term as given by Eq. (5.25)), one has

$$\begin{aligned}&\mathscr{L}_{\text{gauge}}^{(\text{kin.})} + \mathscr{L}_{\text{Higgs}} + \mathscr{L}_{\text{g.f.}} \\ &= -\frac{1}{2}W_{\mu\nu}^- W^{+\mu\nu} - \frac{1}{\xi}(\partial\cdot W^-)(\partial\cdot W^+) + m_W^2 W_\mu^- W^{+\mu} \\ &\quad - \frac{1}{4}Z_{\mu\nu}Z^{\mu\nu} - \frac{1}{2\eta}(\partial\cdot Z)^2 + \frac{1}{2}m_Z^2 Z_\mu Z^\mu \\ &\quad - \frac{1}{4}A_{\mu\nu}A^{\mu\nu} - \frac{1}{2\alpha}(\partial\cdot A)^2 + \cdots\end{aligned} \tag{7.142}$$

[11]Let us remind the reader that a term of the form $\partial_\mu X^\mu$ in a Lagrangian density does not contribute to the action and thus does not influence the dynamical contents (the equations of motion) of the theory.

Here, the mass terms are produced by $\mathscr{L}_{\text{Higgs}}$ while the additional derivative contributions dependent on ξ, η and α are obviously due to $\mathscr{L}_{\text{g.f.}}$. Of course, the part corresponding to the electromagnetic field is the same as in pure QED, as expected.

To sum up the preceding considerations, we have fixed the relevant free-field Lagrangian, which is a prerequisite for defining the perturbation expansion. From (7.142), one can derive equations of motion in a standard manner and, subsequently, the propagators for $W_\mu^\pm$ and Z_μ are determined as the corresponding Green's functions. To this end, we employ the same technique as in the case of covariant photon propagator (7.128) (phrased in a common jargon, one has to "invert the quadratic part of the Lagrangian"). For the W boson, we thus get

$$D_{\mu\nu}^{(\xi)}(q; m_W) = \frac{1}{q^2 - m_W^2 + i\varepsilon}\left[-g_{\mu\nu} + (1-\xi)\frac{q_\mu q_\nu}{q^2 - \xi m_W^2}\right] \tag{7.143}$$

and, similarly, the Z propagator becomes

$$D_{\mu\nu}^{(\eta)}(q; m_Z) = \frac{1}{q^2 - m_Z^2 + i\varepsilon}\left[-g_{\mu\nu} + (1-\eta)\frac{q_\mu q_\nu}{q^2 - \eta m_Z^2}\right] \tag{7.144}$$

The propagators of the unphysical Goldstone bosons follow immediately from (7.140); one has

$$\begin{aligned} D_w^{(\xi)} &= \frac{1}{q^2 - \xi m_W^2 + i\varepsilon} \\ D_z^{(\eta)} &= \frac{1}{q^2 - \eta m_Z^2 + i\varepsilon} \end{aligned} \tag{7.145}$$

for $w^\pm$ and z, respectively.

An astute reader may observe that in the limit $\xi, \eta \to \infty$, one recovers the U-gauge results, namely

$$\lim_{\xi\to\infty} D_{\mu\nu}^{(\xi)}(q; m_W) = \frac{1}{q^2 - m_W^2 + i\varepsilon}\left[-g_{\mu\nu} + \frac{1}{m_W^2} q_\mu q_\nu\right] \tag{7.146}$$

and

$$\lim_{\eta\to\infty} D_{\mu\nu}^{(\eta)}(q; m_Z) = \frac{1}{q^2 - m_Z^2 + i\varepsilon}\left[-g_{\mu\nu} + \frac{1}{m_Z^2} q_\mu q_\nu\right] \tag{7.147}$$

Further, from (7.145), it is obvious that the propagators of $w^\pm$ and z vanish identically for $\xi, \eta \to \infty$. This is gratifying (as a consistency check of our formalism), since the unphysical Goldstone bosons are absent in the U-gauge by definition.

One should also note that the gauge fixing for the electromagnetic field is obviously "decoupled" from the procedure used for the massive vector bosons W and Z, as there is no unphysical Goldstone boson associated with the massless photon. In particular, the covariant term $\frac{1}{2\alpha}(\partial \cdot A)^2$ can either be employed as a part of the scheme (7.135) or added directly to the U-gauge Lagrangian; of course, one is free to use a non-covariant gauge for the photon as well.

The above-described procedure based on adding the terms (7.135) to the GWS gauge invariant Lagrangian defines the class of the so-called ***R*-gauges** (or R_ξ-gauges), where "R" stands for "renormalizable". Such a label refers to the fact that the massive vector boson propagators (7.143) and (7.144) fall off as $1/q^2$ for $q^2 \to \infty$, which in turn means that the theory is of renormalizable type (this is indicated by the usual power-counting analysis based on an evaluation of the index of divergence of a general Feynman graph).[12] From the technical point of view, the formulation of the R-gauges was a real breakthrough as it played a key role in the proof of renormalizability of gauge theories with the Higgs mechanism accomplished first by 't Hooft and Veltman [48]. A more detailed discussion of this problem would go far beyond the scope of this treatment, but at least one important note is in order here. The R-gauges are instrumental in taming the ultraviolet divergences of higher-order Feynman diagrams (that are hard to control within the U-gauge formulation), but there is a price to be paid for that. In particular, the presence of the unphysical fields $w^\pm$ and z requires special care when proving unitarity of the S-matrix and its independence on the gauge-fixing parameters; this, in fact, was the main issue of the pioneering works [48].

In this context, it should be stressed that our discussion of the R-gauges has been incomplete, in the technical sense, since eventually one must also add another set of unphysical fields — the so-called Faddeev–Popov (FP) ghosts. These fictitious particles enter only closed loops of internal lines in Feynman diagrams and are necessary for maintaining the unitarity of the S-matrix in higher orders. Note that they are absent in the U-gauge.

[12]For the calculation of the "index" of a one-particle irreducible Feynman graph (also called "superficial degree of divergence") within a general field theory model, see, e.g., [ItZ] or Appendix G in [Hor].

In the present text, we will not need FP ghosts in any of our calculations and therefore we are not going to pursue this technical issue any further. A detailed discussion of the FP term in the GWS electroweak Lagrangian can be found, e.g., in [BaL].

Finally, let us add a terminological remark. In his original paper [49], 't Hooft proposed the gauge corresponding to $\xi = \eta = 1$ in (7.143) and (7.144). In a sense, this is the simplest possible choice, since the $q_\mu q_\nu$ parts of the W and Z propagators are then absent in analogy with the Feynman gauge for the photon propagator. Thus, such an option is usually called the **'t Hooft–Feynman gauge**. Similarly, for $\xi = \eta = 0$, one gets purely transverse vector boson propagators and this case is therefore referred to as the **'t Hooft–Landau gauge**. Note also that in the 't Hooft–Feynman gauge, the unphysical scalars $w^\pm$ and z have the same masses as $W^\pm$ and Z, respectively, while in the 't Hooft–Landau gauge, they become massless. In the general case, there is a common practice to take the same gauge for W and Z, i.e. set $\xi = \eta$.

7.7 Gauge independence of scattering amplitudes: An example

Having set up the basic framework of the R-gauge formulation of the electroweak standard model, we should now demonstrate, at least on an elementary example, that physical scattering amplitudes are independent of the gauge-fixing parameters ξ, η and α. To this end, we shall first collect the relevant interaction terms contained in the R-gauge GWS Lagrangian. It is clear *a priori* that the total number of the R-gauge interaction vertices must be considerably larger than that within the U-gauge (just because the unphysical scalars $w^\pm$ and z enter the game), so we shall proceed step by step.

The Yang–Mills term $\mathscr{L}_{\text{gauge}}$ (cf. (5.10))

$$\mathscr{L}_{\text{gauge}} = -\frac{1}{4} F^a_{\mu\nu} F^{a\mu\nu} - \frac{1}{4} B_{\mu\nu} B^{\mu\nu} \tag{7.148}$$

involves only the vector boson fields and, consequently, its form in any of the R-gauges is the same as in the U-gauge. Next, let us consider the interactions descending from $\mathscr{L}_{\text{Higgs}}$,

$$\mathscr{L}_{\text{Higgs}} = (D^\mu \Phi)^\dagger (D_\mu \Phi) - \lambda \left(\Phi^\dagger \Phi - \frac{v^2}{2} \right)^2 \tag{7.149}$$

(see (7.133)). When the covariant derivative D_μ is expressed in terms of the physical vector fields $W_\mu^\pm$, Z_μ and A_μ (see (7.7) and (7.8)), one has

$$D_\mu \Phi = \left[\partial_\mu - \frac{1}{2} ig W_\mu^- \tau^- - \frac{1}{2} ig W_\mu^+ \tau^+ - \frac{1}{2} ig(\cos\theta_W Z_\mu + \sin\theta_W A_\mu)\tau^3 \right.$$
$$\left. - \frac{1}{2} ig'(-\sin\theta_W Z_\mu + \cos\theta_W A_\mu) \cdot \mathbb{1} \right] \Phi \tag{7.150}$$

and

$$(D^\mu \Phi)^\dagger = \Phi^\dagger \left[\overleftarrow{\partial}^\mu + \frac{1}{2} ig W^{+\mu} \tau^+ + \frac{1}{2} ig W^{-\mu} \tau^- + \frac{1}{2} ig \right.$$
$$\times (\cos\theta_W Z^\mu + \sin\theta_W A^\mu)\tau^3$$
$$\left. + \frac{1}{2} ig'(-\sin\theta_W Z^\mu + \cos\theta_W A^\mu) \cdot \mathbb{1} \right] \tag{7.151}$$

with $\tau^\pm = \frac{1}{\sqrt{2}}(\tau^1 \pm i\tau^2)$ (see (5.16)). Substituting there the explicit matrix representation

$$\tau^+ = \sqrt{2}\begin{pmatrix} 0 & 1 \\ 0 & 0 \end{pmatrix}, \quad \tau^- = \sqrt{2}\begin{pmatrix} 0 & 0 \\ 1 & 0 \end{pmatrix}, \quad \tau^3 = \begin{pmatrix} 1 & 0 \\ 0 & -1 \end{pmatrix} \tag{7.152}$$

and using (7.131) and (7.132) for Φ and $\Phi^\dagger$, respectively, the term $(D^\mu\Phi)^\dagger(D_\mu\Phi)$ in (7.149) yields a set of interactions involving vector bosons and the scalars $w^\pm$, z, H (including, of course, the couplings WWH, $WWHH$, ZZH and $ZZHH$ encountered earlier in the U-gauge). Further, from the Goldstone potential in (7.149), one obtains a set of purely scalar interactions (including the H self-couplings occurring already in the U-gauge formulation). The scalar — vector and/or pure scalar interactions obtained in this way are trilinear or quadrilinear in the fields involved. Their complete list is rather long and can be found (including explicit formulae for the corresponding Feynman rules), e.g., in [BaL]. Here, we will only summarize, for the purpose of an illustration, the types of trilinear interaction terms appearing in an R-gauge (in addition to those already known from the U-gauge treatment).

In the usual schematic notation, the couplings in question can be grouped naturally in three subsets:

(i) w^-W^+Z, w^+W^-Z, w^+w^-Z, w^-W^+z, w^+W^-z
(ii) $w^-W^+\gamma$, $w^+W^-\gamma$, $w^+w^-\gamma$
(iii) w^-W^+H, w^+W^-H, w^+w^-H, zZH, zzH

There is an obvious mnemonic rule for arriving at such a scheme. One can take the trilinear interactions of vector bosons with themselves or with H, i.e. couplings WWZ, $WW\gamma$, WWH and ZZH, and replace consecutively W and Z by w and z. Note, however, that a coupling W^+W^-z is missing in the above catalogue (it is easy to realize that such terms exactly cancel when working out the expression $(D^\mu\Phi)^\dagger(D_\mu\Phi)$).

As an explicit example of a particular R-gauge interaction term involving vector bosons, let us consider the coupling of the w^+w^- pair to the photon or Z boson. From $\mathscr{L}_{\text{Higgs}}$, one gets, after some algebraic manipulations,

$$\begin{aligned}&\mathscr{L}_{w^+w^-\gamma} + \mathscr{L}_{w^+w^-Z}\\ &\quad = iew^-\overleftrightarrow{\partial}^\mu w^+ A_\mu + i\frac{g}{\cos\theta_W}\left(\frac{1}{2} - \sin^2\theta_W\right) w^-\overleftrightarrow{\partial}^\mu w^+ Z_\mu\end{aligned} \tag{7.153}$$

where we have utilized the familiar relations $g' = g\tan\theta_W$ and $e = g\sin\theta_W$.

An explicit evaluation of the purely scalar interactions is quite straightforward. Using (7.131) and (7.132), one immediately gets

$$\Phi^\dagger\Phi = w^+w^- + \frac{1}{2}[(v+H)^2 + z^2] \tag{7.154}$$

and then

$$\lambda\left(\Phi^\dagger\Phi - \frac{v^2}{2}\right)^2 = \lambda\left(w^+w^- + vH + \frac{1}{2}H^2 + \frac{1}{2}z^2\right)^2 \tag{7.155}$$

From the last expression, the individual scalar interactions can be read off easily; in particular, the interaction w^+w^-H is seen to be described by the Lagrangian

$$\mathscr{L}_{w^+w^-H} = -2\lambda v\; w^+w^-H \tag{7.156}$$

Note that the coupling constant in the last expression can be recast as

$$g_{w^+w^-H} = -\frac{m_H^2}{v} = -\frac{1}{2}g\frac{m_H^2}{m_W} \tag{7.157}$$

when one makes use of the relations $m_H^2 = 2\lambda v^2$ and $m_W = \frac{1}{2}gv$.

Let us now turn to the fermion sector of the GWS standard model. The interactions of vector bosons with fermions obviously remain intact when passing from the U-gauge to the R-gauge. On the other hand, the Yukawa couplings involve the Higgs doublet Φ and thus can produce interactions of fermions with unphysical Goldstone bosons. Below, we shall examine these new fermionic interaction terms in detail. For simplicity, we restrict ourselves to leptons (in the quark sector, one would get similar results, but the flavour mixing must be properly taken into account). Considering an arbitrary lepton type ℓ ($\ell = e, \mu, \tau$) and ignoring the right-handed component of neutrino field, the relevant Yukawa-type Lagrangian is written down as

$$\mathscr{L}^{(\ell)}_{\text{Yukawa}} = -h_\ell \bar{L}^{(\ell)} \Phi \ell_R + \text{h.c.} \tag{7.158}$$

(see (6.82)). Using the representation (7.131) for Φ, (7.158) becomes

$$\begin{aligned}\mathscr{L}^{(\ell)}_{\text{Yukawa}} &= i h_\ell \bar{\nu}_{\ell L} \ell_R w^+ - \frac{i}{\sqrt{2}} h_\ell \bar{\ell}_L \ell_R z - \frac{1}{\sqrt{2}} h_\ell \bar{\ell}_L \ell_R H \\ &\quad - \frac{1}{\sqrt{2}} h_\ell v \bar{\ell}_L \ell_R + \text{h.c.}\end{aligned} \tag{7.159}$$

Now, making an obvious identification $m_\ell = \frac{1}{\sqrt{2}} h_\ell v$ (cf. (6.84)) and using the familiar relation $m_W = \frac{1}{2} g v$, the interaction part of the expression (7.159) is recast as

$$\begin{aligned}&\mathscr{L}_{\nu\ell w} + \mathscr{L}_{\ell\ell z} + \mathscr{L}_{\ell\ell H} \\ &\quad = i \frac{g}{2\sqrt{2}} \frac{m_\ell}{m_W} \bar{\nu}_\ell (1 + \gamma_5) \ell w^+ - i \frac{g}{2\sqrt{2}} \frac{m_\ell}{m_W} \bar{\ell} (1 - \gamma_5) \nu_\ell w^- \\ &\quad\quad - i \frac{g}{2} \frac{m_\ell}{m_W} \bar{\ell} \gamma_5 \ell z - \frac{g}{2} \frac{m_\ell}{m_W} \bar{\ell} \ell H\end{aligned} \tag{7.160}$$

Thus, we have arrived at a complete description of the anticipated R-gauge couplings of $w^\pm$ and z to leptons. Note that we have also recovered (as expected) our earlier U-gauge result for the coupling $\ell\ell H$.

Now, we are ready to examine the gauge independence of an appropriate tree-level scattering amplitude. The particular illustrative example we are going to discuss is the process $e^+e^- \to \mu^+\mu^-$. In the U-gauge, it is described by the Feynman graphs shown in Fig. 7.1 and the relevant R-gauge diagrams are depicted in Fig. 7.2. To begin with, it should be remembered that fixing of the gauge for the photon propagator is always done separately and does not depend on whether we are working in the U-gauge or the R-gauge for W and Z. In fact, the photon-exchange contribution described by the graph in Fig. 7.1(a) is gauge-independent by itself (i.e. does not depend on the gauge-fixing parameter α); as we have noted in the preceding section, this is due to the conservation of the electromagnetic vector current. Further, the contribution of the Higgs boson exchange is the same in all gauges.

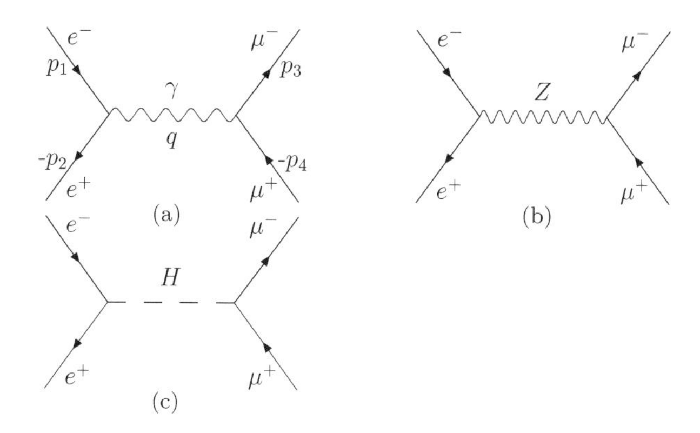

Figure 7.1. Tree diagrams for the process $e^+e^- \to \mu^+\mu^-$ in the U-gauge.

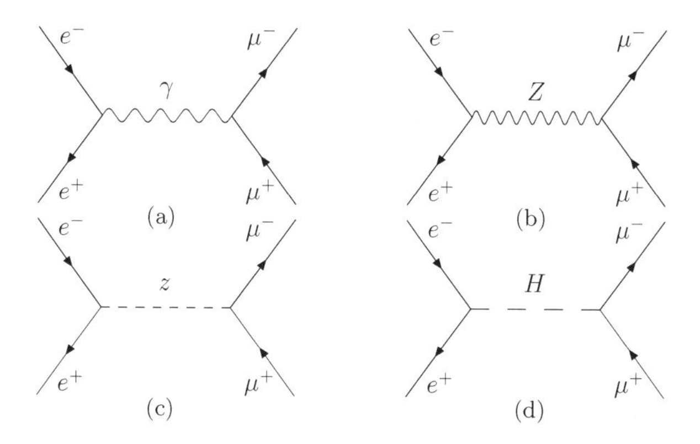

Figure 7.2. Tree diagrams for $e^+e^- \to \mu^+\mu^-$ in an R-gauge.

Thus, it remains to be shown that the sum of the diagrams (b) and (c) in Fig. 7.2 is independent of the relevant gauge-fixing parameter and that it is equal to the U-gauge graph in Fig. 7.1(b).

To see this, let us write down explicitly the matrix elements in question. In an obvious notation, the contributions of the R_ξ-gauge graphs in Fig. 7.2 are given by

$$\begin{aligned} i\mathcal{M}_\xi^{(Z)} &= i^3 \frac{1}{4}\left(\frac{g}{\cos\theta_W}\right)^2 [\bar{v}(p_2)\gamma_\mu(v - a\gamma_5)u(p_1)][\bar{u}(p_3)\gamma_\nu(v - a\gamma_5)v(p_4)] \\ &\quad \times \frac{1}{q^2 - m_Z^2}\left[-g^{\mu\nu} + (1-\xi)\frac{q^\mu q^\nu}{q^2 - \xi m_Z^2}\right] \end{aligned} \tag{7.161}$$

and

$$i\mathcal{M}_\xi^{(z)} = i^3\left(-i\frac{g}{2}\frac{m_e}{m_W}\right)\left(-i\frac{g}{2}\frac{m_\mu}{m_W}\right)[\bar{v}(p_2)\gamma_5 u(p_1)][\bar{u}(p_3)\gamma_5 v(p_4)] \times \frac{1}{q^2 - \xi m_Z^2} \tag{7.162}$$

respectively. Note that the vector and axial coupling parameters v and a appearing in (7.161) are

$$\begin{aligned} v &= \varepsilon_L + \varepsilon_R = -\frac{1}{2} + 2\sin^2\theta_W \\ a &= \varepsilon_L - \varepsilon_R = -\frac{1}{2} \end{aligned} \tag{7.163}$$

(cf. (7.11) and (5.58)). It is convenient to split the amplitude (7.161) as

$$\mathcal{M}_\xi^{(Z)} = \mathcal{M}_{\text{diag.}}^{(Z)} + \mathcal{M}_{\text{long.}}^{(Z)} \tag{7.164}$$

where the two terms in (7.164) correspond to the diagonal and longitudinal part of the Z propagator, respectively, i.e.

$$\mathcal{M}_{\text{diag.}}^{(Z)} = -\frac{1}{4}\left(\frac{g}{\cos\theta_W}\right)^2[\bar{v}(p_2)\gamma_\mu(v - a\gamma_5)u(p_1)][\bar{u}(p_3)\gamma_\nu(v - a\gamma_5)v(p_4)] \times \frac{-g^{\mu\nu}}{q^2 - m_Z^2} \tag{7.165}$$

and

$$\mathcal{M}_{\text{long.}}^{(Z)} = -\frac{1}{4}\left(\frac{g}{\cos\theta_W}\right)^2[\bar{v}(p_2)\not{q}(v - a\gamma_5)u(p_1)][\bar{u}(p_3)\not{q}(v - a\gamma_5)v(p_4)] \times \frac{1-\xi}{(q^2 - m_Z^2)(q^2 - \xi m_Z^2)} \tag{7.166}$$

Obviously, the $\mathcal{M}_{\text{diag.}}^{(Z)}$ is obtained from (7.161) by setting there $\xi = 1$, i.e. it coincides with the $\mathcal{M}_\xi^{(Z)}$ value in the 't Hooft–Feynman gauge:

$$\mathcal{M}_{\text{diag.}}^{(Z)} = \mathcal{M}_{\xi=1}^{(Z)} = \mathcal{M}_{\text{tHF}}^{(Z)} \tag{7.167}$$

Now, taking into account that $q = p_1 + p_2 = p_3 + p_4$, one can employ the equations of motion for the Dirac spinors in (7.166). The vector component of the weak neutral current is conserved, so a non-trivial contribution can only arise from its axial-vector part. In particular, one has

$$\begin{aligned}\bar{v}(p_2)\not{q}\gamma_5 u(p_1) &= -2m_e \bar{v}(p_2)\gamma_5 u(p_1)\\ \bar{u}(p_3)\not{q}\gamma_5 v(p_4) &= +2m_\mu \bar{u}(p_3)\gamma_5 v(p_4)\end{aligned}\tag{7.168}$$

Thus, (7.166) becomes

$$\begin{aligned}\mathcal{M}^{(Z)}_{\text{long.}} &= \frac{g^2}{4m_W^2} m_e m_\mu [\bar{v}(p_2)\gamma_5 u(p_1)][\bar{u}(p_3)\gamma_5 v(p_4)]\\ &\quad\times \frac{m_Z^2(1-\xi)}{(q^2-m_Z^2)(q^2-\xi m_Z^2)}\end{aligned}\tag{7.169}$$

where we have also taken into account (7.163) and the Weinberg mass relation $\cos^2\theta_W = m_W^2/m_Z^2$. With the above results at hand, the sum $\mathcal{M}^{(Z)}_{\text{long.}} + \mathcal{M}^{(z)}_\xi$ can be written as

$$\begin{aligned}\mathcal{M}^{(Z)}_{\text{long.}} + \mathcal{M}^{(z)}_\xi &= \frac{g^2}{4m_W^2} m_e m_\mu [\bar{v}(p_2)\gamma_5 u(p_1)][\bar{u}(p_3)\gamma_5 v(p_4)]\\ &\quad\times \left(\frac{m_Z^2(1-\xi)}{(q^2-m_Z^2)(q^2-\xi m_Z^2)} + \frac{1}{q^2-\xi m_Z^2}\right)\end{aligned}\tag{7.170}$$

It is clear that the ξ-dependent terms drop out from (7.170) and one gets

$$\mathcal{M}^{(Z)}_{\text{long.}} + \mathcal{M}^{(z)}_\xi = \frac{g^2}{4m_W^2}[\bar{v}(p_2)\gamma_5 u(p_1)][\bar{u}(p_3)\gamma_5 v(p_4)]\frac{1}{q^2-m_Z^2}\tag{7.171}$$

However, the last expression is seen to coincide with (7.162) for $\xi = 1$; in other words, (7.171) is equal to the z-exchange contribution in the 't Hooft–Feynman gauge. Then, taking into account also (7.164) and (7.167), one can write

$$\mathcal{M}^{(Z)}_\xi + \mathcal{M}^{(z)}_\xi = \mathcal{M}^{(Z)}_{\xi=1} + \mathcal{M}^{(z)}_{\xi=1} = \mathcal{M}^{(Z)}_{\text{tHF}} + \mathcal{M}^{(z)}_{\text{tHF}}\tag{7.172}$$

and the gauge independence of $\mathcal{M}_\xi(e^+e^- \to \mu^+\mu^-)$ is thereby proved within the class of R_ξ-gauges. The proof of an equivalence of 't Hooft–Feynman gauge and the U-gauge for the considered process goes along the same lines as the above derivation and we leave it to the reader as a straightforward exercise.

Thus, we have been able to prove the gauge independence of the tree-level amplitude in question in an elementary and rather transparent way. In the present context, a cancellation mechanism for the ξ-dependent terms seems to be clear: there is an obvious correlation between the contribution of an unphysical Goldstone boson and that of the longitudinal part of a massive vector boson propagator, which yields the desired compensation (recall that the location of the ξ-dependent extra pole in an R-gauge vector propagator coincides with that of the related unphysical scalar). Technically, an essential ingredient of our calculation has been the "partial conservation" of weak currents (see (7.168)), following simply from the equations of motion for external particles. A generalization of such an analysis to other tree-level scattering amplitudes would be quite straightforward. However, let us emphasize that proving the gauge independence of the S-matrix to all orders of perturbation theory is a highly non-trivial task. In order to accomplish such a goal, some advanced quantum field theory techniques are needed; in particular, one has to employ all relevant Ward identities that express the contents of the original gauge symmetry at quantum level (in fact, our elementary calculation exemplifies how the Ward identities work in the lowest order). Concerning these topics, an interested reader is referred either to the original papers [49, 79] or to the monographs [BaL], [Wei], [Pok].

7.8 Equivalence theorem for longitudinal vector bosons

We have already noted earlier (cf. the end of Section 6.3) that within a gauge theory with the Higgs mechanism, one should expect an intimate dynamical connection between the unphysical Goldstone bosons and longitudinally polarized vector bosons. In a nutshell, such an expectation relies — rather intuitively — on the fact that the physical longitudinal mode of a massive vector boson emerges in place of a would-be Goldstone boson "eaten" by the corresponding (originally massless) gauge field. Now, we are in a position to formulate the relevant "equivalence theorem" in more definite terms. Before doing it, we would like to discuss two instructive examples involving some specific decay and scattering processes.

First, let us consider the decay of a heavy Higgs boson ($m_H \gg m_W$) at rest into a pair of longitudinally polarized vector bosons $W^\pm$. In lowest order, the process is described by the diagram shown in Fig. 7.3. The corresponding amplitude is then

$$\mathcal{M}(H \to W_L^+ W_L^-) = g m_W \varepsilon_L^\mu(k) \varepsilon_{L\mu}(p) \tag{7.173}$$

where we have taken into account the form of the WWH coupling given by (6.76) (plus the fact that the polarization vectors are real). As we know,

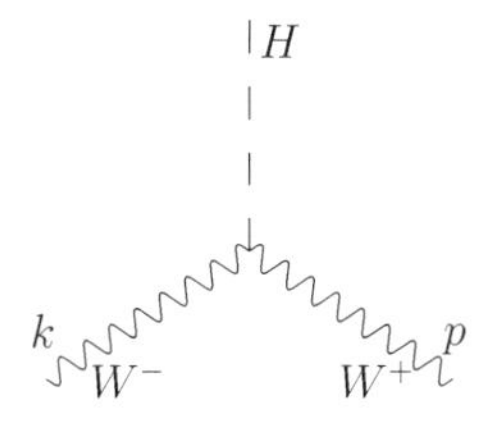

Figure 7.3. Tree-level graph for the decay $H \to W^+W^-$.

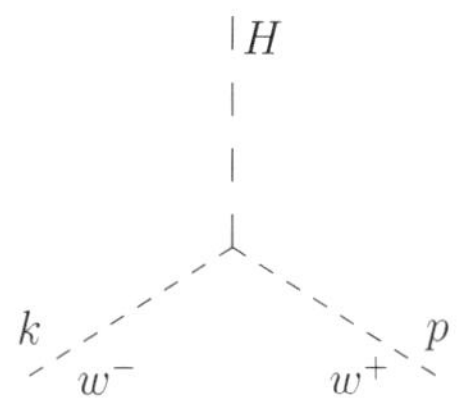

Figure 7.4. Tree-level Feynman graph for the unphysical process $H \to w^+w^-$.

a longitudinal polarization vector can be split as

$$\varepsilon_L^\mu(p) = \frac{1}{m_W} p^\mu + \Delta^\mu(p) \tag{7.174}$$

with $\Delta^\mu(p)$ being of the order $O(m_W/E_W)$ in high-energy limit (cf. (3.29)). In our case, $E_W = \frac{1}{2}m_H$, and thus, $\Delta^\mu(p) \ll 1$ according to the above assumption. Decomposing the polarization vectors in (7.173) according to (7.174), one gets

$$\begin{aligned}
\mathcal{M}(H \to W_L^+ W_L^-) &= g m_W \left(\frac{1}{m_W} k^\mu + \Delta^\mu(k)\right)\left(\frac{1}{m_W} p_\mu + \Delta_\mu(p)\right) \\
&= g m_W \left(\frac{1}{m_W^2} k \cdot p + O(1)\right) \\
&= \frac{g}{2}\frac{m_H^2}{m_W}\left(1 + O\left(\frac{m_W^2}{E_W^2}\right)\right)
\end{aligned} \tag{7.175}$$

Having in mind the envisaged connection between the $W_L^\pm$ and their scalar counterparts $w^\pm$, let us now calculate the amplitude describing formally the unphysical process $H \to w^+w^-$. The corresponding lowest order diagram is shown in Fig. 7.4.

Using our earlier results (7.156) and (7.157), one can write immediately

$$\mathcal{M}(H \to w^+ w^-) = -\frac{g}{2} \frac{m_H^2}{m_W} \tag{7.176}$$

Comparing it with the last line in (7.175), we thus have

$$\mathcal{M}(H \to W_L^+ W_L^-) = -\mathcal{M}(H \to w^+ w^-) \times \left[1 + O\left(\frac{m_W^2}{E_W^2}\right)\right] \tag{7.177}$$

This relation represents a simple explicit example of the equivalence theorem for longitudinal vector bosons and unphysical Goldstone bosons alluded to previously.

As a second example (that may be technically somewhat less trivial), we shall discuss the process $e^+e^- \to W_L^+ W_L^-$. For simplicity, we set $m_e = 0$ throughout the calculation. Then, the relevant U-gauge tree diagrams are those shown in Fig. 7.5 (see p. 126), since the contribution of the Higgs exchange graph vanishes in the considered approximation. In fact, the R-gauge graphs are the same as there is no zWW coupling. Gauge independence of the considered amplitude can be proved readily — it is a matter of a straightforward application of the 't Hooft identity (3.47) for the vertices $WW\gamma$ and WWZ. Further, it is easy to realize that in the chiral limit $m_e \to 0$ we have in mind, the matrix element in question can only be non-vanishing if e^+ and e^- have unlike helicities. For definiteness, let the electron be left-handed and the positron right-handed. Owing to the asymptotic behaviour of longitudinal polarization vectors (cf. (7.174)), contributions of the individual diagrams in Figs. 7.5(a)–(c) diverge in the high-energy limit. As we have already seen in Chapter 5, the leading (quadratic) divergences cancel in their sum (see (5.93)–(5.95)). The residual (linear) divergences are proportional to m_e and therefore entirely disappear in the approximation considered here. Thus, the graphs in Fig. 7.5 yield an amplitude that is asymptotically flat (i.e. satisfies the condition of tree unitarity); as a result of rather long and tedious calculation, one gets

$$\mathcal{M}(e^+e^- \to W_L^+ W_L^-) = \frac{1}{s} \bar{v}_R(l)(\not{p} - \not{r}) u_L(k) \left[e^2 + g^2 \left(-\frac{1}{2} + \sin^2\theta_W\right) \right.$$
$$\left. \times \left(\frac{m_Z^2}{2m_W^2} - 1\right)\right] + O\left(\frac{m_W^2}{s}\right) \tag{7.178}$$

In arriving at the last expression, the computational tricks employed earlier (see, in particular, Chapter 5) are instrumental; note also that another very

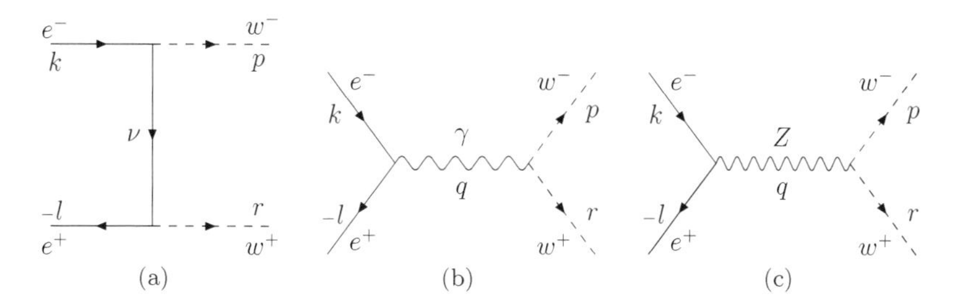

Figure 7.5. Tree-level diagrams for the unphysical process $e^+e^- \to w^+w^-$. The contribution of the diagram (a) vanishes for $m_e = 0$.

useful relation is

$$p \cdot \Delta(p) = -m_W \tag{7.179}$$

where $\Delta(p)$ is the remainder in (7.174) (this is an immediate consequence of the identities $p \cdot \varepsilon_L(p) = 0$ and $p^2 = m_W^2$). With these encouraging remarks, we leave a derivation of (7.178) as a challenge for a seriously interested reader.

Let us now consider the unphysical counterpart of the above process, namely $e^+e^- \to w^+w^-$. The corresponding lowest-order diagrams are depicted in Fig. 7.5.

The relevant interactions of $w^\pm$ are described by the Lagrangian

$$\begin{aligned}\mathscr{L}_{\text{int}}^{(w^\pm)} &= i\frac{g}{2\sqrt{2}}\frac{m_e}{m_W}\bar{\nu}_e(1+\gamma_5)ew^\pm - i\frac{g}{2\sqrt{2}}\frac{m_e}{m_W}\bar{e}(1-\gamma_5)\nu_e w^- \\ &\quad + iew^-\overleftrightarrow{\partial}_\mu w^+ A^\mu + i\frac{g}{\cos\theta_W}\left(\frac{1}{2} - \sin^2\theta_W\right) w^-\overleftrightarrow{\partial}_\mu w^+ Z^\mu\end{aligned} \tag{7.180}$$

(see (7.153) and (7.160)), which means that for $m_e = 0$, only the graphs (b) and (c) contribute. Obviously, the Feynman rules for the interaction vertices $w^+w^-\gamma$ and w^+w^-Z are essentially those of the ordinary scalar QED. Taking into account also the other familiar rules, the amplitude in question can thus be written as

$$\begin{aligned}\mathcal{M}(e^+e^- \to w^+w^-) &= e^2\bar{v}_R(l)\gamma_\mu u_L(k)\frac{-g^{\mu\nu}}{q^2}(p_\nu - r_\nu) \\ &\quad - \left(\frac{g}{\cos\theta_W}\right)^2\left(-\frac{1}{2} + \sin^2\theta_W\right)\left(\frac{1}{2} - \sin^2\theta_W\right) \\ &\quad \times \bar{v}_R(l)\gamma_\mu u_L(k)\frac{-g^{\mu\nu}}{q^2 - m_Z^2}(p_\nu - r_\nu)\end{aligned} \tag{7.181}$$

(we consider again a left-handed electron and right-handed positron). Note that the gauge independence of such a matrix element is proved easily, so in writing down (7.181), we have used the Feynman gauge both for photon and Z. From (7.181), one gets readily

$$\mathcal{M}(e^+e^- \to w^+w^-) = -\frac{1}{s}\bar{v}_R(l)\,(\not{p} - \not{r})\,u_L(k) \times\left[e^2 + \left(\frac{g}{\cos\theta_W}\right)^2\left(\frac{1}{2} - \sin^2\theta_W\right)^2\right] + O\left(\frac{m_W^2}{s}\right) \tag{7.182}$$

where we have set $q^2 = s$. Now, it is clear that when the formula $m_W/m_Z = \cos\theta_W$ is used in (7.178), one has

$$\mathcal{M}(e^+e^- \to W_L^+W_L^-) = -\mathcal{M}\left(e^+e^- \to w^+w^-\right) \times \left[1 + O\Big(\frac{m_W^2}{E_W^2}\Big)\right] \tag{7.183}$$

The analogy between the last relation and Eq. (7.177) is striking. Of course, (7.183) is another example of the equivalence theorem (ET) mentioned above. Let us now formulate a general statement of ET for tree-level matrix elements (for some original papers, see [50]).

Equivalence theorem: Let us consider a process involving, apart from other physical particles, a certain number of longitudinally polarized vector bosons V_L (i.e. $W_L^\pm$ and/or Z_L), with n_1 of them being in the initial state and n_2 in the final state. Let E_V denote generically the vector boson energies; for $E_V \gg m_W$, one then has

$$\begin{aligned}&\mathcal{M}_{fi}\big(V_L(i_1), \ldots, V_L(i_{n_1}), A \to V_L(f_1), \ldots, V_L(f_{n_2}), B\big)\\ &= \mathcal{M}_{fi}\big(\varphi(i_1), \ldots, \varphi(i_{n_1}), A \to \varphi(f_1), \ldots, \varphi(f_{n_2}), B\big)\\ &\times i^{n_1}(-i)^{n_2}\left[1 + O\left(\frac{m_V}{E_V}\right)\right]\end{aligned} \tag{7.184}$$

where φ's stand for the unphysical Goldstone scalar counterparts of V_L's, and A and B symbolize all other incoming and outgoing particles ■

Note that the precise form of the phase factor $i^{n_1}(-i)^{n_2}$ shown in (7.184) is due to the conventional definition of the Goldstone boson fields according to (7.131). Looking back at (7.177) and (7.183), the reader can see immediately that the phase factor contained in the general formula (7.184) is indeed recovered in our previous two examples (where $n_1 = 0$, $n_2 = 2$).

For completeness, a comment is in order here. As we noted, ET in the above form is certainly valid at the tree level. When going to higher orders of perturbation theory, the relation (7.184) gets slightly modified by including a finite renormalization factor $C = 1 + O(g^2)$ (with g denoting generically a gauge coupling constant) *independent of the energies*. Such a generalization of the ET is discussed in detail, e.g., in [80], where also some further references can be found. In fact, it turns out that in a suitable gauge, C can be made equal to unity in all orders.

The ET is a deep general result characteristic of any gauge theory with the Higgs mechanism. Technically, it is a consequence of the gauge symmetry expressed in terms of an appropriate Ward identity. A more detailed commentary concerning the ET proof would go beyond the scope of this text and the interested reader is referred to the literature. In particular, an introductory treatment can be found in [81], together with a comprehensive list of further references. Among other things, the process $e^+e^- \to W_L^+ W_L^-$ is reconsidered in [81] from the point of view of a relevant Ward identity, which provides some insight into the result (7.183) obtained here by means of a straightforward calculation. A brief discussion of the ET can also be found in [PeS] and [Don]; for a more sophisticated survey, see, e.g., [Dob].

It is important to realize that — with the equivalence theorem at hand — general validity of the tree unitarity within the GWS standard electroweak theory becomes quite clear. Indeed, as we know, there are two sources of a possible "bad" high-energy behaviour of the tree-level amplitudes: the U-gauge massive vector boson propagators (that contain a factor of m_V^{-2}) and the polarization vectors of external massive vector bosons (each bringing in a factor of m_V^{-1} in the high-energy limit). Now, one can rely on the gauge independence of physical scattering amplitudes and pass from the U-gauge to an R-gauge, where the "dangerous" parts of vector boson propagators are absent. Subsequently, using ET in the high-energy limit, one can replace the longitudinally polarized vector bosons by the corresponding unphysical Goldstone bosons. However, these are completely innocuous, as the Feynman graphs with external scalars obviously respect the constraints of tree unitarity (once there are no coupling constants with dimension of a negative power of mass). **In this way, one can see that because of the general validity of ET, tree-level unitarity is satisfied in any electroweak theory of renormalizable type.**

Finally, let us add a historical remark. A first proof of the tree unitarity for gauge theories with Higgs mechanism has been given by Bell [82] (at a time when ET has not been known yet) with the help of a different method. Bell's work has actually been a precursor to the papers [31,36,37], where the

GWS electroweak theory has been derived from the constraints of tree-level unitarity (see also [Hor]).

7.9 Effects of the ABJ anomaly

For a complete understanding of structural properties of the electroweak theory, one has to take into account another important concept of quantum field theory, namely the Adler–Bell–Jackiw (ABJ) anomaly [83]. This is a rather subtle phenomenon, which nevertheless plays a substantial role in the discussion of internal consistency and renormalizability of the GWS standard model. The ABJ anomaly reflects the peculiar behaviour of closed fermionic loops involving vector and axial-vector currents; from the technical point of view, it represents a violation of naive Ward identities for such Feynman graphs (for an introduction to the subject, see, e.g., [84] and the monograph [Ber]).

To elucidate the nature of possible effects due to the ABJ anomaly, let us resume the investigation of our recurrent theme — perturbative unitarity or "asymptotic softness" of scattering amplitudes.[13] Throughout the present text, we stressed repeatedly that the tree-level unitarity is a necessary condition for renormalizability in higher orders of perturbation expansion. The crucial point is that power-like growth of a scattering amplitude with energy (for a binary process $1 + 2 \to 3 + 4$) would propagate into higher-order diagrams, leading to an uncontrollable proliferation of divergences and subsequent loss of renormalizability [31]. Such an argument is essentially based on dispersion relations for Feynman diagrams and we shall now recapitulate it briefly.

Using the technique of dispersion relations, a scattering amplitude is evaluated through its imaginary part by means of a Cauchy-type integral and — depending on the asymptotic (high-energy) behaviour of the integrand — one eventually has to employ an appropriate number of subtractions in order to get a finite answer. As regards the imaginary parts, one should remember that these are expressed (via the S-matrix unitarity) in terms of products of the amplitudes in lower perturbative orders. For example, the imaginary part of a one-loop diagram can be represented as a square of a tree-level graph (obtained by cutting the internal lines of the closed loop), etc. All this means that the leading asymptotic energy dependence of a higher-order graph is given, roughly

[13]The reader may find it useful to look back into Section 3.1; in what follows, we are going to slightly generalize our earlier considerations.

speaking, by a product of contributions from lower orders. Thus, if the tree-level unitarity is violated, one can expect that the power-like growth of a considered scattering amplitude will get worse at higher orders (barring some accidental cancellations). Consequently, an indefinitely growing number of subtractions is needed to make the dispersion integrals convergent. In fact, the subtractions are tantamount to the renormalization counterterms and their infinite number means that the theory is not renormalizable in the usual perturbative sense.

On the other hand, if the tree unitarity holds, one can expect (naively) that the scattering amplitude in question remains sufficiently "soft" for $E \to \infty$ even at higher orders of perturbation theory[14]; in simple terms, the idea is (having in mind, e.g., a binary process) that through the successive multiplication of asymptotically flat matrix elements of lower order, one should get a result with the same high-energy behaviour. However, there is a snag. When calculating the full one-loop amplitude, one must perform — apart from algebraic manipulations — an integration over an energy variable in the relevant dispersion relation, and it might happen (in principle at least) that the result would behave differently than the basic tree-level amplitude. In particular, the real part of a one-loop amplitude could pick up a contribution, scaling as a positive power of energy for $E \to \infty$. This would mean that an originally expected chain of well-behaved (i.e. asymptotically soft) perturbative iterations breaks down: at the one-loop level and higher, one would face a rapid violation of unitarity, ending up with non-renormalizable perturbation series. In other words, while it seems to be true beyond any reasonable doubt that the tree-level unitarity is a *necessary* condition for renormalizability, one cannot be sure whether it is a *sufficient* condition as well. **It turns out that the tree unitarity indeed does not, in general, guarantee renormalizability.** The presaged "pathological" behaviour of one-loop scattering amplitudes is rather exceptional, but it does occur within some field theory models. As we shall explain below, its source is just the ABJ anomaly. Later on, we will also show how this potential problem is avoided within the GWS standard model.[15]

---

[14]As usual, E is a generic notation for a relevant energy variable, e.g., the total centre-of-mass energy of the considered process, $E_{\mathrm{c.m.}} = s^{1/2}$.

[15]For the purpose of the preceding heuristic discussion, we have invoked the method of dispersion relations, but we shall not pursue it any further. In our subsequent calculations, we simply utilize some particular results of a direct evaluation of one-loop

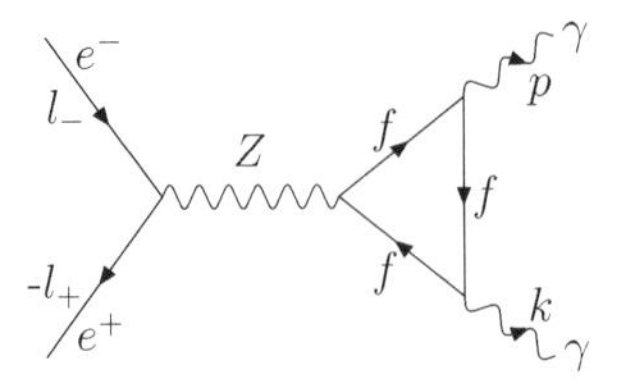

Figure 7.6. "Anomalous" contribution to the process $e^+e^- \to \gamma\gamma$ involving fermionic triangle loop. Adding an analogous diagram with crossed external photon lines is tacitly assumed.

As an illustrative example, let us consider the e^+e^- annihilation into two photons and, for definiteness, we shall first work in the U-gauge. As regards the high-energy behaviour of the corresponding amplitude, such a process is completely innocuous at the tree level, where it is represented by the familiar QED diagrams (these, of course, do not involve any "dangerous" components since the massless photons can only have transverse polarizations). In the one-loop approximation, there are many diagrams that contribute to the process in question and most of them simply reproduce the decent behaviour of tree-level matrix element (modified only by some logarithmic corrections). However, there is one exception, namely the graph shown in Fig. 7.6. The corresponding matrix element can be written as

$$\begin{aligned} i\mathcal{M}_\triangle = i^8(-1)\frac{1}{4}\left(\frac{g}{\cos\theta_W}\right)^2 \bar{v}(l_+)\gamma_\lambda(v_e - a_e\gamma_5)u(l_-) \\ \times \frac{-g^{\lambda\alpha} + m_Z^{-2}q^\lambda q^\alpha}{q^2 - m_Z^2} \cdot (-a_f)Q_f^2 e^2 T_{\alpha\mu\nu}(k,p)\varepsilon^{*\mu}(k)\varepsilon^{*\nu}(p) \end{aligned} \tag{7.185}$$

where the coupling parameters for neutral currents have the usual meaning (cf. (5.58) or (7.163)), Q_f is the charge factor for the fermion circulating in the loop (e.g., $Q_e = -1$, etc.) and $T_{\alpha\mu\nu}(k,p)$ stands for the triangle loop itself. Before specifying its form, note that in (7.185), we have already taken into account that only the axial-vector (A) part of the weak neutral current entering the triangle (in the vertex attached to the Z propagator) can give a non-vanishing contribution. The reason is that the electromagnetic currents

Feynman graphs. For the dispersion-relation approach to the ABJ anomaly in the present context, see, e.g., [Hor].

appearing in the other two vertices are of pure vector (V) nature and in combination with the vector part of the weak neutral current, one would get a VVV triangle as in spinor QED — but this is known to vanish identically according to the Furry's theorem.[16] Thus, the triangle loop appearing in Fig. 7.6 is of the VVA type and it is represented formally as

$$T_{\alpha\mu\nu}(k,p) = \int \frac{d^4l}{(2\pi)^4} \mathrm{Tr}\left(\frac{1}{\not{l} - \not{k} - m_f}\gamma_\mu \frac{1}{\not{l} - m_f}\gamma_\nu \frac{1}{\not{l} + \not{p} - m_f}\gamma_\alpha\gamma_5\right) + [(k,\mu) \leftrightarrow (p,\nu)] \tag{7.186}$$

where we have included the crossing of the external photon lines ("Bose symmetrization") indicated in Fig. 7.6. The usual factor of (-1) associated with any purely fermionic closed loop has already been incorporated into the overall factor in the expression (7.185) and the meaning of the integration variable (loop momentum l) is obvious. At first sight, the integral in (7.186) has an ultraviolet divergence and must be defined properly. A brief discussion of this issue, together with a succinct summary of basic properties of $T_{\alpha\mu\nu}$ can be found in Appendix E; below, we will only utilize some key relations that are substantial for the understanding of the high-energy behaviour of the matrix element (7.185).

Obviously, the only potentially dangerous term is that involving the longitudinal part of the Z boson propagator (because of the factor m_Z^{-2}). Denoting the corresponding contribution to (7.185) as $\mathcal{M}_\triangle^{(\mathrm{long.})}$, one has

$$\mathcal{M}_\triangle^{(\mathrm{long.})} = -i\frac{1}{4}\left(\frac{g}{\cos\theta_W}\right)^2 m_Z^{-2} a_f Q_f^2 e^2 \bar{v}(l_+)\not{q}(v_e - a_e\gamma_5)u(l_-) \times \frac{1}{s - m_Z^2} q^\alpha T_{\alpha\mu\nu}(k,p)\varepsilon^{*\mu}(k)\varepsilon^{*\nu}(p) \tag{7.187}$$

where we have also set $s = q^2$. Taking into account that $q = l_- + l_+$ and employing the equations of motion for the Dirac spinors, one gets

$$\begin{aligned} \bar{v}(l_+)\not{q}u(l_-) &= 0 \\ \bar{v}(l_+)\not{q}\gamma_5 u(l_-) &= -2m_e\bar{v}(l_+)\gamma_5 u(l_-) \end{aligned} \tag{7.188}$$

[16] Let us remind the reader that purely fermionic closed loops with an (arbitrary) odd number of vertices are discarded within spinor QED because to any such graph, one can add the contribution of its counterpart with a reverse orientation of internal lines (i.e. with fermion circulating inside the loop in opposite direction) and this is exactly opposite to the original one. For a triangle graph, reverting the loop orientation is tantamount to the crossing of two external photon lines attached to its vertices.

and (7.187) thus becomes

$$\mathcal{M}_{\triangle}^{(\text{long.})} = \frac{i}{4}\left(\frac{g}{\cos\theta_W}\right)^2 a_f Q_f^2 e^2 \frac{m_e}{m_Z^2}\frac{1}{s-m_Z^2}\bar{v}(l_+)\gamma_5 u(l_-) \times q^\alpha T_{\alpha\mu\nu}(k,p)\varepsilon^{*\mu}(k)\varepsilon^{*\nu}(p) \tag{7.189}$$

where we have already set $a_e = -1/2$ according to (7.163). It means that one factor of m_Z^{-1} is effectively compensated by the electron mass factorized from the four-divergence of the axial-vector part of the weak neutral current and the matrix element in question can therefore grow at worst linearly in the high-energy limit. Now, we assume that $T_{\alpha\mu\nu}(k,p)$ is defined in such a way that

$$k^\mu T_{\alpha\mu\nu}(k,p) = 0, \quad p^\nu T_{\alpha\mu\nu}(k,p) = 0 \tag{7.190}$$

i.e. the vector Ward identities (cf. Appendix E) are imposed in order to maintain electromagnetic gauge invariance. Then

$$q^\alpha T_{\alpha\mu\nu}(k,p) = 2m_f T_{\mu\nu}(k,p) + \frac{1}{2\pi^2}\epsilon_{\mu\nu\rho\sigma}k^\rho p^\sigma \tag{7.191}$$

where $T_{\mu\nu}(k,p)$ is obtained from (7.186) by replacing $\gamma_\alpha\gamma_5$ with γ_5 (let us stress that the integral defining the $T_{\mu\nu}$ is perfectly convergent). The relation (7.191) represents an "anomalous axial-vector Ward identity". The first contribution in its right-hand side is usually called the "normal term" and the second one is the famous **ABJ anomaly**. It is clear that the two contributions in (7.191) have substantially different impact on the high-energy behaviour of the expression (7.189). The normal term is proportional to the fermion mass and thus it compensates the remaining factor of m_Z^{-1}; in other words, this yields a contribution that is asymptotically flat for $E \to \infty$ (and proportional to $m_e m_f/m_Z^2$). However, the ABJ anomaly represents a "hard" contribution that does not contain any compensating mass factor and one is thus indeed left with a result for (7.189) that is linearly divergent in the limit $E \to \infty$. Explicitly, the leading term in (7.189) (and, consequently, in (7.185)) has the form

$$\mathcal{M}_{\triangle\text{anomaly}}^{(\text{long.})} = \frac{i}{4}\left(\frac{g}{\cos\theta_W}\right)^2 a_f Q_f^2 e^2 \frac{m_e}{m_Z^2}\frac{1}{s-m_Z^2}\bar{v}(l_+)\gamma_5 u(l_-) \times \frac{1}{2\pi^2}\epsilon_{\mu\nu\rho\sigma}k^\rho p^\sigma \varepsilon^{*\mu}(k)\varepsilon^{*\nu}(p) \tag{7.192}$$

i.e. the high-energy asymptotics can be written schematically as

$$\mathcal{M}_\triangle \simeq a_f Q_f^2 \, O\left(\frac{m_e}{m_Z^2}E\right) \tag{7.193}$$

where we have singled out only the coefficients that depend on the characteristics of the fermion circulating in the triangle loop. Let us emphasize that it is the *real part* of the triangle subgraph that yields the observed bad high-energy behaviour of (7.185); the corresponding imaginary part is sufficiently "soft" (for a more detailed discussion of this point and for further explicit formulae, see, e.g., [Hor]).

The lesson to be learned from the considered example is as follows. If one considers, e.g., the GWS electroweak theory with the fermion sector restricted to a single lepton type, the tree-level unitarity surely holds, but at the one-loop (and higher) level, one observes a rapid violation of unitarity induced by an effect of the ABJ anomaly. In fact, the coefficient in (7.193) is the same for all lepton species (note that $a_e = a_\mu = a_\tau = -\frac{1}{2}$ and $Q_e = Q_\mu = Q_\tau = -1$) and this means that adding more "standard" leptons to a single generation does not make the situation any better. Thus, **the GWS model for a leptonic world would not be renormalizable**; in particular, the original Weinberg model [40] certainly suffers from such an "anomaly disease" and — as we will show later in this section — this is cured only when the quark sector is taken into account properly.

In any case, the problem described above is characteristic of the U-gauge (since only there one encounters vector boson propagators containing pieces proportional to m_V^{-2}). Irrespective of its ultimate solution within the full Standard Model, it is interesting to know how the ABJ anomaly can manifest itself in an R-gauge, where all propagators behave properly. To examine this, let us consider again the process $e^+e^- \to \gamma\gamma$. Obviously, in an R_ξ-gauge, there is no power-like growth of the scattering amplitude in question for $E \to \infty$, even at the level of individual Feynman graphs. Instead, the crucial point now is gauge invariance, i.e. the independence of the S-matrix element on the gauge parameter ξ (order by order in perturbation theory). For tree graphs, the problem is trivial, but it becomes rather subtle already at the one-loop level. Motivated by our previous experience, we shall focus our attention on the "suspect" diagrams depicted in Fig. 7.7.[17]

[17]Of course, there are many other one-loop Feynman graphs contributing to the considered process, which depend on the gauge parameter ξ. It can be shown that the sum

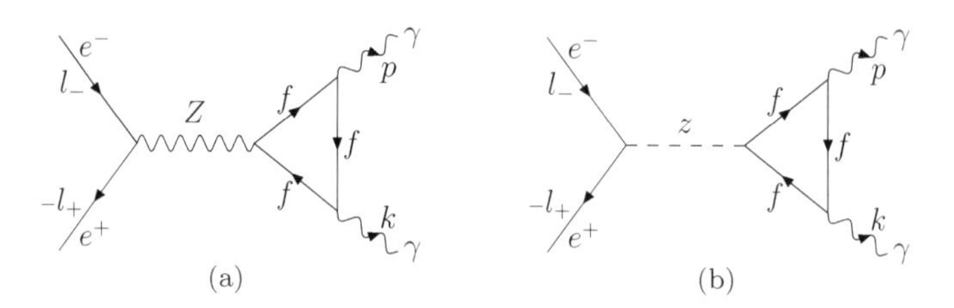

Figure 7.7. Contributions of triangle fermion loops to the process $e^+e^- \to \gamma\gamma$ in an R-gauge: (a) exchange of the Z boson and (b) exchange of its unphysical counterpart z.

Normally, one would expect a cancellation of the ξ-dependent part of the graph in Fig. 7.7(a) against the contribution of Fig. 7.7(b), similar to that in the tree-level example discussed in Section 7.7. However, the ABJ anomaly may violate such a mechanism, as it represents an extra contribution to naive Ward identities (that simply "copy" classical relations for current divergences). We are going to show that such an effect really occurs. For the purpose of our discussion, let us denote the contributions of Fig. 7.7(a) and (b) (including the crossing of external photon lines) simply as $\mathcal{M}_\xi^{(Z)}$ and $\mathcal{M}_\xi^{(z)}$, respectively, and identify the fermion inside the triangle loops with a lepton, i.e. $f = e, \mu$ or τ. According to the rules established in Sections 7.6 and 7.7, one has

$$i\mathcal{M}_\xi^{(Z)} = i^8(-1)\frac{1}{4}\left(\frac{g}{\cos\theta_W}\right)^2(-a_f)Q_f^2e^2\bar{v}(l_+)\gamma_\lambda(v_e - a_e\gamma_5)u(l_-)$$
$$\times\frac{-g^{\lambda\alpha} + (1-\xi)q^\lambda q^\alpha(q^2 - \xi m_Z^2)^{-1}}{q^2 - m_Z^2}T_{\alpha\mu\nu}(k,p)\varepsilon^{*\mu}(k)\varepsilon^{*\nu}(p) \tag{7.194}$$

and

$$i\mathcal{M}_\xi^{(z)} = i^8(-1)\left(-i\frac{g}{2}\frac{m_e}{m_W}\right)\left(-i\frac{g}{2}\frac{m_f}{m_W}\right)Q_f^2e^2\bar{v}(l_+)\gamma_5u(l_-)$$
$$\times\frac{1}{q^2 - \xi m_Z^2}T_{\mu\nu}(k,p)\varepsilon^{*\mu}(k)\varepsilon^{*\nu}(p) \tag{7.195}$$

where the used symbols have the same meaning as before (note that in writing (7.194), we have automatically discarded a term that would

of all one-loop graphs, except those in Fig. 7.7, is ξ-independent, but a straightforward proof based on an explicit diagram calculations is tedious.

correspond to a loop of the VVV type). Now, it is convenient to compare $\mathcal{M}_\xi^{(Z)} + \mathcal{M}_\xi^{(z)}$ with a "reference value", corresponding, e.g., to the 't Hooft–Feynman gauge (for which $\xi = 1$). The expression (7.194) is naturally split as

$$\mathcal{M}_\xi^{(Z)} = \mathcal{M}_{\text{diag.}}^{(Z)} + \mathcal{M}_{\text{long.}}^{(Z)} \tag{7.196}$$

in correspondence with the structure of Z boson propagator (cf. (7.164)). Obviously, $\mathcal{M}_{\text{diag.}}^{(Z)}$ coincides with $\mathcal{M}_{\xi=1}^{(Z)}$, i.e.

$$\mathcal{M}_{\text{diag.}}^{(Z)} = \mathcal{M}_{tHF}^{(Z)} \tag{7.197}$$

Proceeding in the usual way, $\mathcal{M}_{\text{long.}}^{(Z)}$ is recast as

$$\begin{aligned}\mathcal{M}_{\text{long.}}^{(Z)} &= -i\frac{1}{2}\left(\frac{g}{\cos\theta_W}\right)^2 a_f Q_f^2 e^2 a_e m_e \frac{1-\xi}{(q^2-\xi m_Z^2)(q^2-m_Z^2)} \\ &\quad \times \bar{v}(l_+)\gamma_5 u(l_-)\big[2m_f T_{\mu\nu}(k,p) + \mathcal{A}_{\mu\nu}(k,p)\big]\varepsilon^{*\mu}(k)\varepsilon^{*\nu}(p)\end{aligned} \tag{7.198}$$

where we have also used the identity (7.191) and denoted

$$\mathcal{A}_{\mu\nu}(k,p) = \frac{1}{2\pi^2}\epsilon_{\mu\nu\rho\sigma}k^\rho p^\sigma \tag{7.199}$$

$\mathcal{M}_{\text{long.}}^{(Z)}$ is thus divided into its "normal part" (corresponding to $2m_f T_{\mu\nu}$) and a contribution of the ABJ anomaly. It is easy to show that by adding $\mathcal{M}_\xi^{(z)}$ shown in (7.195) to the normal part of (7.198), one recovers the z-exchange contribution in the 't Hooft–Feynman gauge (the reader is recommended to verify this explicitly). Taking into account also (7.197), our results can be summarized as follows:

$$\mathcal{M}_\xi^{(Z)} + \mathcal{M}_\xi^{(z)} = \mathcal{M}_{tHF}^{(Z)} + \mathcal{M}_{tHF}^{(z)} + \mathcal{M}_{\text{anomaly}}^{(Z)} \tag{7.200}$$

where

$$\begin{aligned}\mathcal{M}_{\text{anomaly}}^{(Z)} &= \frac{i}{4}\left(\frac{g}{\cos\theta_W}\right)^2 a_f Q_f^2 e^2 m_e \frac{1-\xi}{(q^2-\xi m_Z^2)(q^2-m_Z^2)} \\ &\quad \times \bar{v}(l_+)\gamma_5 u(l_-)\mathcal{A}_{\mu\nu}(k,p)\varepsilon^{*\mu}(k)\varepsilon^{*\nu}(p)\end{aligned} \tag{7.201}$$

In other words, the sum $\mathcal{M}_\xi^{(Z)} + \mathcal{M}_\xi^{(z)}$ would be gauge-independent, were it not for the ABJ anomaly term in (7.198). **The anomaly effect destroys gauge invariance** of the matrix element in question at the one-loop level and this, of course, is a fatal blow to the internal consistency of the GWS electroweak theory involving just one fermion species. Note that the resulting anomalous contribution in (7.201) is non-vanishing for any $\xi \neq 1$ and, as regards its dependence on the properties of the fermion inside the triangle loop, this is carried by an overall factor $a_f Q_f^2$ — the same as in our previous result concerning the violation of perturbative unitarity in the U-gauge (cf. (7.193)). Thus, in analogy with the observation following the relation (7.193), we can also conclude that the GWS theory of leptons would be internally inconsistent (irrespective of the number of lepton flavours), since the gauge independence of the S-matrix would be lost because of the ABJ anomaly.

An upshot of the preceding discussion is as follows. If the spectrum of elementary fermions is reduced to leptons alone, the GWS electroweak theory suffers from serious problems due to the ABJ anomaly. In the U-gauge, such a model is non-renormalizable, even though the tree-level unitarity is satisfied. In renormalizable R_ξ-gauges, some particular one-loop scattering amplitudes depend explicitly on the gauge-fixing parameter ξ, i.e. there is a manifest violation of gauge invariance. Both difficulties are of similar technical origin, but the second issue is in fact more fundamental: while one can imagine a perfectly consistent quantum field theory model that is not perturbatively renormalizable,[18] the loss of gauge independence makes the considered perturbative approximation (and thereby the whole perturbation expansion) totally meaningless. Anyway, within the GWS theory, both problems are two sides of the same coin, which is the anomalous Ward identity for a triangle fermion loop.

It is gratifying that within the full GWS standard model, there is a natural way out of the difficulties described above: it turns out that the ABJ anomaly effects due to the lepton triangle loops are exactly cancelled by an analogous contribution coming from quarks. Let us now show how such a simple mechanism works in the example discussed above. As we have seen, the anomaly coefficient corresponding to the VVA triangle loop made of a fermion f is $a_f Q_f^2$, with a_f and Q_f having the usual meaning

[18] It would only mean that an infinite number of renormalization counterterms is needed and, consequently, such a theory has less predictive power in comparison with models renormalizable in the conventional sense.

explained above. According to the familiar rules for the weak neutral current couplings, one has

$$a_f = T_{3L}^{(f)} \tag{7.202}$$

where $T_{3L}^{(f)}$ is the weak isospin of f_L. Thus, for u-quark ($Q_u = \frac{2}{3}$) and d-quark ($Q_d = -\frac{1}{3}$), one gets

$$a_u Q_u^2 = +\frac{1}{2}\left(\frac{2}{3}\right)^2, \quad a_d Q_d^2 = -\frac{1}{2}\left(-\frac{1}{3}\right)^2 \tag{7.203}$$

(obviously, the result (7.203) is valid for any up-type and down-type quark flavours). Now, it is important to realize that any quark can occur in three colour "copies" (more precisely, a quark with a given flavour can exist in three states distinguished by the colour quantum number). Colour charge is substantial for strong interactions (described by quantum chromodynamics), but does not play any dynamical role in electroweak interactions. This means that the contribution of the considered triangle loop for a given quark flavour should be simply multiplied by the number of colours $N_c = 3$. Thus, restricting ourselves to the first generation of fermions (ν_e, e, u, d), the relevant results can be summarized as follows. Lepton contribution to the full coefficient of the ABJ anomaly is

$$a_{\mathrm{ABJ}}^{(\mathrm{leptons})} = -\frac{1}{2}(-1)^2 = -\frac{1}{2} \tag{7.204}$$

while the quark loops yield

$$a_{\mathrm{ABJ}}^{(\mathrm{quarks}\, u,d)} = 3 \times \left[\frac{1}{2}\left(\frac{2}{3}\right)^2 - \frac{1}{2}\left(-\frac{1}{3}\right)^2\right] = +\frac{1}{2} \tag{7.205}$$

In this straightforward way, it is seen that the lepton and quark anomalies indeed compensate each other. Obviously, for the other two generations of elementary fermions, the result must be the same, since the pattern of the relevant parameters a_f and Q_f repeats itself.

The algebraic condition for the anomaly cancellation can be put in a more elegant form. To see this, let us start with a general expression for the sum of the ABJ anomaly coefficients (e.g. within the first generation) that can be written as

$$a_{\mathrm{ABJ}} = \sum_f a_f Q_f^2 = T_{3L}^{(\nu)} Q_\nu^2 + T_{3L}^{(e)} Q_e^2 + N_c (T_{3L}^{(u)} Q_u^2 + T_{3L}^{(d)} Q_d^2) \tag{7.206}$$

(for the sake of full symmetry, we have also included formally the neutrino with $Q_\nu = 0$, though its contribution vanishes). Utilizing the known values of the weak isospin for leptons and quarks, (7.206) is recast as

$$a_{\mathrm{ABJ}} = \frac{1}{2}(Q_\nu^2 - Q_e^2) + \frac{1}{2}N_c(Q_u^2 - Q_d^2) \tag{7.207}$$

However, electric charges of two fermions belonging to the same isospin doublet differ by one unit, so the last expression becomes

$$a_{\mathrm{ABJ}} = \frac{1}{2}\left[Q_\nu + Q_e + N_c(Q_u + Q_d)\right] \tag{7.208}$$

Thus, the ABJ anomaly coefficient in question is seen to be proportional to the sum of all fermion charges, with each quark taken in N_c colour mutations. The condition of vanishing of the ABJ anomaly in the considered example thus reads

$$\sum_f Q_f = 0 \tag{7.209}$$

where the sum in (7.209) extends over all fermion species, including the colour factor N_c for quarks. From the above discussion, it is clear that for achieving such a cancellation of anomalies, it is essential that there are just three colours, $N_c = 3$. Equally obvious is the fact that the presence of two quark flavours in each generation (with electric charges $+\frac{2}{3}$ and $-\frac{1}{3}$, respectively) is necessary for this purpose. Historically, this observation provided (among other things) strong motivation for the t-quark searches after the discoveries of τ-lepton in 1975 and b-quark in 1977: at that time, the top was desperately needed to make the third generation complete. As we noted earlier in this chapter, this superheavy quark has been directly observed only in mid-1990s. The simple algebraic condition (7.209) represents indeed a highly remarkable result: it provides the only known successful theoretical relation *between leptons and quarks*, which otherwise form entirely independent sectors of the spectrum of elementary fermions within SM.

In fact, within the GWS standard model, one can find many other examples of scattering amplitudes that could be affected by the ABJ anomaly effects. The point is that any triangular fermion loop of the type VVA yields the anomaly, irrespective of the assignments of the attached vector boson lines. Moreover, ABJ anomalies also occur in AAA fermion triangles (i.e. in those made of three axial-vector currents). The potentially anomalous cases can be simply classified in terms of labels for the external vector boson lines entering vertices of the triangle loops in question. Thus, apart from the $Z\gamma\gamma$ case discussed previously, the other relevant configurations can be $ZZ\gamma, ZZZ, ZWW$ and γWW (for example, the configuration $ZZ\gamma$ can occur in the amplitude for the process $e^+e^- \to Z\gamma$, etc.). Obviously, for combinations $Z\gamma\gamma, ZZ\gamma$ and γWW, only the VVA anomalies contribute, while in cases labelled as ZZZ and ZWW, both VVA and AAA anomalies can play a role. When the whole collection of triangle graphs is analyzed, a result that emerges is remarkably simple: **it turns out that satisfying the relation (7.209) already suffices for the elimination of all anomalies enumerated here** (a proof of this statement is left as a challenge for a seriously interested reader). Therefore, we can conclude that the GWS standard model is completely free of ABJ anomalies, which means that it is an internally consistent and perturbatively renormalizable theory of electroweak interactions.

The possibility of a mutual compensation of ABJ triangle anomalies due to different fermion species has been first observed — within the $SU(2) \times U(1)$ electroweak theory — by Bouchiat, Iliopoulos and Meyer [85]. Thus, the anomaly cancellation mechanism outlined above should perhaps be appropriately called the "BIM mechanism". Note also that almost simultaneously with the work [85], the same issue was analyzed independently in [86] and [87].

7.10 Synopsis of the GWS standard model

We have already described in detail all relevant parts of the Glashow–Weinberg–Salam standard model of electroweak interactions. For the reader's convenience, we shall now summarize the whole construction as well as the corresponding interaction Lagrangian in the physical U-gauge.

As for the particle contents of the GWS standard model, there are

(i) three generations of spin-1/2 elementary fermions, i.e. six leptons (e, ν_e, μ, ν_μ, τ, ν_τ) and six quarks (d, u, s, c, b, t);
(ii) four spin-1 bosons, namely the massive intermediate vector bosons $W^\pm$, Z^0 and massless photon γ;
(iii) a spin-0 Higgs boson H.

The fermions are conventionally considered as constituting the "matter", while the spin-1 bosons are "carriers of electroweak force" (since they mediate electroweak interactions). The Higgs boson is intimately related to the mechanism of mass generation for the other particles.

The fundamental dynamical principle is that of gauge invariance (i.e. local internal symmetry); the relevant symmetry group is non-Abelian, namely $SU(2) \times U(1)$. $SU(2)$ is referred to as "weak isospin" subgroup and the factor $U(1)$ corresponds to "weak hypercharge". Within this framework, $W^{\pm}$, Z^0 and γ mediating electroweak interactions are quanta of physical vector fields that are made of the four original Yang–Mills fields associated with the four generators of $SU(2) \times U(1)$. Basic building blocks of the fermion sector are left-handed $SU(2)$ doublets

$$L^{(e)} = \begin{pmatrix} \nu_{eL} \\ e_L \end{pmatrix}, \quad L^{(\mu)} = \begin{pmatrix} \nu_{\mu L} \\ \mu_L \end{pmatrix}, \quad L^{(\tau)} = \begin{pmatrix} \nu_{\tau L} \\ \tau_L \end{pmatrix}$$

$$L_0^{(d)} = \begin{pmatrix} u_{0L} \\ d_{0L} \end{pmatrix}, \quad L_0^{(s)} = \begin{pmatrix} c_{0L} \\ s_{0L} \end{pmatrix}, \quad L_0^{(b)} = \begin{pmatrix} t_{0L} \\ b_{0L} \end{pmatrix} \tag{7.210}$$

and right-handed singlets

$$e_R, \quad \mu_R, \quad \tau_R \tag{7.211}$$

$$d_{0R}, \quad u_{0R}, \quad s_{0R}, \quad c_{0R}, \quad b_{0R}, \quad t_{0R}$$

where $e_L = \frac{1}{2}(1 - \gamma_5)e$, $e_R = \frac{1}{2}(1 + \gamma_5)e$, etc. Note that we have not included here the right-handed components of neutrino fields; for the purpose of this overview, we simply ignore neutrino masses (a more detailed discussion concerning this issue can be found in Section 6.6). The quark variables in (7.210) and (7.211) carry the label "0" as they represent a set of "primordial" fields (or "protofields") that have yet to be transformed into the physical ones, carrying definite masses.[19] For implementing the Higgs mechanism, which yields mass terms of vector bosons, one makes use of an $SU(2)$ doublet of complex scalar fields

$$\Phi = \begin{pmatrix} \varphi^+ \\ \varphi^0 \end{pmatrix} \tag{7.212}$$

Furthermore, in order to generate masses of all fermions (in particular, the Dirac masses of both down- and up-type quarks), one has to employ both (7.212) and a conjugate doublet

$$\widetilde{\Phi} = i\tau_2 \Phi^* \tag{7.213}$$

where τ_2 is the second Pauli matrix.

[19]In comparison with the notation introduced in Sections 7.3 and 7.4, we have changed the symbols for quark doublets; such a slight modification makes the subsequent formulae more compact. The logic of the labelling employed in (7.210) should be obvious.

The GWS gauge invariant Lagrangian can be written as consisting of four parts:

$$\mathscr{L}_{\text{GWS}} = \mathscr{L}_{\text{gauge}} + \mathscr{L}_{\text{fermion}} + \mathscr{L}_{\text{Higgs}} + \mathscr{L}_{\text{Yukawa}} \tag{7.214}$$

and the individual terms in (7.214) are consecutively defined below.

First, $\mathscr{L}_{\text{gauge}}$ is the pure Yang–Mills Lagrangian corresponding to the local symmetry $SU(2) \times U(1)$:

$$\mathscr{L}_{\text{gauge}} = -\frac{1}{4} F^a_{\mu\nu} F^{a\mu\nu} - \frac{1}{4} B_{\mu\nu} B^{\mu\nu} \tag{7.215}$$

where

$$\begin{aligned} F^a_{\mu\nu} &= \partial_\mu A^a_\nu - \partial_\nu A^a_\mu + g\epsilon^{abc} A^b_\mu A^c_\nu \\ B_{\mu\nu} &= \partial_\mu B_\nu - \partial_\nu B_\mu \end{aligned} \tag{7.216}$$

with A^a_μ, $a = 1, 2, 3$ and B_μ being the $SU(2)$ and $U(1)$ gauge fields, respectively, ϵ^{abc} stands for the totally antisymmetric Levi-Civita symbol (structure constants of the $SU(2)$ algebra) and g is the $SU(2)$ gauge coupling constant.

Next, $\mathscr{L}_{\text{fermion}}$ comprises kinetic terms for leptons and quarks, and their interactions with gauge fields. Making use of the doublets and singlets (7.210) and (7.211), it can be written as

$$\begin{aligned} \mathscr{L}_{\text{fermion}} = &\sum_{\ell=e,\mu,\tau} i\bar{L}^{(\ell)}\gamma^\mu \left(\partial_\mu - igA^a_\mu \frac{\tau^a}{2} - iY^{(\ell)}_L g' B_\mu\right) L^{(\ell)} \\ &+ \sum_{q=d,s,b} i\bar{L}^{(q)}_0 \gamma^\mu \left(\partial_\mu - igA^a_\mu \frac{\tau^a}{2} - iY^{(q)}_L g' B_\mu\right) L^{(q)}_0 \\ &+ \sum_{\ell=e,\mu,\tau} i\bar{\ell}_R \gamma^\mu \left(\partial_\mu - iY^{(\ell)}_R g' B_\mu\right) \ell_R \\ &+ \sum_{q=d,u,s,c,b,t} i\bar{q}_{0R} \gamma^\mu \left(\partial_\mu - iY^{(q)}_R g' B_\mu\right) q_{0R} \end{aligned} \tag{7.217}$$

where we have introduced standard covariant derivatives. Apart from the parameter g that has already appeared in (7.216), there is another independent coupling constant g' associated with the $U(1)$ subgroup. The weak hypercharge assignments for the matter fields are conventionally defined by

$$Q = T_3 + Y \tag{7.218}$$

with T_3 being the third component of weak isospin and Q the corresponding particle charge (in units of positron charge). Taking into account that

$$\begin{aligned} Q_e &= Q_\mu = Q_\tau = -1 \\ Q_d &= Q_s = Q_b = -\frac{1}{3} \\ Q_u &= Q_c = Q_t = +\frac{2}{3} \end{aligned} \tag{7.219}$$

the relation (7.218) yields

$$\begin{aligned} Y_L^{(\ell)} &= -\frac{1}{2}, \quad \ell = e, \mu, \tau \\ Y_L^{(q)} &= +\frac{1}{6}, \quad q = d, s, b \\ Y_R^{(q)} &= -\frac{1}{3}, \quad q = d, s, b \\ Y_R^{(q)} &= +\frac{2}{3}, \quad q = u, c, t \end{aligned} \tag{7.220}$$

$\mathscr{L}_{\text{Higgs}}$ has the form

$$\begin{aligned} \mathscr{L}_{\text{Higgs}} = \Phi^\dagger \left(\overleftarrow{\partial}_\mu + igA_\mu^a \frac{\tau^a}{2} + \frac{i}{2} g' B_\mu \right) \left(\partial^\mu - igA^{b\mu} \frac{\tau^b}{2} - \frac{i}{2} g' B^\mu \right) \Phi \\ - \lambda \left(\Phi^\dagger \Phi - \frac{v^2}{2} \right)^2 \end{aligned} \tag{7.221}$$

where, in accordance with the rule (7.218), we have set $Y_\Phi = +\frac{1}{2}$ in the relevant covariant derivatives. λ is a coupling constant for the Higgs scalar self-interaction and the parameter v ("vacuum expectation value" or "vacuum shift" of the Higgs field) provides an overall scale for particle masses. Working out the scalar "potential" in (7.221), one can see immediately that λv^2 is the coefficient of a wrong-sign mass term for Φ.

Finally, the Yukawa-type term reads

$$\begin{aligned} \mathscr{L}_{\text{Yukawa}} = &- \sum_{\ell = e, \mu, \tau} h_\ell \bar{L}^{(\ell)} \Phi \ell_R + \text{h.c.} - \sum_{\substack{q = d, s, b \\ q' = d, s, b}} h_{qq'} \bar{L}_0^{(q)} \Phi q'_{0R} + \text{h.c.} \\ &- \sum_{\substack{q = d, s, b \\ q' = u, c, t}} \tilde{h}_{qq'} \bar{L}_0^{(q)} \widetilde{\Phi} q'_{0R} + \text{h.c.} \end{aligned} \tag{7.222}$$

with h_ℓ, $h_{qq'}$, and $\tilde{h}_{qq'}$ being essentially arbitrary (real) coupling constants; the overall minus sign is purely conventional. If one takes into account

(7.220) as well as

$$Y_{\widetilde{\Phi}} = -Y_{\Phi} = -\frac{1}{2} \tag{7.223}$$

the $U(1)$ invariance of (7.222) can be checked easily (its symmetry with respect to $SU(2)$ is obvious).

Now, the complex doublet (7.212) embodies four real scalars and three of them are would-be Goldstone bosons associated with a spontaneously broken $SU(2)$ symmetry of the potential in $\mathscr{L}_{\text{Higgs}}$. These unphysical scalars can be eliminated by means of an appropriate choice of gauge; such a procedure is formally equivalent to an $SU(2)$ transformation within the Lagrangian (7.214) and amounts to replacing Φ by

$$\Phi_U = \begin{pmatrix} 0 \\ \dfrac{1}{\sqrt{2}}(v+H) \end{pmatrix} \tag{7.224}$$

where H denotes the physical Higgs boson. Note also that (7.213) then immediately yields

$$\widetilde{\Phi}_U = \begin{pmatrix} \dfrac{1}{\sqrt{2}}(v+H) \\ 0 \end{pmatrix} \tag{7.225}$$

In what follows, we shall describe how the contents of the original Lagrangian (7.214) is disentangled in terms of physical fields. We are going to concentrate first on the structural aspects and a detailed form of the interaction Lagrangian will be summarized later on. Let us start with the leptonic part of $\mathscr{L}_{\text{fermion}}$. A_μ^a and B_μ are primordial gauge fields without a direct particle contents, but the physical vector fields can be obtained from them by means of appropriate linear combinations. In particular, $W_\mu^\pm$ defined by

$$W_\mu^\pm = \frac{1}{\sqrt{2}}\left(A_\mu^1 \mp i A_\mu^2\right) \tag{7.226}$$

are coupled to weak charged currents (with coupling constant being proportional to g) and Z_μ and A_μ introduced via a real orthogonal transformation

$$\begin{aligned} A_\mu^3 &= \cos\theta_W Z_\mu + \sin\theta_W A_\mu \\ B_\mu &= -\sin\theta_W Z_\mu + \cos\theta_W A_\mu \end{aligned} \tag{7.227}$$

represent the Z boson field and the electromagnetic four-potential, respectively. The mixing embodied in (7.227) represents the mathematical basis of the concept of "electroweak unification" within the GWS theory. The parameter θ_W is usually called the Weinberg angle or simply "weak mixing angle". The requirement that A_μ be coupled with equal strength to the left-handed and right-handed leptons (in other words, the current coupled to A_μ be a pure vector) leads to the condition $\tan\theta_W = g'/g$, i.e.

$$\cos\theta_W = \frac{g}{\sqrt{g^2+g'^2}}, \quad \sin\theta_W = \frac{g'}{\sqrt{g^2+g'^2}} \tag{7.228}$$

and, subsequently, the electromagnetic coupling constant is expressed as

$$e = \frac{gg'}{\sqrt{g^2+g'^2}} \tag{7.229}$$

One thus arrives at a relation between e and g, namely

$$e = g\sin\theta_W \tag{7.230}$$

Note that (7.230) is sometimes called the "unification condition" (for an $SU(2)\times U(1)$ electroweak theory). Z_μ defined by (7.227) is then coupled to a weak neutral current whose structure is fully determined in terms of the parameter $\sin^2\theta_W$ (that has to be fixed by experiments). As regards the electroweak interactions of quarks, we shall discuss them a bit later; now, let us come back to $\mathscr{L}_{\text{gauge}}$.

Using the linear transformations (7.226) and (7.227) in (7.215), one gets kinetic terms for the vector fields $W_\mu^\pm$, Z_μ and A_μ, namely

$$\mathscr{L}_{\text{gauge}}^{(\text{kin})} = -\frac{1}{2}W_{\mu\nu}^- W^{+\mu\nu} - \frac{1}{4}Z_{\mu\nu}Z^{\mu\nu} - \frac{1}{4}A_{\mu\nu}A^{\mu\nu} \tag{7.231}$$

where $W_{\mu\nu}^- = \partial_\mu W_\nu^- - \partial_\nu W_\mu^-$, etc. and a set of trilinear and quadrilinear vector boson self-interactions. These are of the following types: $WW\gamma$, WWZ, $WWWW$, $WWZZ$, $WWZ\gamma$ and $WW\gamma\gamma$. Note that other types, such as, $Z\gamma\gamma$, ZZZ, and $ZZZZ$, are automatically excluded.

Next, we proceed to $\mathscr{L}_{\text{Higgs}}$. Substituting (7.224) into (7.221), one identifies readily free Lagrangian for the Higgs boson H, with a mass given by

$$m_H^2 = 2\lambda v^2 \tag{7.232}$$

As the most important item, mass terms for $W^\pm$ and Z are obtained from the Higgs mechanism. To that end, one has to work out the relevant quadratic form in variables A_μ^a and B_μ (which is induced by the vacuum shift v in (7.224)): A_μ^1 and A_μ^2 are replaced by $W_\mu^\pm$ defined according to (7.226) and a mass matrix for A_μ^3 and B_μ is diagonalized in a straightforward way. Taking into account the normalization of (7.231), the vector boson masses in question can then be identified as

$$m_W = \frac{1}{2} g v, \qquad m_Z = \frac{1}{2}(g^2 + g'^2)^{1/2} v \tag{7.233}$$

Thus, in view of (7.228), one has

$$m_W = m_Z \cos\theta_W \tag{7.234}$$

It should be stressed that Z_μ carrying the mass shown in (7.233) is given precisely by the expression following from (7.227), i.e.

$$Z_\mu = \cos\theta_W A_\mu^3 - \sin\theta_W B_\mu \tag{7.235}$$

(with $\cos\theta_W$ and $\sin\theta_W$ taking on the values (7.228)). Similarly, the orthogonal combination

$$A_\mu = \sin\theta_W A_\mu^3 + \cos\theta_W B_\mu \tag{7.236}$$

corresponds to the massless photon.[20] Invoking the familiar formula $G_F/\sqrt{2} = g^2/(8m_W^2)$ and using (7.233), one finds out immediately that v is simply related to the Fermi constant:

$$v = (G_F\sqrt{2})^{-1/2} \doteq 246 \text{ GeV} \tag{7.237}$$

Further, utilizing the unification condition (7.230) and the relation between e and the fine structure constant α, $\alpha = e^2/(4\pi)$, the mass formulae (7.233) and (7.234) can be recast in a form most suitable for practical purposes, namely

$$m_W = \left(\frac{\pi\alpha}{G_F\sqrt{2}}\right)^{1/2} \frac{1}{\sin\theta_W}, \quad m_Z = \left(\frac{\pi\alpha}{G_F\sqrt{2}}\right)^{1/2} \frac{1}{\sin\theta_W \cos\theta_W} \tag{7.238}$$

Apart from the above-mentioned mass terms, $\mathscr{L}_{\text{Higgs}}$ also yields a set of interactions involving $W^\pm$, Z and H. Schematically, the relevant couplings are WWH, ZZH, $WWHH$, $ZZHH$, HHH and $HHHH$.

[20] In other words, it is seen that the fields diagonalizing gauge boson mass terms in $\mathscr{L}_{\text{Higgs}}$ coincide with those descending from the analysis of interactions in the leptonic sector of $\mathscr{L}_{\text{fermion}}$. Such a result is gratifying (as it manifests the internal consistency of the electroweak theory), but one should also keep in mind that we have actually anticipated it by choosing carefully the relevant hypercharge values according to (7.218).

Last but not least, let us consider the term $\mathscr{L}_{\text{Yukawa}}$. As for the leptonic part of (7.222), this is worked out in a straightforward way. Substituting there (7.224), one gets readily mass terms for charged leptons, with

$$m_\ell = \frac{1}{\sqrt{2}} h_\ell v \tag{7.239}$$

and pure scalar Yukawa interactions of the type $\ell\ell H$, whose strength is — in view of (7.239) — obviously proportional to m_ℓ / v. In the quark sector, one gets two different (in general, non-diagonal) 3×3 mass matrices for primordial fields (corresponding to the matrices of coupling constants in (7.222)); in particular, the original Yukawa interactions involving Φ yield the down-type quark masses while the up-type quarks gain masses from interactions with $\widetilde{\Phi}$ displayed in (7.225). The quark mass matrices are diagonalized by means of appropriate biunitary transformations (involving independent rotations of left-handed and right-handed fields). Owing to the simple structure of Φ_U and $\widetilde{\Phi}_U$, Higgs boson interactions are diagonalized simultaneously with mass terms and one thus arrives at the same pattern of coupling constants as in the case of leptons: the strength of a coupling qqH ($q = d, u, s, c, b, t$) is proportional to m_q / v.

Having diagonalized the mass matrices in question, one can return to the quark sector of $\mathscr{L}_{\text{fermion}}$. Performing the relevant unitary rotations of the primordial fields appearing in (7.217), the interaction Lagrangian is recast in terms of variables corresponding to mass eigenstates. In this context, one has to keep in mind that transformations of the up-type and down-type quarks are completely independent, e.g., for the left-handed fields, one can write symbolically

$$\begin{pmatrix} d_L \\ s_L \\ b_L \end{pmatrix} = \mathcal{U} \begin{pmatrix} d_{0L} \\ s_{0L} \\ b_{0L} \end{pmatrix}, \quad \begin{pmatrix} u_L \\ c_L \\ t_L \end{pmatrix} = \widetilde{\mathcal{U}} \begin{pmatrix} u_{0L} \\ c_{0L} \\ t_{0L} \end{pmatrix} \tag{7.240}$$

where $\mathcal{U}$ and $\widetilde{\mathcal{U}}$ are in general different unitary 3×3 matrices. Thus, an essentially arbitrary unitary matrix

$$\widetilde{\mathcal{U}}\mathcal{U}^\dagger = V = \begin{pmatrix} V_{ud} & V_{us} & V_{ub} \\ V_{cd} & V_{cs} & V_{cb} \\ V_{td} & V_{ts} & V_{tb} \end{pmatrix} \tag{7.241}$$

shows up in the resulting weak interaction of charged currents. V is the celebrated Cabibbo–Kobayashi–Maskawa (CKM) matrix and can be ultimately parametrized in terms of just four physically relevant parameters — three "Cabibbo-like" angles and one $\mathcal{CP}$ — violating phase (in case of two generations of quarks one would end up with a real orthogonal 2×2 matrix described in terms of the Cabibbo angle). In this way, one arrives at a natural view of the origin of flavour mixing and $\mathcal{CP}$ violation: both these phenomena are intimately related to the diagonalization of quark matrices descending from the most general Yukawa couplings compatible with electroweak symmetry. Furthermore, one observes that *three* is just the minimum number of fermion generations for which $\mathcal{CP}$ violation can occur within the SM scheme. On the other hand, neutral quark currents, which are obviously flavour-diagonal in the primordial basis, remain diagonal even after the transformation to the fields with definite masses. In this way, SM provides a natural explanation for the conspicuous absence of strangeness-changing (more generally, flavour-changing) weak neutral currents.

Note finally that massive neutrinos and their eventual mixings can be incorporated quite naturally into the SM scheme. Introducing also the right-handed (singlet) components of neutrino fields from the very beginning, the technique used for quarks can be generalized in a straightforward way to the lepton sector of $\mathscr{L}_{\text{Yukawa}}$; one thus gets Dirac mass terms for neutrinos (along with the corresponding Higgs boson Yukawa interactions) and a leptonic analogue of the CKM matrix.[21]

Let us now summarize the explicit form of the SM interaction Lagrangian in the U-gauge. This can be written as

$$\begin{aligned}\mathscr{L}_{\text{int}}^{(\text{GWS})} &= \sum_f Q_f e \bar{f}\gamma^\mu f A_\mu + \mathscr{L}_{\text{CC}} + \mathscr{L}_{\text{NC}} - ig(W_\mu^0 W_\nu^- \overleftrightarrow{\partial}^\mu W^{+\nu} \\ &\quad + W_\mu^- W_\nu^+ \overleftrightarrow{\partial}^\mu W^{0\nu} + W_\mu^+ W_\nu^0 \overleftrightarrow{\partial}^\mu W^{-\nu}) \\ &\quad - g^2 \left[\frac{1}{2}(W^- \cdot W^+)^2 - \frac{1}{2}(W^-)^2(W^+)^2 + (W^0)^2(W^- \cdot W^+)\right. \\ &\quad \left. - (W^- \cdot W^0)(W^+ \cdot W^0)\right] \\ &\quad + g m_W W_\mu^- W^{+\mu} H + \frac{1}{2\cos\theta_W} g m_Z Z_\mu Z^\mu H\end{aligned}$$

[21]In fact, when introducing neutrino mass terms, one has more possibilities than in the case of quarks; neutrinos are electrically neutral and this opens the possibility that they could be Majorana particles. As we said earlier, in the present text, we do not pursue the issue of neutrino masses in detail, although this currently represents one of the hot topics of particle physics. The interested reader is referred, e.g., to [Vog].

$$+\frac{1}{4}g^2W_\mu^-W^{+\mu}H^2+\frac{1}{8}\frac{g^2}{\cos^2\theta_W}Z_\mu Z^\mu H^2$$
$$-\sum_f\frac{1}{2}g\frac{m_f}{m_W}\bar{f}fH-\frac{1}{4}g\frac{m_H^2}{m_W}H^3-\frac{1}{32}g^2\frac{m_H^2}{m_W^2}H^4 \tag{7.242}$$

where the indicated sums run over all elementary fermions (leptons and quarks) and the relevant charge factors Q_f are displayed in (7.219). In the self-interactions of vector bosons we have used, for mnemonic convenience, the notation

$$W_\mu^0=\cos\theta_W Z_\mu+\sin\theta_W A_\mu \tag{7.243}$$

(of course, W_μ^0 coincides with A_μ^3 used before). The term $\mathscr{L}_{\mathrm{CC}}$ describes the interactions of weak charged currents and vector bosons $W^\pm$:

$$\mathscr{L}_{\mathrm{CC}}=\frac{g}{2\sqrt{2}}\sum_{\ell=e,\mu,\tau}\bar{\nu}_\ell\gamma^\lambda(1-\gamma_5)\ell W_\lambda^+$$
$$+\frac{g}{2\sqrt{2}}(\bar{u},\ \bar{c},\ \bar{t})\gamma^\lambda(1-\gamma_5)V_{\mathrm{CKM}}\begin{pmatrix}d\\ s\\ b\end{pmatrix}W_\lambda^+ +\mathrm{h.c.} \tag{7.244}$$

where V_{CKM} is the CKM unitary matrix (7.241). $\mathscr{L}_{\mathrm{NC}}$ stands for the interaction of weak neutral currents and Z:

$$\mathscr{L}_{\mathrm{NC}}=\frac{g}{\cos\theta_W}\sum_f\left(\varepsilon_L^{(f)}\bar{f}_L\gamma^\lambda f_L+\varepsilon_R^{(f)}\bar{f}_R\gamma^\lambda f_R\right)Z_\lambda \tag{7.245}$$

where

$$\begin{aligned}\varepsilon_L^{(f)}&=T_{3L}^{(f)}-Q_f\sin^2\theta_W\\ \varepsilon_R^{(f)}&=T_{3R}^{(f)}-Q_f\sin^2\theta_W\end{aligned} \tag{7.246}$$

with $T_{3L}^{(f)}=+\frac{1}{2}$ for $f=\nu_e,\nu_\mu,\nu_\tau,u,c,t$, $T_{3L}^{(f)}=-\frac{1}{2}$ for $f=e,\mu,\tau,d,s,b$ and $T_{3R}^{(f)}=0$ for any f.

The neutral current interaction may alternatively be written in the form

$$\mathscr{L}_{\mathrm{NC}}=\frac{g}{2\cos\theta_W}\sum_f\bar{f}\gamma^\lambda(v_f-a_f\gamma_5)fZ_\lambda \tag{7.247}$$

with

$$\begin{aligned}v_f&=\varepsilon_L^{(f)}+\varepsilon_R^{(f)}\\ a_f&=\varepsilon_L^{(f)}-\varepsilon_R^{(f)}\end{aligned} \tag{7.248}$$

i.e.

$$
\left.\begin{aligned}
v_f &= -\frac{1}{2} - 2Q_f \sin^2\theta_W \\
a_f &= -\frac{1}{2}
\end{aligned}\right\} \quad \text{for } f = e, \mu, \tau, d, s, b
$$
$$
\left.\begin{aligned}
v_f &= +\frac{1}{2} - 2Q_f \sin^2\theta_W \\
a_f &= +\frac{1}{2}
\end{aligned}\right\} \quad \text{for } f = \nu_e, \nu_\mu, \nu_\tau, u, c, t
\tag{7.249}
$$

As for the self-interactions of vector bosons, the compact form shown in (7.242) is worked out readily by using the definition (7.243) and one can thus identify the individual couplings $WW\gamma$, WWZ, $WW\gamma\gamma$, $WWWW$, $WWZZ$ and $WWZ\gamma$. For completeness, let us reiterate the important relations

$$
\begin{aligned}
e/g &= \sin\theta_W \\
m_W/m_Z &= \cos\theta_W \\
G_F/\sqrt{2} &= g^2/(8m_W^2)
\end{aligned}
\tag{7.250}
$$

In summarizing the electroweak standard model, we should also count the number of free parameters involved in its Lagrangian. First of all, there are coupling constants g, g', λ and the mass scale v (obviously, in view of (7.250), these basic four parameters can be traded, e.g., for α, $\sin^2\theta_W$, m_Z, m_H or α, G_F, m_Z, and m_H). The remaining free parameters come from the Yukawa sector. Taking neutrinos as massless for the moment, one has three masses of charged leptons, six quark masses and four parameters of the CKM mixing matrix. Thus, within the original version of SM, one has $4 + 3 + 6 + 4 = 17$ free parameters. If one also allows for neutrino masses and mixings, one has seven additional parameters (three neutrino masses and four parameters of a leptonic CKM-like mixing matrix). Thus, the total number of free parameters in a "realistic" present-day variant of electroweak SM is 24.

Finally, the reader may find it instructive to see all the interactions contained in (7.242) depicted, schematically, as the corresponding Feynman-graph vertices. Such a collection is displayed in Fig. 7.8. In this context, it is useful to realize that despite the common label "Standard Model" used for the GWS electroweak theory, detailed experimental tests are currently not available for all interaction vertices shown here. The present-day situation can be roughly summarized as follows. Interactions of W and Z with

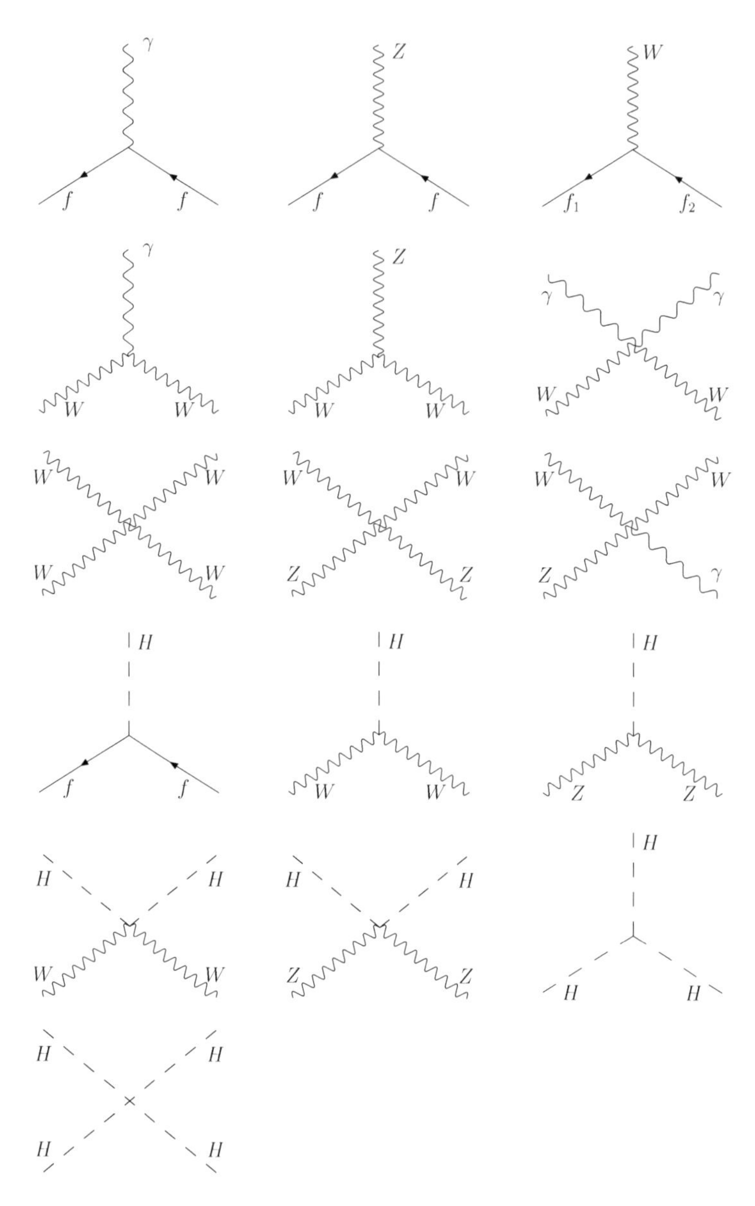

Figure 7.8. All types of interactions contained in the electroweak SM Lagrangian. The set of vertices displayed here corresponds to the physical U-gauge. The label f refers generally to any relevant fermion species. In the CC weak interaction vertex, f_1 and f_2 denote symbolically either ν_ℓ, ℓ or a pair of quarks with charges differing by one unit.

fermions (both leptons and quarks) are tested with good accuracy. Similarly, the couplings $WW\gamma$ and WWZ have been already tested well at the facility Large Electron Positron Collider (LEP) at CERN. In contrast to this, the quartic self-interactions of vector bosons are tested rather poorly (some events corresponding to the WW scattering have been detected only recently at LHC). Concerning the Higgs boson interactions, there are some experimental data for the couplings WWH, ZZH, ttH, bbH and $\tau\tau H$. Couplings of H to light fermions (including c-quark) as well as the quartic interactions $WWHH$ and $ZZHH$ are at present experimentally inaccessible. Higgs boson self-interactions HHH and $HHHH$ are still entirely untested, but represent great challenge for the forthcoming collider experiments. The point is that such measurements would be crucial for a definitive identification of the currently known Higgs-like particle H as the Higgs boson of SM. Thus, taking into account the full number of elementary fermions (and the related number of the elements of the CKM matrix), one may say that, roughly, about one-third of the interaction vertices in Fig. 7.8 remain untested up to now.

Anyway, in view of the stunning phenomenological success of the GWS theory since the 1970s until present day, the label "Standard Model" is well justified and understandable (though the term "Standard Theory" would perhaps be more pertinent) and it seems to be clear that SM will remain a "textbook" effective theory of electroweak interactions valid up to energy scale of $O(100~\text{GeV})$. An excellent survey of the SM physics (in particular, with regard to the performance of the LEP at CERN) can be found in [Ven]. Further, as indicated above, during the past two decades, great progress was made in the experiments on several facilities, most prominent being LHC at CERN. These achievements are covered in considerable detail, e.g., in [Alt], [Lan] and, of course, the full overview of the current data can be found in [6]. It turns out that up to now all available experimental results confirm the validity of SM (though some "smoking guns" occur, encouraging the permanent quest for new physics beyond SM). Some popular theory models going far beyond the present-day SM are described, e.g., in [Lan] and [Pal].

Problems

7.1 Evaluate lepton and hadron decay widths of the W boson within SM (in the tree approximation). Needless to say, one may assume complete hadronization of final-state quarks, so that the hadronic width is to be calculated by summing the W decay rates involving all relevant quark–antiquark pairs. Note that to a good accuracy, one may neglect masses of quarks u, d, s, c, and b (as well as the lepton masses), since these are much smaller than m_W. Show that

$$\Gamma(W \to \text{leptons}) = 3\Gamma_0, \quad \Gamma(W \to \text{quarks}) = 6\Gamma_0$$

where

$$\Gamma_0 = \frac{1}{6\pi\sqrt{2}} G_F m_W^3$$

(in this way, one sees that the ratio of the hadronic and leptonic widths is 2:1 in good agreement with experimental data). Using the known values of G_F and m_W, one thus gets

$$\Gamma_W = \Gamma(W \to \text{all}) = 9\Gamma_0 \doteq 2.1\,\text{GeV}$$

Hint: For the calculation of the full decay width into quarks, the unitarity of the CKM-mixing matrix is to be utilized. Further, one should not forget to include the colour factor $N_c = 3$.

7.2 Evaluate lepton and hadron decay widths of the Z boson within SM (in tree approximation). Neglecting the relevant fermion masses, show first that for an individual decay $Z \to f\bar{f}$), one gets

$$\Gamma(Z \to f\bar{f}) = \frac{G_F m_Z^3}{6\pi\sqrt{2}} \left(v_f^2 + a_f^2\right)$$

with the coupling factors v_f and a_f being given by (7.249) (cf. also Problem 5.2). Concerning the inclusive decay rates, one is supposed to recover the formulae (see also [Pas])

$$\Gamma(Z \to \text{leptons}) = \frac{G_F m_Z^3}{3\pi\sqrt{2}} \left(\frac{3}{2} - 3\sin^2\theta_W + 6\sin^4\theta_W\right)$$

$$\Gamma(Z \to \text{quarks}) = \frac{G_F m_Z^3}{3\pi\sqrt{2}} \left(\frac{15}{4} - 7\sin^2\theta_W + \frac{22}{3}\sin^4\theta_W\right)$$

so that

$$\Gamma_Z = \Gamma(Z \to \text{all}) = \frac{G_F m_Z^3}{3\pi\sqrt{2}} \left(\frac{21}{4} - 10\sin^2\theta_W + \frac{40}{3}\sin^4\theta_W\right)$$

Employing the numerical value $\sin^2\theta_W \approx 0.23$, one may then check that the hadronic and leptonic decays constitute roughly 70% and 30% of the full Z width, respectively. Note that utilizing also the known values of G_F and m_Z, one is led to an approximate prediction for Γ_Z, which reads $\Gamma_Z \approx 2.4\,\text{GeV}$ (an attentive reader may thus observe readily that the results for Γ_W and Γ_Z correspond to mean lifetimes of W and Z of the order of 10^{-25} s).

7.3 Calculate the width (and mean lifetime) of the top quark. What is its dominant decay mode?

7.4 Concerning the decay $t \to b + W^+$, it is also interesting to consider the production of longitudinal and transverse W separately. Evaluate the ratio of the decay rates in question and show that

$$\frac{\Gamma(t \to b\, W_L)}{\Gamma(t \to b\, W_T)} = \frac{m_t^2}{2m_W^2}$$

in the approximation $m_b = 0$.

7.5 Compute the electron energy spectrum in the b-quark decay $b \to c + e^- + \bar{\nu}_e$. Similarly, calculate the energy spectrum for positron in $c \to s + e^+ + \bar{\nu}_e$.
Hint: For such low-energy processes, one can employ an effective four-fermion Lagrangian involving charged $V-A$ currents. Throughout the calculation, neglect the electron mass.

7.6 Show that the SM tree-level amplitude for the process $d + \bar{s} \to W^+W^-$ behaves well in the high-energy limit.

7.7 Compute the cross-section of the process $e^+e^- \to \mu^+\mu^-$ in the vicinity of the Z resonance, i.e. for $s = E_{\text{c.m.}}^2$ close to m_Z^2. Calculate also the corresponding forward–backward asymmetry A_{FB} (for relevant definitions, see Problem 5.6 at the end of Chapter 5). Throughout the calculation, employ the Breit–Wigner form for denominator of the Z boson propagator, i.e. replace the expression $(q^2 - m_Z^2)^{-1}$ with $(q^2 - m_Z^2 + im_Z\Gamma)^{-1}$, where Γ stands for the Z total width (concerning this, see, e.g., [PeS], Section 7.3). Show that the interference cross-section $\sigma_{\gamma Z}$ vanishes for $s = m_Z^2$ and σ_Z (i.e. the Z-exchange contribution) becomes

$$\begin{aligned}
\sigma_Z\Big|_{s=m_Z^2} &= 12\pi \frac{\Gamma(Z \to e^+e^-)\Gamma(Z \to \mu^+\mu^-)}{m_Z^2\Gamma^2} \\
&= \frac{12\pi}{m_Z^2}\,\text{BR}\left(Z \to e^+e^-\right)\text{BR}\left(Z \to \mu^+\mu^-\right)
\end{aligned}$$

Using the familiar formula

$$\sigma_\gamma \doteq \frac{4\pi\alpha^2}{3s}$$

for the photon-exchange contribution at high energy, evaluate the ratio σ_Z/σ_γ for $s = m_Z^2$.

Further, one can define the cross-section for Z boson production in e^+e^- annihilation as

$$\sigma(e^+e^- \to Z) = \sum_f \sigma_Z\left(e^+e^- \to f\bar{f}\right)$$

where the sum runs over all fermions for which the decay channel $Z \to f\bar{f}$ is open. Using the preceding results, it is easy to see that

$$\sigma(e^+e^- \to Z) = \frac{12\pi}{m_Z^2}\mathrm{BR}(Z \to e^+e^-)$$

What is the numerical value of the ratio $\sigma(e^+e^- \to Z)/\sigma_\gamma(s = m_Z^2)$? If the luminosity of an electron–positron collider is $10^{32}\mathrm{cm}^{-2}\mathrm{s}^{-1}$, how many Z bosons are then produced within one year?
Hint: Remember that one year has approximately $\pi \times 10^7$ s.

7.8 Imagine that you are in the position of a particle physics aficionado who on July 4, 2012 reads in news headlines about the discovery of a Higgs-like boson with mass of 125 GeV. Would you then be able to predict, at least roughly, its lifetime?

7.9 Once again, suppose that the SM Higgs boson has the mass $m_H \approx$ 125 GeV. In analogy with contents of Problem 7.7, perform an analysis of the cross-section for $\sigma(e^+e^- \to f\bar{f})$ in the vicinity of the Higgs boson resonance, i.e. for $s = E_{\mathrm{c.m.}}^2$ close to m_H^2. Evaluate the ratios $\sigma(e^+e^- \to H)/\sigma_\gamma$ and $\sigma(e^+e^- \to H)/\sigma_Z$, with σ_γ and σ_Z taken at $s = m_H^2$.

7.10 Consider the production of a pair of Higgs bosons in electron–positron annihilation, i.e. the process $e^+e^- \to HH$. Identify the corresponding tree diagrams, single out the one giving the dominant contribution, and evaluate the cross-section as a function of the collision energy in the c.m. system. For an explicit numerical illustration, choose, e.g., the energy $E_{\mathrm{c.m.}} = s^{1/2} = 500$ GeV as a reference point. It should be obvious *a priori* that the resulting value of such a tree-level cross-section must be extremely small because of the suppression factor m_e/m_W due to the eeH coupling. Indeed, the reader is supposed to find out, by means of an explicit calculation, that the value in question is of the order of 10^{-24} barn (i.e. yoctobarn). Actually, the considered process is a curious example of a situation where the higher-order (one-loop) diagrams give much larger contribution than the tree-level ones.

The interested reader is encouraged to figure out what the relevant SM one-loop diagrams could be. It turns out that, at the one-loop level, the relevant cross-section may be of the order of 10^{-17} barn (i.e. 10^{-2} femtobarn). For details, see, e.g., [88], [89] and references therein.

7.11 Show that amplitudes of the processes $W^-W^+ \to H\gamma$ and $W^-W^+ \to HZ$ satisfy the condition of tree unitarity.

7.12 The observation of the rare decay $H \to \gamma\gamma$ was one of the first experimental signals marking the discovery of the Higgs boson. As we know, within the SM Lagrangian there is no direct $H\gamma\gamma$ interaction, so that the process in question can only occur at the one-loop (and higher) level. The contribution of the relevant Feynman diagrams is free of the UV divergences, as one may anticipate in view of the perturbative renormalizability of SM. The reader is encouraged to demonstrate by means of an explicit calculation that the contribution of a purely fermionic triangle loop for $H \to \gamma\gamma$ is indeed UV finite (to this end, one may utilize the elementary techniques displayed in Appendix E). Obviously, there are other two relevant loops that any observant reader is supposed to draw readily: a triangle and a bubble made of W boson internal lines, involving the $WW\gamma$ and $WW\gamma\gamma$ couplings, respectively. Their sum is UV finite as well, but the corresponding calculation is much more laborious; this may be left as a challenge for truly hard-working SM aficionados. Anyway, one may find a lot of detailed information concerning the decay process in question in [Gun].

Epilogue

The saga of the standard model of electroweak interactions is undoubtedly one of the most fascinating chapters of modern physics history. The road to the final form of SM had been a remarkable interplay of bold theoretical hypotheses and brilliant experiments that gradually confirmed the theory. Particularly impressive is the way how several new particles were successfully predicted: intermediate vector bosons W and Z, the fourth quark c (actually, also two extra quarks b and t) and the enigmatic Higgs boson. In view of more than four decades of doubts and conceptual disputes concerning the nature of the electroweak symmetry breaking, the ultimate observation of the solitary scalar particle endowed with properties of the Higgs boson was perhaps one of the most astonishing discoveries of particle physics ever made. Thus, the GWS model of electroweak unification, which was a highly speculative construction at the beginning of the 1970s, has finally become a widely recognized physically realistic theory of natural phenomena at a fundamental level. As we know, there are still some long-standing fundamental questions on the interface between particle physics and cosmology, which the standard model is not able to answer; another challenge is understanding a deeper unification of fundamental interactions (the time-honoured subject of "grand unification"). Consequently, there is a lot of activity in the quest for physics beyond SM, both in theory and experiment. Anyway, the present-day SM represents one of the greatest achievements of modern physics and for many years to come will certainly stay with us as a robust reference theory for evaluating the results of forthcoming experiments.

Appendix A

Dirac Equation and its Solutions

The use of Dirac equation in physics is twofold. Either it is treated as relativistic quantum-mechanical equation for a spin-1/2 particle, or it describes a classical bispinor field (that is subsequently quantized in terms of spin-1/2 particles and their antiparticles).[1] In what follows, we consider the case of free particles (fields) — this is just what is needed for the purposes of perturbative quantum field theory (i.e. for the Feynman diagram calculations).

The Dirac equation is written in the familiar covariant form as

$$(i\gamma^\mu \partial_\mu - m)\psi(x) = 0 \tag{A.1}$$

where m is a mass parameter and the coefficients γ^μ, $\mu = 0, 1, 2, 3$ are 4×4 matrices satisfying anticommutation relations

$$\{\gamma^\mu, \gamma^\nu\} \equiv \gamma^\mu\gamma^\nu + \gamma^\nu\gamma^\mu = 2g^{\mu\nu} \tag{A.2}$$

Here, $g^{\mu\nu}$ denotes a metric tensor in the flat four-dimensional space–time; in our conventions, this is taken to be

$$g^{\mu\nu} = g_{\mu\nu} = \begin{pmatrix} 1 & 0 & 0 & 0 \\ 0 & -1 & 0 & 0 \\ 0 & 0 & -1 & 0 \\ 0 & 0 & 0 & -1 \end{pmatrix} \tag{A.3}$$

(of course, the mixed components are $g^\mu_\nu = \delta^\mu_\nu$). Needless to say, multiplication by the 4×4 unit matrix in the right-hand side of (A.2) is tacitly

[1] A terminological remark is perhaps in order here. In general, bispinor (or Dirac spinor) is a quantity that transforms according to the four-dimensional representation $(\frac{1}{2}, 0) \oplus (0, \frac{1}{2})$ of the Lorentz group, i.e. it behaves as a direct sum of two inequivalent two-component Weyl spinors. An elementary introduction to the theory of relativistic spinors can be found in most of the textbooks on field theory, see, e.g., [Ryd].

understood. Thus, (A.2) means that $\gamma^\mu\gamma^\nu = -\gamma^\nu\gamma^\mu$ for $\mu \neq \nu$, $(\gamma^0)^2 = 1$ and $(\gamma^j)^2 = -1$ for $j = 1, 2, 3$ (we denote the 4×4 unit matrix simply as 1).

It is natural to also introduce matrices γ_μ, defined by lowering formally the Lorentz labels of the γ^μ, i.e.

$$\gamma_\mu = g_{\mu\nu}\gamma^\nu \tag{A.4}$$

Using (A.3), one thus has

$$\gamma_0 = \gamma^0, \quad \gamma_j = -\gamma^j \quad \text{for } j = 1, 2, 3 \tag{A.5}$$

A standard ingredient of the relevant notation is the "slash" symbol $\not{a}$ defined for any four-vector a as

$$\not{a} = a_\mu\gamma^\mu = a^\mu\gamma_\mu \tag{A.6}$$

Employing this, the Dirac equation can be recast as

$$(i\not{\partial} - m)\psi(x) = 0 \tag{A.7}$$

with $\not{\partial} = \gamma^\mu\partial_\mu = \gamma_\mu\partial^\mu$.

In general, Dirac matrices must have some specific properties under hermitian conjugation; for our conventional choice of the $g^{\mu\nu}$, one has[2]

$$\gamma_0^\dagger = \gamma_0, \qquad \gamma_j^\dagger = -\gamma_j \tag{A.8}$$

Obviously, this can be written compactly as

$$\gamma_\mu^\dagger = \gamma_0\gamma_\mu\gamma_0 \tag{A.9}$$

In this context, it is natural to introduce another standard symbol, namely that of Dirac conjugation: for a ψ, the conjugate spinor $\bar{\psi}$ is defined by

$$\bar{\psi} = \psi^\dagger\gamma_0 \tag{A.10}$$

A simple consequence of such a definition is that the $\bar{\psi}(x)$ satisfies the equation

$$\bar{\psi}(x)(i\overleftarrow{\not{\partial}} + m) = 0 \tag{A.11}$$

if $\psi(x)$ is a solution of (A.1).

For various purposes, it is highly useful to introduce an additional matrix denoted as γ_5, which anticommutes with all γ^μ, $\mu = 0, 1, 2, 3$. In view

[2]For the metric with opposite signature, i.e. for $g^{\mu\nu} = \text{diag}(-1, 1, 1, 1)$, the γ^j, $j = 1, 2, 3$ would be hermitian and γ^0 antihermitian.

of the anticommutativity of different γ^μ, it is obvious that the product $\gamma^0\gamma^1\gamma^2\gamma^3$ has the desired property; conventionally, we shall define γ_5 as

$$\gamma_5 = i\gamma^0\gamma^1\gamma^2\gamma^3 \tag{A.12}$$

Then

$$\gamma_5^\dagger = \gamma_5, \quad (\gamma_5)^2 = 1 \tag{A.13}$$

Taking into account (A.5), one can also recast (A.12) as

$$\gamma_5 = -i\gamma_0\gamma_1\gamma_2\gamma_3 \tag{A.14}$$

Let us stress that we do not introduce two different symbols γ_5 and γ^5, as it would make little practical sense.

Having displayed basic definitions and some elementary facts concerning the gamma matrices, we should also recall that γ^μ are simply related to the matrices β, α^j, $j = 1, 2, 3$ introduced originally by Dirac; one has

$$\gamma^0 = \beta, \quad \gamma^j = \beta\alpha^j \tag{A.15}$$

or, inverting the last relation,

$$\alpha^j = \gamma^0\gamma^j \tag{A.16}$$

The "old" matrices β and α^j are all hermitian and appear in the Schrödinger-like form of the Dirac equation (in which Lorentz covariance is not "manifest"), namely

$$i\partial_0\psi(x) = \left(-i\vec{\alpha}\cdot\vec{\nabla} + \beta m\right)\psi(x) \tag{A.17}$$

Under a Lorentz transformation of space–time coordinates $x' = \Lambda x$, the Dirac spinor ψ in (A.1) is transformed as

$$\psi'(x') = S(\Lambda)\psi(x) \tag{A.18}$$

where $S(\Lambda)$ is a non-singular 4×4 matrix fulfilling the condition[3]

$$S^{-1}(\Lambda)\gamma^\mu S(\Lambda) = \Lambda^\mu_\nu\gamma^\nu \tag{A.19}$$

An explicit general form of the $S(\Lambda)$ can be found in any textbook on relativistic quantum theory, but it will not be needed for our present

[3]Note that (A.19) reflects the relativistic covariance of Dirac equation, which means that if ψ is a solution of Eq. (A.1), ψ' defined by (A.18) satisfies the same equation, written in primed coordinates.

purposes. We recall here at least some of its important properties. First, it holds[4]

$$S^{-1} = \gamma_0 S^\dagger \gamma_0 \tag{A.20}$$

Its immediate consequence is a simple transformation law for conjugate Dirac spinor: if $\psi'(x') = S\psi(x)$, then

$$\bar{\psi}'(x') = \bar{\psi}(x) S^{-1} \tag{A.21}$$

Further, note that the $S(\Lambda)$ has a particularly simple form for the space inversion $\mathcal{P}$, i.e. for $x' = x_\mathcal{P} = (x^0, -\vec{x})$ (obviously, the corresponding Λ is $\Lambda_\mathcal{P} = \text{diag}(1, -1, -1, -1)$). In that case, one has

$$\psi_\mathcal{P}(x_\mathcal{P}) = \gamma_0 \psi(x) \tag{A.22}$$

Finally, let us add that the covariance relation (A.19) has a simple counterpart involving the γ_5, namely

$$S^{-1}(\Lambda)\gamma_5 S(\Lambda) = \det \Lambda \gamma_5 \tag{A.23}$$

(remember that $\det \Lambda = +1$ for Lorentz boosts and spatial rotations, while $\det \Lambda = -1$ for space inversion).

There are infinitely many realizations of the anticommutation relations (A.2) in terms of 4×4 matrices, but it turns out that they are all equivalent. If γ^μ and $\gamma^{\mu\prime}$ are two sets satisfying (A.2), then there is a non-singular matrix U such that $\gamma^{\mu\prime} = U\gamma^\mu U^{-1}$; moreover, if both γ^μ and $\gamma^{\mu\prime}$ have the above-mentioned hermiticity properties, U is unitary (for a proof of this non-trivial statement, see, e.g., [Mes]). For an illustration, let us display three frequently used representations. The standard (or Dirac) representation reads

$$\gamma^0 = \begin{pmatrix} \mathbb{1} & 0 \\ 0 & -\mathbb{1} \end{pmatrix}, \quad \gamma^j = \begin{pmatrix} 0 & \sigma_j \\ -\sigma_j & 0 \end{pmatrix} \tag{A.24}$$

where $\mathbb{1}$ is the 2×2 unit matrix and σ_j, $j = 1, 2, 3$ are Pauli matrices

$$\sigma_1 = \begin{pmatrix} 0 & 1 \\ 1 & 0 \end{pmatrix}, \quad \sigma_2 = \begin{pmatrix} 0 & -i \\ i & 0 \end{pmatrix}, \quad \sigma_3 = \begin{pmatrix} 1 & 0 \\ 0 & -1 \end{pmatrix} \tag{A.25}$$

Consequently,

$$\gamma_5 = \begin{pmatrix} 0 & \mathbb{1} \\ \mathbb{1} & 0 \end{pmatrix} \tag{A.26}$$

[4]Thus, (A.20) indicates that, in general, $S(\Lambda)$ is not unitary. In fact, $S(\Lambda)$ is unitary for spatial rotations and hermitian for pure Lorentz boosts.

and

$$\alpha^j = \begin{pmatrix} 0 & \sigma_j \\ \sigma_j & 0 \end{pmatrix} \tag{A.27}$$

Next, the so-called chiral (also spinor or Weyl) representation is defined as

$$\gamma^0_{\text{chiral}} = \begin{pmatrix} 0 & \mathbb{1} \\ \mathbb{1} & 0 \end{pmatrix}, \quad \gamma^j_{\text{chiral}} = \begin{pmatrix} 0 & -\sigma_j \\ \sigma_j & 0 \end{pmatrix} \tag{A.28}$$

Then

$$(\gamma_5)_{\text{chiral}} = \begin{pmatrix} \mathbb{1} & 0 \\ 0 & -\mathbb{1} \end{pmatrix} \tag{A.29}$$

Finally, the Majorana representation

$$\begin{aligned} \gamma^0_{\text{Majorana}} &= \begin{pmatrix} 0 & \sigma_2 \\ \sigma_2 & 0 \end{pmatrix}, \quad \gamma^1_{\text{Majorana}} = \begin{pmatrix} i\sigma_3 & 0 \\ 0 & i\sigma_3 \end{pmatrix} \\ \gamma^2_{\text{Majorana}} &= \begin{pmatrix} 0 & -\sigma_2 \\ \sigma_2 & 0 \end{pmatrix}, \quad \gamma^3_{\text{Majorana}} = \begin{pmatrix} -i\sigma_1 & 0 \\ 0 & -i\sigma_1 \end{pmatrix} \end{aligned} \tag{A.30}$$

consists of purely imaginary matrices (it means that Dirac equation has only real coefficients in such a representation). In fact, the explicit form (A.30) corresponds to simple expressions made of standard Dirac matrices (A.24), namely

$$\begin{aligned} \gamma^0_{\text{Majorana}} &= \gamma^0\gamma^2, \quad \gamma^1_{\text{Majorana}} = -\gamma^1\gamma^2 \\ \gamma^2_{\text{Majorana}} &= -\gamma^2, \quad \gamma^3_{\text{Majorana}} = \gamma^2\gamma^3 \end{aligned} \tag{A.31}$$

For completeness, let us specify the equivalence transformations between the above-mentioned representations. One has

$$\gamma^\mu_{\text{chiral}} = U\gamma^\mu U^{-1} \tag{A.32}$$

where

$$U = U^\dagger = U^{-1} = \frac{1}{\sqrt{2}}(\gamma_0 + \gamma_5) = \frac{1}{\sqrt{2}} \begin{pmatrix} \mathbb{1} & \mathbb{1} \\ \mathbb{1} & -\mathbb{1} \end{pmatrix} \tag{A.33}$$

(the gamma matrices in the last expression are taken in the standard representation) and

$$\gamma^\mu_{\text{Majorana}} = V\gamma^\mu V^{-1} \tag{A.34}$$

where

$$V = V^\dagger = V^{-1} = \frac{1}{\sqrt{2}}\gamma^0(1 + \gamma^2) = \frac{1}{\sqrt{2}} \begin{pmatrix} \mathbb{1} & \sigma_2 \\ \sigma_2 & -\mathbb{1} \end{pmatrix} \tag{A.35}$$

In the present text, we employ only the standard representation (A.24).

Dirac matrices are endowed with many remarkable properties that hold independently of a specific representation. Some of the relevant identities are summarized below. Let us start with a series of "sandwich" relations

$$\begin{aligned}\gamma_\alpha \gamma^\alpha &= 4 \\ \gamma_\alpha \gamma_\mu \gamma^\alpha &= -2\gamma_\mu \\ \gamma_\alpha \gamma_\mu \gamma_\nu \gamma^\alpha &= 4g_{\mu\nu} \\ \gamma_\alpha \gamma_\mu \gamma_\nu \gamma_\rho \gamma^\alpha &= -2\gamma_\rho \gamma_\nu \gamma_\mu\end{aligned} \tag{A.36}$$

etc., which follow easily from the basic anticommutation relation (A.2) (when checking the above identities, don't forget that $g^\alpha_\alpha = 4$).

Further, there is a set of formulae for traces of products of gamma matrices. First of all, trace of the product of an arbitrary odd number of γ^μ's is identically zero, i.e.

$$\mathrm{Tr}(\gamma_{\mu_1} \dots \gamma_{\mu_{2k+1}}) = 0 \tag{A.37}$$

(for proving this, the existence of the fully anticommuting γ_5 satisfying $(\gamma_5)^2 = 1$ is instrumental). For products involving an even number of γ^μ's, one has, in particular,

$$\begin{aligned}\mathrm{Tr}(\gamma_\mu \gamma_\nu) &= 4g_{\mu\nu} \\ \mathrm{Tr}(\gamma_\mu \gamma_\nu \gamma_\rho \gamma_\sigma) &= 4(g_{\mu\nu} g_{\rho\sigma} - g_{\mu\rho} g_{\nu\sigma} + g_{\mu\sigma} g_{\nu\rho})\end{aligned} \tag{A.38}$$

These relations, as well as their eventual extensions for longer chains of Dirac matrices, can be derived systematically by using (A.2) and the cyclicity property of traces, i.e. $\mathrm{Tr}(AB) = \mathrm{Tr}(BA)$. Of course, the universal factor 4 appearing in (A.38) is due to the trace of 4×4 unit matrix (acting in the four-dimensional space of Dirac spinors).[5]

Similarly, there is a series of formulae for traces involving also γ_5. In particular,

$$\begin{aligned}\mathrm{Tr}(\gamma_5) &= 0 \\ \mathrm{Tr}(\gamma_\mu \gamma_\nu \gamma_5) &= 0 \\ \mathrm{Tr}(\gamma_\mu \gamma_\nu \gamma_\rho \gamma_\sigma \gamma_5) &= 4i\epsilon_{\mu\nu\rho\sigma}\end{aligned} \tag{A.39}$$

[5]Note also that traces of products of Dirac matrices behave, in general, as tensors under Lorentz transformations (this is a simple consequence of trace cyclicity and the covariance relation (A.19)). At the same time, they consist of pure numbers and therefore can depend only on components of metric tensor. Such an argument provides a useful additional insight into the algebraic structure of (A.38).

where $\epsilon_{\mu\nu\rho\sigma}$ is the totally antisymmetric Levi-Civita symbol; in our conventions, $\epsilon_{0123} = +1$. Again, there is a tensor argument for the algebraic structure of the last relation in (A.39). Taking into account (A.19) together with (A.23), it is seen that the trace in question is a (purely numerical) pseudotensor under Lorentz transformations; however, the only numerical four-index pseudotensor in four space–time dimensions is just the Levi-Civita symbol. Thus, the last trace in (A.39) can only be proportional to the $\epsilon_{\mu\nu\rho\sigma}$ (for the same reason, the first two traces must vanish as there is no possibility to make a pseudoscalar or a two-index pseudotensor out of $\epsilon_{\mu\nu\rho\sigma}$ and the metric tensor). Traces of longer chains of the type (A.39) can be expressed as linear combinations of appropriate products of $g_{\alpha\beta}$ and $\epsilon_{\mu\nu\rho\sigma}$ (one such example is shown in (A.50)).

Finally, for the sake of completeness, one should mention another general trace identity, namely

$$\mathrm{Tr}(\gamma_\alpha\gamma_\beta\ldots\gamma_\tau\gamma_\omega) = \mathrm{Tr}(\gamma_\omega\gamma_\tau\ldots\gamma_\beta\gamma_\alpha)$$

In this context, let us list some general relations for products of two Levi-Civita tensors, which are highly useful in calculations involving Lorentz pseudotensors. The "master formula" reads

$$\epsilon^{\iota\kappa\lambda\mu}\epsilon_{\rho\sigma\tau\omega} = -\begin{vmatrix} \delta^\iota_\rho & \delta^\iota_\sigma & \delta^\iota_\tau & \delta^\iota_\omega \\ \delta^\kappa_\rho & \delta^\kappa_\sigma & \delta^\kappa_\tau & \delta^\kappa_\omega \\ \delta^\lambda_\rho & \delta^\lambda_\sigma & \delta^\lambda_\tau & \delta^\lambda_\omega \\ \delta^\mu_\rho & \delta^\mu_\sigma & \delta^\mu_\tau & \delta^\mu_\omega \end{vmatrix} \tag{A.40}$$

(note that such a result is quite natural *a priori*, since the product of two pseudotensors must be a true tensor and the determinant maintains automatically the required antisymmetry). Contractions of Lorentz indices in the left-hand side of (A.40) yield

$$\epsilon^{\iota\kappa\lambda\omega}\epsilon_{\rho\sigma\tau\omega} = -\begin{vmatrix} \delta^\iota_\rho & \delta^\iota_\sigma & \delta^\iota_\tau \\ \delta^\kappa_\rho & \delta^\kappa_\sigma & \delta^\kappa_\tau \\ \delta^\lambda_\rho & \delta^\lambda_\sigma & \delta^\lambda_\tau \end{vmatrix} \tag{A.41}$$

and, in particular,

$$\epsilon^{\iota\kappa\tau\omega}\epsilon_{\rho\sigma\tau\omega} = -2\begin{vmatrix} \delta^\iota_\rho & \delta^\iota_\sigma \\ \delta^\kappa_\rho & \delta^\kappa_\sigma \end{vmatrix} = -2(\delta^\iota_\rho\delta^\kappa_\sigma - \delta^\iota_\sigma\delta^\kappa_\rho) \tag{A.42}$$

From (A.42), one then gets readily

$$\epsilon^{\iota\sigma\tau\omega}\epsilon_{\rho\sigma\tau\omega} = -6\delta^\iota_\rho \tag{A.43}$$

There is another useful relation (of a completely different type), which is worth mentioning here:

$$g_{\lambda\mu}\epsilon_{\nu\rho\sigma\tau} - g_{\lambda\nu}\epsilon_{\mu\rho\sigma\tau} + g_{\lambda\rho}\epsilon_{\mu\nu\sigma\tau} - g_{\lambda\sigma}\epsilon_{\mu\nu\rho\tau} + g_{\lambda\tau}\epsilon_{\mu\nu\rho\sigma} = 0 \tag{A.44}$$

(note that this identity comes out easily when working out the expression $\text{Tr}(\gamma_\lambda\gamma_\mu\gamma_\nu\gamma_\rho\gamma_\sigma\gamma_\tau\gamma_5)$ by using (A.2), the γ_5 anticommutativity and trace cyclicity).

A highly useful technical device of "diracology" is a special basis in the 16-dimensional space of all 4×4 matrices, which is made of appropriate products of the γ^μ's. The "canonical" choice is

$$\Gamma_S = 1, \quad \Gamma_V = \gamma_\mu, \quad \Gamma_T = \sigma_{\mu\nu}, \quad \Gamma_A = \gamma_5\gamma_\mu, \quad \Gamma_P = \gamma_5 \tag{A.45}$$

with $\mu, \nu = 0, 1, 2, 3$; the $\sigma_{\mu\nu}$ is defined as

$$\sigma_{\mu\nu} = \frac{i}{2}[\gamma_\mu, \gamma_\nu] \tag{A.46}$$

Total number of the matrices (A.45) can be checked immediately; one gets $1 + 4 + 6 + 4 + 1 = 16$ (obviously, there are only six linearly independent matrices $\sigma_{\mu\nu}$ because of antisymmetry, $\sigma_{\mu\nu} = -\sigma_{\nu\mu}$). The indices S, V, T, A, P stand for scalar, vector, tensor, axial-vector (pseudovector) and pseudoscalar, respectively, and they refer to the transformation properties of bilinear quantities obtained by sandwiching the matrices (A.45) between Dirac spinors. In particular, let $\psi_1 = \psi_1(x)$ and $\psi_2 = \psi_2(x)$ be two Dirac spinors; then the expressions

$$\bar{\psi}_1\psi_2, \quad \bar{\psi}_1\gamma_\mu\psi_2, \quad \bar{\psi}_1\sigma_{\mu\nu}\psi_2, \quad \bar{\psi}_1\gamma_5\gamma_\mu\psi_2, \quad \bar{\psi}_1\gamma_5\psi_2 \tag{A.47}$$

behave consecutively as a scalar, vector, antisymmetric tensor, axial vector and pseudoscalar under a Lorentz transformation. This is proved easily if one takes into account the transformation laws (A.18) and (A.21) and the relations (A.19) and (A.23).

It is not difficult to see that the matrices Γ_j, $j = S, V, T, A, P$ have the properties

$$(\Gamma_j)^2 = \pm 1, \quad \text{Tr}(\Gamma_j\Gamma_k) = 0 \quad \text{for } j \neq k \tag{A.48}$$

Obviously, the relations (A.48) are instrumental for calculating the expansion coefficients of a general matrix in the basis (A.45). As a simple application, one can derive the following formula for the product of three Dirac matrices:

$$\gamma_\lambda\gamma_\mu\gamma_\nu = (g_{\lambda\mu}g_{\nu\rho} - g_{\lambda\nu}g_{\mu\rho} + g_{\lambda\rho}g_{\mu\nu})\gamma^\rho + i\epsilon_{\lambda\mu\nu\rho}\gamma_5\gamma^\rho \tag{A.49}$$

(proving the last identity is left to the reader as an instructive exercise). Note that using this and the identity (A.44), one easily obtains the trace formula

$$\mathrm{Tr}(\gamma_\lambda\gamma_\mu\gamma_\nu\gamma_\rho\gamma_\sigma\gamma_\tau\gamma_5) = 4i\big(g_{\lambda\mu}\epsilon_{\nu\rho\sigma\tau} - g_{\lambda\nu}\epsilon_{\mu\rho\sigma\tau} + g_{\mu\nu}\epsilon_{\lambda\rho\sigma\tau} + g_{\sigma\tau}\epsilon_{\lambda\mu\nu\rho} - g_{\rho\tau}\epsilon_{\lambda\mu\nu\sigma} + g_{\rho\sigma}\epsilon_{\lambda\mu\nu\tau}\big) \tag{A.50}$$

When calculating scattering cross-sections and decay probabilities within perturbative quantum field theory, one often encounters products of two traces of the type (A.38) and/or (A.39), contracted over Lorentz indices. Here are some practical formulae that improve greatly the efficiency of algebraic manipulations:

$$\begin{aligned}
\mathrm{Tr}(\not{a}\gamma^\mu\not{b}\gamma^\nu)\cdot\mathrm{Tr}(\not{c}\gamma_\mu\not{d}\gamma_\nu) &= 32\big[(a\cdot c)(b\cdot d) + (a\cdot d)(b\cdot c)\big] \\
\mathrm{Tr}(\not{a}\gamma^\mu\not{b}\gamma^\nu\gamma_5)\cdot\mathrm{Tr}(\not{c}\gamma_\mu\not{d}\gamma_\nu\gamma_5) &= 32\big[(a\cdot c)(b\cdot d) - (a\cdot d)(b\cdot c)\big] \\
\mathrm{Tr}(\not{a}\gamma^\mu\not{b}\gamma^\nu)\cdot\mathrm{Tr}(\not{c}\gamma_\mu\not{d}\gamma_\nu\gamma_5) &= 0
\end{aligned} \tag{A.51}$$

Note that these relations can be obtained in a straightforward way by using (A.38), (A.39) and the identity (A.42). Let us also give an analogous formula involving $\sigma_{\mu\nu}$:

$$\mathrm{Tr}(\not{a}\sigma^{\alpha\beta}\not{b}\sigma^{\mu\nu})\cdot\mathrm{Tr}(\not{c}\sigma_{\alpha\beta}\not{d}\sigma_{\mu\nu}) = 128\big[2(a\cdot c)(b\cdot d) + 2(a\cdot d)(b\cdot c) - (a\cdot b)(c\cdot d)\big]$$

(needless to say, a derivation of the last relation is much more tedious than in the preceding case).

For completeness, we list some useful identities for Pauli matrices:

$$\begin{aligned}
\sigma_j\sigma_k &= \delta_{jk}\cdot\mathbb{1} + i\epsilon_{jkl}\sigma_l \\
\mathrm{Tr}(\sigma_j\sigma_k) &= 2\delta_{jk} \\
\sum_i(\sigma_i)_{ab}(\sigma_i)_{cd} &= 2\delta_{ad}\delta_{bc} - \delta_{ab}\delta_{cd}
\end{aligned} \tag{A.52}$$

Let us now proceed further to summarize some essential properties of solutions of the free-particle Dirac equation (A.1). We consider plane waves, i.e. the solutions corresponding to a definite energy and momentum. There are two independent Ansätze for such a solution, namely

$$\begin{aligned}
\psi_+(x) &= u(p)\,\mathrm{e}^{-ipx} \\
\psi_-(x) &= v(p)\,\mathrm{e}^{ipx}
\end{aligned} \tag{A.53}$$

with $px = p_0x_0 - \vec{p}\cdot\vec{x}$, where we take $p_0 > 0$ by definition. Substituting (A.53) into (A.1), one gets

$$(\not{p} - m)u(p) = 0 \tag{A.54}$$

and

$$(\not{p} + m)v(p) = 0 \tag{A.55}$$

Obviously, p must then satisfy $p^2 = m^2$. Further, taking into account (A.17), it becomes clear that ψ_+ corresponds to a positive energy $E = p_0 = \sqrt{\vec{p}^2 + m^2}$ while ψ_- carries negative energy $-\sqrt{\vec{p}^2 + m^2}$. It is useful to know that solutions of (A.54) and (A.55) are interrelated through the operation of charge conjugation defined in terms of the matrix $C = i\gamma^2\gamma^0$ (in standard representation): if $u(p)$ is a solution of (A.54), then

$$u_c(p) = C\bar{u}(p)^T \tag{A.56}$$

(with T denoting matrix transposition) satisfies Eq. (A.55).

A frequently used set of $u(p)$ and $v(p)$, corresponding to the standard representation of Dirac matrices in (A.54) and (A.55), can be described explicitly as

$$u^{(r)}(p) = \sqrt{E+m}\begin{pmatrix} \chi^{(r)} \\ \dfrac{\vec{\sigma}\cdot\vec{p}}{E+m}\chi^{(r)} \end{pmatrix}, \quad r = 1, 2 \tag{A.57}$$

and

$$v^{(r)}(p) = \pm\sqrt{E+m}\begin{pmatrix} \dfrac{\vec{\sigma}\cdot\vec{p}}{E+m}\chi^{(r)} \\ \chi^{(r)} \end{pmatrix}, \quad r = 1, 2 \tag{A.58}$$

where $E = \sqrt{\vec{p}^2 + m^2}$ and

$$\chi^{(1)} = \begin{pmatrix} 1 \\ 0 \end{pmatrix}, \quad \chi^{(2)} = \begin{pmatrix} 0 \\ 1 \end{pmatrix} \tag{A.59}$$

The upper and lower signs in (A.58) holds for $r = 1$ and $r = 2$, respectively; note that the form of the $v^{(r)}(p)$ (including the overall sign) is determined by the charge-conjugation transformation (A.56). It is important to stress that the solutions (A.57) and (A.58) are normalized according to

$$\begin{aligned} \bar{u}(p)u(p) &= 2m \\ \bar{v}(p)v(p) &= -2m \end{aligned} \tag{A.60}$$

(such a normalization is most convenient for the discussion of high-energy behaviour of scattering amplitudes represented by Feynman graphs).

Let us also remark that in the non-relativistic limit, i.e. for $|\vec{p}| \ll m$, the lower two components of the bispinor (A.57) become negligible and the Dirac particle is thus effectively described by means of a two-component spinor, proportional to $\chi^{(r)}$; this is the main advantage of working in the standard representation.

The index $r = 1, 2$ in (A.57) and (A.58) labels spin degrees of freedom. In particular, $u^{(r)}(p)$ corresponds to positive-energy solution with spin up ($r = 1$) or down ($r = 2$) along the third axis of the coordinate system in the particle rest frame. In fact, spin states of a Dirac particle can be described, quite generally, in an elegant covariant way. We are now going to summarize briefly the contents of such a formalism as well as some relevant formulae. Unless stated otherwise, we assume explicitly that $m \neq 0$.

A basic notion is that of the "spin four-vector". For a given four-momentum p, one defines $s^\mu = s^\mu(p)$, $\mu = 0, 1, 2, 3$, so that

$$s^\mu p_\mu = 0 \tag{A.61}$$

and the s^μ behaves as a space-like Lorentz four-vector; its normalization is conveniently fixed by

$$s^2 = -1 \tag{A.62}$$

(concerning the terminology, let us add that the spin four-vector s is sometimes also called "polarization vector"). A remark is in order here. For a conceptual construction of the spin vector $s = s(p)$, one can start in the particle rest frame, where $p = p^{(0)} = (m, 0, 0, 0)$ and take $s = s^{(0)} = (0, \vec{s})$, with $\vec{s}$ being a unit vector in three-dimensional space ($\vec{s}$ is to be understood as the spin direction in the rest frame). Passing from $p^{(0)}$ to an arbitrary p, $p^2 = m^2$, $s(p)$ is defined by means of the corresponding Lorentz transformation of the $s^{(0)}$. Of course, given a spatial direction $\vec{s}$ (in the rest frame), there are two independent states of a Dirac particle, characterized by the spin projection pointing up or down along the $\vec{s}$. Alternatively, one can say that these two states correspond to the opposite directions $\vec{s}$ and $-\vec{s}$ (i.e. to the spin parallel with either $\vec{s}$ or $-\vec{s}$). In a general reference frame, this means that for a given four-momentum p (and, say, for a positive energy), there are two independent states corresponding to the spin four-vectors s and $-s$, respectively.

Covariant description of the spin states in question can be formulated as follows. With an $s = s(p)$ at hand, one considers solutions of (A.54) and (A.55) satisfying[6]

$$\begin{aligned} \gamma_5 \not{s} u(p,s) &= u(p,s) \\ \gamma_5 \not{s} v(p,s) &= v(p,s) \end{aligned} \tag{A.63}$$

In other words, $u(p,s)$ or $v(p,s)$ is obtained from an arbitrary solution of (A.54) or (A.55) by means of the projector

$$P_+(s) = \frac{1}{2}(1 + \gamma_5 \not{s}) \tag{A.64}$$

(the reader is recommended to verify explicitly that the $P_+(s)$ is indeed a projector, i.e. that it holds $(P_+(s))^2 = P_+(s)$). Similarly, using the projector

$$P_-(s) = \frac{1}{2}(1 - \gamma_5 \not{s}) \tag{A.65}$$

one obtains the remaining independent spin states, corresponding to the spin vector $-s$. It can be shown that the four spinors $u(p, \pm s)$ and $v(p, \pm s)$ determine a complete system of solutions of the Dirac equation for a free particle. For routine Feynman diagram calculations of cross-sections or decay probabilities (employing the familiar "trace techniques"), explicit expressions for $u(p,s)$ and $v(p,s)$ are not necessary; one really needs only the combinations like $u(p,s)\bar{u}(p,s)$, etc. The relevant results are

$$\begin{aligned} u(p,s)\bar{u}(p,s) &= (\not{p} + m)\frac{1 + \gamma_5 \not{s}}{2} \\ v(p,s)\bar{v}(p,s) &= (\not{p} - m)\frac{1 + \gamma_5 \not{s}}{2} \end{aligned} \tag{A.66}$$

The corresponding formulae for $u(p,-s)\bar{u}(p,-s)$ and $v(p,-s)\bar{v}(p,-s)$ are obtained from (A.66) trivially by replacing s with $-s$ there. Summing the expressions (A.66) over the individual spin states (or "polarizations"), one gets

$$\begin{aligned} \sum_{\text{spin}} u(p,s)\bar{u}(p,s) &= \not{p} + m \\ \sum_{\text{spin}} v(p,s)\bar{v}(p,s) &= \not{p} - m \end{aligned} \tag{A.67}$$

There is an important particular example of the spin vector that deserves special attention. The specific spin states we have in mind

[6]It should be noted that $\not{p}$ and $\gamma_5 \not{s}$ commute; it is a simple consequence of the relation $s \cdot p = 0$.

correspond to **helicity** or "longitudinal polarization", described in terms of an $s(p)$, whose spatial part is directed along the three-momentum $\vec{p}$. For reasons that are explained below, we denote the spin four-vector $s = (s^0, \vec{s})$ having $\vec{s}$ parallel to $\vec{p}$ as $s_R(p)$, indicating thus that it corresponds to right-handed particle (with positive helicity); similarly, the left-handed state (with negative helicity) is described by $s_L(p) = -s_R(p)$. It is not difficult to find an explicit form of the $s_R(p)$. Using the Ansatz $s_R(p) = (s^0, \lambda\vec{p})$ with $\lambda > 0$ and taking into account the general relations $s \cdot p = 0$, $s^2 = -1$, one readily gets

$$s_R^{\mu}(p) = \left(\frac{|\vec{p}|}{m}, \frac{E}{m}\frac{\vec{p}}{|\vec{p}|}\right) \tag{A.68}$$

Let us now comment on the connection between the above formal description of helicity and its straightforward physical definition (which may be more familiar to an average reader). The helicity of a particle with definite momentum is generally defined as the projection of spin on the direction of motion. Thus, for a Dirac particle with momentum $\vec{p}$, helicity is identified with an eigenvalue of the 4×4 matrix

$$h(p) = \frac{\vec{\Sigma} \cdot \vec{p}}{|\vec{p}|} \tag{A.69}$$

where

$$\vec{\Sigma} = \begin{pmatrix} \vec{\sigma} & 0 \\ 0 & \vec{\sigma} \end{pmatrix} \tag{A.70}$$

(strictly speaking, (A.69) represents the spin projection up to a factor of 1/2, since the spin matrix for a Dirac particle is $\frac{1}{2}\vec{\Sigma}$). Now, the crucial observation is that

$$\gamma_5 \not{s}_R(p) u(p) = \frac{\vec{\Sigma} \cdot \vec{p}}{|\vec{p}|} u(p) \tag{A.71}$$

for any $u(p)$ satisfying Eq. (A.54). It is clear that the identity (A.71) establishes the aforementioned equivalence between the two descriptions of helicity. A proof of (A.71) is not difficult; apart from some straightforward algebraic manipulations, one has to take into account the identity

$$\vec{\Sigma} = \gamma_5 \vec{\alpha} \tag{A.72}$$

that obviously holds in the standard representation (in fact, defining generally $\Sigma^j = \frac{1}{2}\epsilon^{jkl}\sigma^{kl}$, (A.72) is valid in any representation). Note that an identity analogous to (A.71) can also be derived for solutions of (A.55); however, when considering the helicity of a $v(p)$, one should not forget that the corresponding plane wave carries momentum $-\vec{p}$ (and negative energy).

For completeness, let us now discuss briefly the case of a massless particle. Obviously — the expression (A.68) makes no sense for $m = 0$; more generally, one can verify directly — starting from the basic requirements — that a space-like longitudinal spin four-vector simply cannot be constructed in the massless case. Nevertheless, the definition of helicity based on (A.69) is still applicable. Moreover, making use of the identity (A.72), the characterization of the helicity states is greatly simplified: it turns out that helicity is essentially reduced to **chirality**, which is an eigenvalue of γ_5. In particular, for $u(p)$, helicity coincides with chirality, while for $v(p)$, helicity is equal to chirality taken with minus sign. Thus, for $m = 0$, the left-handed and right-handed spinors $u_{L,R}(p)$ and $v_{L,R}(p)$ satisfy

$$\begin{aligned}\gamma_5 u_L(p) &= -u_L(p)\\ \gamma_5 u_R(p) &= u_R(p)\end{aligned} \tag{A.73}$$

and

$$\begin{aligned}\gamma_5 v_L(p) &= v_L(p)\\ \gamma_5 v_R(p) &= -v_R(p)\end{aligned} \tag{A.74}$$

These relations can be recast in terms of appropriate projectors, namely

$$\begin{aligned}\frac{1}{2}(1-\gamma_5)u_L(p) &= u_L(p)\\ \frac{1}{2}(1+\gamma_5)u_R(p) &= u_R(p)\end{aligned} \tag{A.75}$$

and

$$\begin{aligned}\frac{1}{2}(1+\gamma_5)v_L(p) &= v_L(p)\\ \frac{1}{2}(1-\gamma_5)v_R(p) &= v_R(p)\end{aligned} \tag{A.76}$$

Similarly, as in the massive case, for practical calculations, one needs combinations like $u_L(p)\bar{u}_L(p)$, etc. For that purpose, one cannot simply take the limit $m \to 0$ in (A.66) since it does not exist. On the other hand, the summed expression (A.67) is safe in the massless limit and one has

$$\sum_{\text{spin}} u(p)\bar{u}(p) = u_L(p)\bar{u}_L(p) + u_R(p)\bar{u}_R(p) = \not{p}$$
$$\sum_{\text{spin}} v(p)\bar{v}(p) = v_L(p)\bar{v}_L(p) + v_R(p)\bar{v}_R(p) = \not{p} \tag{A.77}$$

The desired "anatomy" of the relations (A.77) can be obtained from (A.75), (A.76) and (A.77) by means of simple algebraic tricks (among other things, one has to utilize obvious relations like $\frac{1}{2}(1-\gamma_5)u_R = 0$, etc.). Leaving a detailed derivation to the interested reader, we give here only the result:

$$u_L(p)\bar{u}_L(p) = \not{p}\,\frac{1+\gamma_5}{2}, \quad u_R(p)\bar{u}_R(p) = \not{p}\,\frac{1-\gamma_5}{2}$$
$$v_L(p)\bar{v}_L(p) = \not{p}\,\frac{1-\gamma_5}{2}, \quad v_R(p)\bar{v}_R(p) = \not{p}\,\frac{1+\gamma_5}{2} \tag{A.78}$$

When summarizing important properties of plane-wave solutions of the Dirac equation, one should also mention the so-called Gordon identity that represents a practically useful decomposition of the current $\bar{u}(p)\gamma_\mu u(p')$ into "convective" and "spin" parts. To arrive at such a result, one can start with the identity

$$\bar{u}(p)\big[(\not{p}-m)\gamma_\mu + \gamma_\mu(\not{p}'-m)\big]u(p') = 0 \tag{A.79}$$

that obviously holds for solutions of Eq. (A.54) (we suppress here the spin labels, since these are irrelevant in the present context). Decomposing the matrix products in (A.79) into anticommutators and commutators, employing the basic relation (A.2) and the definition (A.46), one readily gets

$$\bar{u}(p)\gamma_\mu u(p') = \frac{1}{2m}\bar{u}(p)\big[(p_\mu + p'_\mu) + i\sigma_{\mu\nu}(p^\nu - p'^\nu)\big]u(p') \tag{A.80}$$

In fact, it is easy to realize that there are three additional identities of such a type, involving one or two spinors v instead of u. Such generalizations of (A.80) are derived by modifying appropriately the "master identity" (A.79): since a $v(p)$ satisfies Eq. (A.55), it is sufficient to change the sign of the corresponding four-momentum whenever v stands in place of u. Thus, it becomes clear that the resulting Gordon identities are obtained in the same manner — simply by changing signs of the relevant four-momenta in (A.80).

To conclude this appendix, let us now recapitulate briefly some basic relations concerning the quantized free Dirac field. This is represented by a four-component spinor operator in the Fock space written as

$$\psi(x) = \sum_{\pm s} \int \frac{d^3p}{(2\pi)^{3/2}(2p_0)^{1/2}} \left[b(p,s)u(p,s)\,\mathrm{e}^{-ipx} + d^+(p,s)v(p,s)\,\mathrm{e}^{ipx} \right]$$

$$\bar{\psi}(x) = \sum_{\pm s} \int \frac{d^3p}{(2\pi)^{3/2}(2p_0)^{1/2}} \left[b^+(p,s)\bar{u}(p,s)\,\mathrm{e}^{ipx} + d(p,s)\bar{v}(p,s)\,\mathrm{e}^{-ipx} \right] \tag{A.81}$$

Here, $b(p,s)$, $d(p,s)$ are annihilation operators of the particle and antiparticle, respectively, and $b^+(p,s)$, $d^+(p,s)$ are the corresponding creation operators. Of course, the annihilation and creation operators are related through hermitian conjugation, i.e. $b^+(p,s) = b^\dagger(p,s)$, $d^+(p,s) = d^\dagger(p,s)$. The four-momenta in (A.81) are on the mass shell, i.e. one takes everywhere $p_0 = E(p) = \sqrt{\vec{p}^2 + m^2}$.

The field operators satisfy equal-time (E.T.) anticommutation relations

$$\begin{aligned} \{\psi_a(x), \psi_b(y)\}_{\text{E.T.}} &= 0 \\ \{\psi_a(x), \psi_b^\dagger(y)\}_{\text{E.T.}} &= \delta_{ab}\delta^3(\vec{x} - \vec{y}) \end{aligned} \tag{A.82}$$

that yield the algebra of creation and annihilation operators

$$\begin{aligned} \{b(p,s), b(p',s')\} = 0, &\quad \{d(p,s), d(p',s')\} = 0 \\ \{b(p,s), b^+(p',s')\} &= \delta_{ss'}\delta^3(\vec{p} - \vec{p}\,') \\ \{d(p,s), d^+(p',s')\} &= \delta_{ss'}\delta^3(\vec{p} - \vec{p}\,') \\ \{b(p,s), d(p',s')\} = 0, &\quad \{b(p,s), d^+(p',s')\} = 0 \end{aligned} \tag{A.83}$$

(other anticommutators are obtained by means of hermitian conjugation). Note that a passage from (A.82) to (A.83) (and vice versa) is guaranteed, among other things, by the choice of the normalization factor $(2\pi)^{-3/2}(2p_0)^{-1/2}$ introduced in the definition (A.81). It should be emphasized that the anticommutation relations (A.83) imply a specific normalization of one-particle states; in particular, defining $|p,s\rangle = b^+(p,s)|0\rangle$, etc. (with $|0\rangle$ being the Fock vacuum state), one has

$$\langle p,s|p',s'\rangle = \delta_{ss'}\delta^3(\vec{p} - \vec{p}\,') \tag{A.84}$$

Although such a non-covariant normalization is not universally accepted in the current literature, we stick to this convention (essentially corresponding to [BjD]) throughout the present text.

Appendix B

Scattering Amplitudes, Cross-Sections and Decay Rates

Let us start with definition of the Lorentz invariant scattering (or decay) amplitude $\mathcal{M}_{fi}$ in terms of an S-matrix element $S_{fi} = \langle f|S|i\rangle$. This reads

$$S_{fi} = \delta_{fi} + (2\pi)^4\delta^4(P_f - P_i)(i\mathcal{M}_{fi})\prod_{f,i}\frac{1}{(2\pi)^{3/2}(2E_{f,i})^{1/2}} \tag{B.1}$$

where P_f and P_i denote the total four-momenta of the final and initial particles, respectively. The normalization factors under the product symbol correspond to the conventional choice, exemplified by the formula (A.81) for quantized Dirac field; such a choice means that these factors have the same form for bosons and fermions. Note also that our sign convention for $\mathcal{M}_{fi}$ differs from that adopted in some standard textbooks, e.g., the definition used in [BjD] is obtained from (B.1) by replacement $i\mathcal{M}_{fi} \to -i\mathcal{M}_{fi}$.

In the context of practical calculations, $\mathcal{M}_{fi}$ is often called simply "matrix element" (for a given process). Within perturbation theory, this is evaluated by means of the relevant covariant Feynman rules; in particular, the contributions of external Dirac particles are represented by the corresponding spinors u or v, etc.[1] Knowing the matrix element $\mathcal{M}_{fi}$, one can compute physically observable quantities for the considered process. In particular, the differential cross-section for a reaction $1 + 2 \to 3 + 4 + \cdots + n$ is given by the general formula

$$d\sigma = \frac{1}{|\vec{v}_1 - \vec{v}_2|}\frac{1}{2E_1}\frac{1}{2E_2}|\mathcal{M}_{fi}|^2(2\pi)^4\delta^4\left(p_1 + p_2 - \sum_{j=3}^{n} p_j\right)$$
$$\times \frac{d^3p_3}{(2\pi)^3 2E_3}\cdots\frac{d^3p_n}{(2\pi)^3 2E_n}K \tag{B.2}$$

[1]In other words, the Eq. (B.1) means that the non-covariant normalization factors do not enter the routine Feynman diagram calculations.

(irrespectively of whether the particles $1, \ldots, n$ are bosons or fermions). $\vec{v}_1$ and $\vec{v}_2$ denote velocities of the initial particles (we assume that vectors $\vec{v}_1$ and $\vec{v}_2$ are parallel and have opposite directions), $p_j = (E_j, \vec{p}_j)$, $j = 1, 2, \ldots, n$ are four-momenta, i.e. $E_j = \sqrt{\vec{p}_j^{\,2} + m_j^2}$, and K is a combinatorial ("statistical") factor, which is different from 1 only when some of the final-state particles are identical:

$$K = \prod_{r=1}^{k} \frac{1}{n_r!} \tag{B.3}$$

where n_r is the number of identical particles of the rth kind in the final state $|f\rangle$ (of course, $n_1 + \cdots + n_k = n - 2$).

It is worth noting here that with the result (B.2) at hand, one can determine the dimension of the matrix element $\mathcal{M}_{fi}$ on quite general grounds. The argument goes as follows. The dimension of the left-hand side of (B.2) is (length)^2, i.e. (mass)$^{-2}$ in the system of units where $\hbar = c = 1$. Thus,

$$[d\sigma] = M^{-2} \tag{B.4}$$

with M being an arbitrary mass and in the right-hand side of (B.2), one has

$$M^{-1} \cdot M^{-1} \cdot \left[|\mathcal{M}_{fi}|^2\right] \cdot M^{-4} \cdot (M^2)^{n-2} = \left[|\mathcal{M}_{fi}|^2\right] \cdot M^{2n-10} \tag{B.5}$$

(recall that dimension of $\delta^4(P_f - P_i)$ is M^{-4} !). Comparing (B.4) and (B.5), one gets immediately the desired result:

$$[\mathcal{M}_{fi}] = M^{4-n} \tag{B.6}$$

In particular, (B.6) shows that the matrix element for an arbitrary *binary* process $1 + 2 \to 3 + 4$ is *dimensionless*; this simple observation is quite useful in estimating the high-energy behaviour of scattering amplitudes.

Before proceeding further, let us recall briefly some elementary kinematics. Considering a binary process and using the above notation for the corresponding four-momenta, one defines the Lorentz invariant Mandelstam variables as

$$\begin{aligned} s &= (p_1 + p_2)^2 = (p_3 + p_4)^2 \\ t &= (p_1 - p_3)^2 = (p_2 - p_4)^2 \\ u &= (p_1 - p_4)^2 = (p_2 - p_3)^2 \end{aligned} \tag{B.7}$$

(in writing (B.7), we have taken into account explicitly the four-momentum conservation $p_1 + p_2 = p_3 + p_4$). It is not difficult to show that s, t, and u

satisfy the identity

$$s + t + u = \sum_{j=1}^{4} m_j^2 \tag{B.8}$$

One should also note that s has a simple physical meaning: it coincides with the square of total centre-of-mass (c.m.) energy of the colliding particles. This is obvious, since $(p_1 + p_2)^2$ has the same value in any Lorentz frame and in the c.m. system, one has $p_1 + p_2 = (E_1 + E_2, \vec{0})$ by definition. Thus, one has

$$s^{1/2} = \sqrt{|\vec{p}_{\text{c.m.}}|^2 + m_1^2} + \sqrt{|\vec{p}_{\text{c.m.}}|^2 + m_2^2} \tag{B.9}$$

with $\vec{p}_{\text{c.m.}}$ standing for the c.m. momentum of one of the colliding particles (one can take, e.g., $\vec{p}_{\text{c.m.}} = \vec{p}_{1\text{c.m.}} = -\vec{p}_{2\text{c.m.}}$). An explicit formula for the $|\vec{p}_{\text{c.m.}}|$ then follows easily from (B.9); one gets

$$|\vec{p}_{\text{c.m.}}| = \left[\frac{\lambda(s, m_1^2, m_2^2)}{4s}\right]^{1/2} \tag{B.10}$$

where

$$\lambda(x, y, z) = x^2 + y^2 + z^2 - 2xy - 2xz - 2yz \tag{B.11}$$

Integrating over an appropriate part of the phase space of final states in (B.2), one can derive special formulae that are suitable for practical applications. The most frequently used result is the expression for angular distribution of final-state particles in a binary process $1 + 2 \to 3 + 4$, considered in the c.m. frame. Below, we quote the standard formula for the corresponding differential cross-section (its derivation is rather straightforward and can be found in many places, see, e.g., Appendix C in [Hor]). Assuming that the final-state particles are not identical, one has

$$\frac{d\sigma}{d\Omega_{\text{c.m.}}} = \frac{1}{64\pi^2} \frac{1}{s} \frac{|\vec{p}\,'_{\text{c.m.}}|}{|\vec{p}_{\text{c.m.}}|} |\mathcal{M}_{fi}|^2 \tag{B.12}$$

where the $|\vec{p}_{\text{c.m.}}|$ has been defined in (B.9) and $|\vec{p}\,'_{\text{c.m.}}|$ has an analogous meaning for the final-state particles; therefore,

$$|\vec{p}\,'_{\text{c.m.}}| = \left[\frac{\lambda(s, m_3^2, m_4^2)}{4s}\right]^{1/2} \tag{B.13}$$

The $d\Omega_{\text{c.m.}}$ is an element of solid angle corresponding to the direction of $\vec{p}\,'_{\text{c.m.}}$; in spherical coordinates, this has the standard form

$$d\Omega_{\text{c.m.}} = \sin\vartheta_{\text{c.m.}} d\vartheta_{\text{c.m.}} d\varphi_{\text{c.m.}}$$

If particles 3 and 4 were identical, the right-hand side of (B.12) would include the combinatorial factor $K = 1/2$.

For an elastic scattering process (where the final particles are the same as those in the initial state), the formula (B.12) gets simplified: in such a case, one has, obviously, $|\vec{p}\,'_{\text{c.m.}}| = |\vec{p}_{\text{c.m.}}|$ and (B.12) thus becomes

$$\left.\frac{d\sigma}{d\Omega_{\text{c.m.}}}\right|_{\text{elast.}} = \frac{1}{64\pi^2}\frac{1}{s}|\mathcal{M}_{fi}|^2 \tag{B.14}$$

There is another frequently occurring situation, where this kind of simplification is relevant. Considering a general binary process $1 + 2 \to 3 + 4$ in the high-energy limit, i.e. for $s^{1/2} \gg m_j$, $j = 1, \ldots, 4$, the particle masses can be safely neglected in kinematical relations and thus $|\vec{p}\,'_{\text{c.m.}}|/|\vec{p}_{\text{c.m.}}| \doteq 1$ with good accuracy. In fact, when all particles involved in a given process are taken as effectively massless, one can further streamline the cross-section calculations by introducing a suitable new kinematical variable. To comply with a traditional notation, let us label the four-momenta of particles 1, 2, 3, 4 consecutively as k, p, k', p' and define

$$y = \frac{p \cdot q}{p \cdot k} \tag{B.15}$$

where $q = k - k'$ (obviously, y is Lorentz invariant and dimensionless). Then it is not difficult to see that for $m_j = 0$, $j = 1, \ldots, 4$, the Mandelstam variables t and u can be expressed in terms of s and y as

$$\begin{aligned} t &= -sy \\ u &= -s(1-y) \end{aligned} \tag{B.16}$$

Moreover, in such a case, y is related simply to the scattering angle in the c.m. system:

$$y = \frac{1}{2}(1 - \cos\vartheta_{\text{c.m.}}) \tag{B.17}$$

($\vartheta_{\text{c.m.}}$ is defined here as the angle between $\vec{p}\,'_{\text{c.m.}}$ and $\vec{p}_{\text{c.m.}}$). From (B.17), it is then clear that for vanishing masses, y takes on values between 0 and 1. The above observations make the practical importance of the kinematical variable y obvious. Thus, it is also desirable to have an expression for differential cross-section, written directly with respect to y. This is achieved easily. Assuming that the matrix element squared $|\mathcal{M}_{fi}|^2$ does not depend on the polar angle φ (which is usually the case), the integration of (B.12) over φ is done trivially and, taking into account (B.17), one readily gets

$$\frac{d\sigma}{dy} = \frac{1}{16\pi}\frac{1}{s}|\mathcal{M}_{fi}|^2 \tag{B.18}$$

When writing (B.18), it is assumed implicitly that the expression for $|\mathcal{M}_{fi}|^2$ has been recast in terms of s and y by using (B.16) (this is certainly possible when considering scattering of unpolarized particles, i.e. when $|\mathcal{M}_{fi}|^2$ is summed over the relevant spin states). Let us emphasize again that (B.18) is valid as an approximate formula in the high-energy limit or as an exact formula in a strictly massless case.

Another important case that deserves a separate treatment is the scattering on a fixed target, i.e. in the *laboratory* (rest) system of one of the initial particles. In particular, let us consider a binary process $1+2 \to 3+4$ in the rest frame of particle 2. One can calculate, e.g., the angular distribution of the particle 3 with respect to the direction of incident particle 1. We shall not present here a derivation of the formula in question from the basic relation (B.2) (though it is not a difficult task) and quote only the final result for the corresponding differential cross-section:

$$\frac{d\sigma}{d\Omega} = \frac{1}{64\pi^2} \frac{1}{|\vec{p}_1| m_2} |\mathcal{M}_{fi}|^2 \frac{|\vec{p}_3|}{E_1 + m_2 - \frac{|\vec{p}_1| E_3}{|\vec{p}_3|} \cos\vartheta} \tag{B.19}$$

where $d\Omega$ is an element of solid angle along the direction of $\vec{p}_3$, ϑ is the angle between $\vec{p}_3$ and $\vec{p}_1$, and the meaning of the other symbols should be obvious (note that for brevity we drop everywhere the labels referring explicitly to the laboratory frame and write simply $d\Omega$, $d\vartheta$ instead of $d\Omega_{\text{lab.}}$, $d\vartheta_{\text{lab.}}$, etc.). Of course, for an evaluation of the right-hand side of (B.19), one has to take into account the relevant energy–momentum constraints. Making use of the definition $p_2 = (m_2, \vec{0})$, one has $\vec{p}_4 = \vec{p}_1 - \vec{p}_3$ and the energy conservation then yields

$$\sqrt{|\vec{p}_3|^2 + m_3^2} + \sqrt{|\vec{p}_1|^2 - 2|\vec{p}_1||\vec{p}_3| \cos\vartheta + |\vec{p}_3|^2 + m_4^2} = E_1 + m_2 \tag{B.20}$$

Equation (B.20) can be explicitly solved for $|\vec{p}_3|$; for a general combination of $m_1, \ldots, m_4$, the result is quite a complicated function of the scattering angle ϑ, but it can be considerably simplified when some of the masses vanish. This, of course, is of practical interest, since such a configuration occurs in some familiar physical processes, e.g., in the Compton scattering $\gamma + e^- \to \gamma + e^-$ or in the elastic scattering of (quasi)massless neutrino on a charged lepton. Thus, let us consider the case of an elastic scattering with $m_1 = m_3 = 0$ and $m_2 = m_4 = m$. We shall label p_1, p_2, p_3, p_4 consecutively as k, p, k', p' and denote the energies $E(k)$, $E(k')$ simply as E, E', respectively (thus, $E = |\vec{k}|$ and $E' = |\vec{k}'|$). The relation (B.20) then becomes

$$E' + \sqrt{E'^2 - 2EE' \cos\vartheta + E^2 + m^2} = E + m \tag{B.21}$$

and this is easily reduced to

$$\frac{1}{E'} - \frac{1}{E} = \frac{1}{m}(1 - \cos\vartheta) \tag{B.22}$$

From (B.22), one gets immediately

$$E' = \frac{E}{1 + \frac{E}{m}(1 - \cos\vartheta)} \tag{B.23}$$

(note that the last result is precisely the famous Compton relation for the change of frequency of a photon scattered off a free electron). Using all kinematical relations shown above, the formula (B.19) is recast, after some simple manipulations, as

$$\frac{d\sigma}{d\Omega} = \frac{1}{64\pi^2}\frac{1}{m^2}|\mathcal{M}_{fi}|^2\left(\frac{E'}{E}\right)^2 \tag{B.24}$$

Thus, we have arrived at the desired result: Eq. (B.24) represents a relatively simple formula for angular distribution of elastically scattered particles in the laboratory frame, which is applicable whenever the incident particle is much lighter than the target (so that its mass can be safely neglected).

Next, let us turn to the decay processes. In general, we consider a particle with the mass M decaying in its rest system into a number (n) of lighter particles. The differential probability of such a decay per unit of time (the differential decay rate) is given by

$$dw = \frac{1}{2M}|\mathcal{M}_{fi}|^2(2\pi)^4\delta^4\Big(P - \sum_{j=1}^{n} p_j\Big)\frac{d^3p_1}{(2\pi)^3 2E_1}\cdots\frac{d^3p_n}{(2\pi)^3 2E_n}K \tag{B.25}$$

where $\mathcal{M}_{fi}$ is the corresponding Lorentz invariant matrix element, P denotes the four-momentum of the decaying particle, i.e. (in the rest frame) $P = (M, \vec{0})$, $p_j = (E_j, \vec{p}_j)$ for $j = 1, \ldots, n$ are the four-momenta of the decay products and K stands for the combinatorial factor defined in (B.3). The simplest configuration is a two-body decay, i.e. $n = 2$ in (B.25). In such a case, the phase-space integration is particularly simple and one can thus derive easily the formulae of immediate practical interest. Below, we summarize some relevant results (their detailed derivation can be found in many places, see, e.g., Appendix C in [Hor]). For definiteness, we assume that the decay products 1 and 2 are not identical particles, i.e. we set $K = 1$ in (B.25).

Thus, we start with the elementary 2-body differential decay rate

$$dw = \frac{1}{2M}|\mathcal{M}_{fi}|^2(2\pi)^4\delta^4(P - p_1 - p_2)\frac{d^3p_1}{(2\pi)^3 2E_1}\frac{d^3p_2}{(2\pi)^3 2E_2} \tag{B.26}$$

When it makes sense to consider an angular distribution of the decay products (e.g., when the decaying particle is polarized, defining thus a preferred direction in space), one can just integrate over the magnitudes of the final-state momenta (with the energy–momentum constraint defined by the delta function in (B.26)) and express the decay rate in question as

$$dw = \frac{1}{2M}|\mathcal{M}_{fi}|^2 d(\mathrm{LIPS}_2) \tag{B.27}$$

where LIPS_2 is an acronym for "2-body Lorentz Invariant Phase Space"; its element $d(\mathrm{LIPS}_2)$ is given by

$$d(\mathrm{LIPS}_2) = \frac{|\vec{p}|}{M}\frac{d\Omega}{16\pi^2} \tag{B.28}$$

with $\vec{p}$ denoting the momentum of a decay product (one can take, e.g., $\vec{p} = \vec{p}_1 = -\vec{p}_2$) and $d\Omega$ stands for an element of the solid angle along the direction of $\vec{p}$. Obviously, $|\vec{p}|$ can be calculated by means of the formula (B.10) with $s = M^2$ and one thus has

$$|\vec{p}| = \frac{1}{2M}\left[\lambda(M^2, m_1^2, m_2^2)\right]^{1/2} \tag{B.29}$$

It is useful to note that the expression (B.11) for $\lambda(M^2, m_1^2, m_2^2)$ can be recast, after some simple manipulations, as

$$\lambda(M^2, m_1^2, m_2^2) = \left[M^2 - (m_1 + m_2)^2\right]\left[M^2 - (m_1 - m_2)^2\right] \tag{B.30}$$

When the initial and final particles are unpolarized, the quantity $|\mathcal{M}_{fi}|^2$ is summed (and averaged) over the relevant spin states; it is easy to realize that the result can only depend on M^2, m_1^2, m_2^2. The expression (B.27) can then be integrated trivially over the angles (one thus gets just a multiplicative factor of 4π) and the resulting decay rate (decay width) Γ becomes

$$\Gamma = \frac{1}{2M}\overline{|\mathcal{M}_{fi}|^2}\mathrm{LIPS}_2 \tag{B.31}$$

where $\overline{|\mathcal{M}_{fi}|^2}$ stands for the spin-averaged matrix element squared and

$$\mathrm{LIPS}_2 = \frac{1}{4\pi}\frac{|\vec{p}|}{M} = \frac{1}{8\pi}\left[1 - \frac{(m_1 + m_2)^2}{M^2}\right]^{1/2}\left[1 - \frac{(m_1 - m_2)^2}{M^2}\right]^{1/2} \tag{B.32}$$

(in writing the last expression, we have utilized the relations (B.29) and (B.30)). For completeness, let us display two frequently used particular forms of (B.32):

(i) For $m_1 = m_2 = m$, (B.32) is reduced to

$$\text{LIPS}_2 \bigg|_{m_1=m_2=m} = \frac{1}{8\pi}\sqrt{1-\frac{4m^2}{M^2}} \tag{B.33}$$

(ii) For $m_1, m_2 \ll M$, one has the approximate relation

$$\text{LIPS}_2 \bigg|_{m_1,m_2\ll M} \doteq \frac{1}{8\pi} \tag{B.34}$$

Finally, we shall discuss some general properties of relativistic scattering amplitudes. In particular, below, we briefly summarize basic formulae concerning the partial-wave expansion (usually called the Jacob–Wick expansion). More details can be found, e.g., [ItZ]. First, let us consider the elastic scattering of particles 1 and 2; as a reference frame, we always use the corresponding c.m. system, but we suppress the label c.m. in what follows. Initial and final states of both particles are characterized by definite momenta ($\vec{p}_1 = -\vec{p}_2 = \vec{p}$, $\vec{p}_1' = -\vec{p}_2' = \vec{p}'$) and helicities (denoted as h_1, h_2, h_1', h_2'); note that $|\vec{p}| = |\vec{p}'|$ for elastic scattering. We identify the third axis of our coordinate system with the direction of $\vec{p}$. For a scattering amplitude f, normalized with respect to the differential cross-section in such a way that

$$\frac{d\sigma}{d\Omega} = |f|^2 \tag{B.35}$$

one can write the Jacob–Wick expansion

$$f_{h'h}(s,\Omega) = \sum_j (2j+1) f_{h'h}^{(j)}(s) \mathscr{D}_{\lambda'\lambda}^{(j)}(\Omega) \tag{B.36}$$

where we denote collectively $h \equiv (h_1, h_2)$, $h' \equiv (h_1', h_2')$, the angles $\Omega = (\vartheta, \phi)$ define the direction of $\vec{p}'$ and $\mathscr{D}_{\lambda'\lambda}^{(j)}(\Omega)$ are Wigner functions (known from the theory of angular momentum as matrix elements of finite rotations, cf., e.g., [Sak]). The indices λ and λ' are given by $\lambda = h_1 - h_2$, $\lambda' = h_1' - h_2'$. Some basic properties of the $\mathscr{D}$-functions are summarized at the end of this appendix. The coefficients $f^{(j)}$ are the partial-wave amplitudes; the label j stands for the total angular momentum characterizing an individual partial wave. The sum in (B.36) runs over all non-negative integer or half-integer values of j, depending on whether there is an even or odd number of fermions among the particles 1 and 2. An $f^{(j)}$ has the form

$$f_{h'h}^{(j)}(s) = \frac{1}{2i|\vec{p}|}\left(S_{h'h}^{(j)} - 1\right) \tag{B.37}$$

where $S_{h'h}^{(j)}$ is an element of the S-matrix in the angular momentum basis.[2] The crucial point is that the S-matrix is unitary; this implies an important constraint for the $f^{(j)}$, namely

$$|f^{(j)}(s)| \leq \frac{1}{|\vec{p}|} \tag{B.38}$$

(here and in what follows, we usually suppress the indices h, h').

From Eq. (B.36), one can obtain easily a corresponding expansion for the Lorentz invariant matrix element $\mathcal{M}$ entering the cross-section formula (B.12). Indeed, rescaling the f normalized according to (B.35) so as to get an $\mathcal{M}$ satisfying (B.12), and taking into account that $|\vec{p}| = |\vec{p}\,'|$ for elastic scattering, one can write

$$\mathcal{M}(s, \Omega) = 16\pi \sum_j (2j+1)\mathcal{M}^{(j)}(s)\mathscr{D}_{\lambda'\lambda}^{(j)}(\Omega) \tag{B.39}$$

with

$$\mathcal{M}^{(j)}(s) = \frac{s^{1/2}}{4i|\vec{p}|}\left(S^{(j)} - 1\right) \tag{B.40}$$

This yields a unitarity bound for $\mathcal{M}^{(j)}(s)$, namely

$$\left|\mathcal{M}^{(j)}(s)\right| \leq \frac{s^{1/2}}{2|\vec{p}|} \tag{B.41}$$

In high-energy limit or for massless particles, one has $|\vec{p}| \doteq \frac{1}{2}\sqrt{s}$, and (B.41) then simplifies to

$$\left|\mathcal{M}^{(j)}(s)\right| \leq 1 \tag{B.42}$$

Now, in hindsight, it becomes clear that the choice of the overall factor 16π in (B.39) has been convenient, as it leads to the simple constraint (B.42).

A remark on practical evaluation of the partial-wave amplitudes is in order here. For a given physical process, the matrix element $\mathcal{M}(s, \Omega)$ can be calculated, e.g., within standard covariant perturbation theory (i.e. by means of Feynman diagrams). Taking into account an appropriate orthogonality relation for the Wigner $\mathscr{D}$-functions (see (B.56)), one can evaluate an $\mathcal{M}^j(s)$ by means of the angular integration

$$\mathcal{M}^{(j)}(s) = \frac{1}{16\pi}\int \mathcal{M}(s, \Omega)\mathscr{D}_{\lambda'\lambda}^{(j)*}(\Omega)\frac{d\Omega}{4\pi} \tag{B.43}$$

[2]In such a basis, the S-matrix has a block-diagonal form; for a fixed j, $S^{(j)}$ is a finite dimensional matrix living in a subspace spanned by the helicity states.

In the particular case where $\lambda' = \lambda = 0$ (i.e. for $h_1 = h_2$, $h_1' = h_2'$), the $\mathscr{D}$-functions are reduced to Legendre polynomials (see (B.54)) and the formula (B.43) then becomes

$$\mathcal{M}^{(j)}(s) = \frac{1}{32\pi}\int_{-1}^{1} \mathcal{M}(s,\vartheta)P_j(\cos\vartheta)d(\cos\vartheta) \tag{B.44}$$

For an inelastic process $1+2 \to 3+4$, one can also write a partial-wave expansion in the form (B.36) or (B.39); however, in such a case, only the purely non-diagonal S-matrix elements are involved. Instead of (B.40), one then has

$$\mathcal{M}^{(j)}_{\text{inel.}}(s) = \frac{s^{1/2}}{4i|\vec{p}|}S^{(j)}_{\text{inel.}} \tag{B.45}$$

where the symbol $S^{(j)}_{\text{inel.}}$ again represents collectively elements of the relevant unitary matrix and the index "inel." denotes the inelastic channel $1+2 \to 3+4$. In high-energy limit, the relation (B.45) implies the bound

$$\left|\mathcal{M}^{(j)}_{\text{inel.}}(s)\right| \le \frac{1}{2} \tag{B.46}$$

The constraints for partial-wave amplitudes following from S-matrix unitarity can also be easily converted into inequalities for partial cross-sections (i.e. for cross-sections corresponding to the individual partial waves). Using the expansion (B.39) in the formula (B.12) for differential cross-section, integrating (B.12) over the angles and utilizing the orthogonality relation (B.56), one gets

$$\sigma(s) = \sum_j \sigma^{(j)}(s) \tag{B.47}$$

(for a given set of the initial and final helicities), where

$$\sigma^{(j)}(s) = \frac{16\pi}{s}(2j+1)\left|\mathcal{M}^{(j)}(s)\right|^2 \tag{B.48}$$

For elastic scattering, the inequality (B.41) then implies a bound for the partial cross-sections (B.48), namely

$$\sigma^{(j)}(s) \le (2j+1)\frac{4\pi}{|\vec{p}|^2} \tag{B.49}$$

which in the high-energy limit becomes

$$\sigma^{(j)}(s) \le (2j+1)\frac{16\pi}{s} \tag{B.50}$$

In the case of an inelastic process, it is easy to derive analogous inequalities; in high-energy limit (or for massless particles), one gets

$$\sigma^{(j)}_{\text{inel.}}(s) \leq (2j+1)\frac{4\pi}{s} \tag{B.51}$$

We close this appendix by collecting some important formulae for the Wigner $\mathscr{D}$-functions appearing in the Jacob–Wick expansion. For a non-negative integer or half-integer j, one defines

$$\mathscr{D}^{(j)}_{m'm}(\Omega) = \mathrm{e}^{im\varphi} d^{(j)}_{m'm}(\vartheta) \tag{B.52}$$

where the indices m and m' may only take on values $-j, -j+1, \ldots, j-1, j$, and the functions $d^{(j)}_{m'm}(\vartheta)$ are given by the general formula

$$d^{(j)}_{m'm}(\vartheta) = \sum_k (-1)^{k-m+m'} \frac{\left[(j+m)!(j-m)!(j+m')!(j-m')!\right]^{1/2}}{(j+m-k)!k!(j-k-m')!(k-m+m')!} \times \left(\cos\frac{\vartheta}{2}\right)^{2j-2k+m-m'} \left(\sin\frac{\vartheta}{2}\right)^{2k-m+m'} \tag{B.53}$$

where the sum runs over the integers k such that the arguments of all factorials in (B.53) are non-negative.

When $m = m' = 0$, for an arbitrary integer $l \geq 0$, one has

$$\mathscr{D}^{(l)}_{00}(\Omega) = P_l(\cos\vartheta) \tag{B.54}$$

where P_l is Legendre polynomial.

As another example, let us show the explicit form of the functions $d^{(j)}_{m'm}(\vartheta)$ for $j = 1$:

$$\begin{aligned}
d^{(1)}_{11}(\vartheta) &= d^{(1)}_{\text{-1-1}}(\vartheta) = \frac{1}{2}(1+\cos\vartheta) \\
d^{(1)}_{00}(\vartheta) &= \cos\vartheta \\
d^{(1)}_{\text{1-1}}(\vartheta) &= d^{(1)}_{\text{-11}}(\vartheta) = \frac{1}{2}(1-\cos\vartheta) \\
d^{(1)}_{10}(\vartheta) &= -d^{(1)}_{01}(\vartheta) = d^{(1)}_{\text{0-1}}(\vartheta) = -d^{(1)}_{\text{-10}}(\vartheta) = \frac{1}{\sqrt{2}}\sin\vartheta
\end{aligned} \tag{B.55}$$

An orthogonality relation for the $\mathscr{D}$-functions reads:

$$\int \mathscr{D}^{(j_1)*}_{m'_1 m_1}(\Omega)\mathscr{D}^{(j_2)*}_{m'_2 m_2}(\Omega)\frac{d\Omega}{4\pi} = \frac{1}{2j_1+1}\delta_{j_1 j_2}\delta_{m_1 m_2} \tag{B.56}$$

Appendix C

Beta Decay of Polarized Neutron

Our starting point is the beta-decay matrix element

$$\mathcal{M} = C_V (U_p^\dagger U_n)\big[\bar{u}_e(1+\alpha_V\gamma_5)\gamma_0 v_\nu\big] \\ + C_A (U_p^\dagger \sigma_j U_n)\big[\bar{u}_e(1+\alpha_A\gamma_5)\gamma_5\gamma^j v_\nu\big] \quad \text{(C.1)}$$

(cf. (1.111)), where the neutron is assumed to be polarized along the third axis. It is easy to see that such an assumption can be technically implemented by writing

$$U_n U_n^\dagger = 2M\frac{1+\sigma_3}{2} \quad \text{(C.2)}$$

where M denotes, in accordance with conventions in Chapter 1, the average nucleon mass.

First, we are going to calculate the matrix element (C.1) squared, summing eventually over the spin states of $p, e, \bar{\nu}$. Employing some familiar properties of the Dirac matrices and introducing spinor traces in the usual way, $|\mathcal{M}|^2$ can be written as

$$|\mathcal{M}|^2 = C_V^2 \operatorname{Tr}\big(U_p U_p^\dagger U_n U_n^\dagger\big)\cdot \operatorname{Tr}\big[(1-\alpha_V\gamma_5)u\bar{u}(1+\alpha_V\gamma_5)\gamma_0 v\bar{v}\gamma_0\big] \\ + C_A C_V \operatorname{Tr}\big(U_p U_p^\dagger U_n U_n^\dagger \sigma_j\big)\cdot \operatorname{Tr}\big[(\alpha_A-\gamma_5)u\bar{u}(1+\alpha_V\gamma_5)\gamma_0 v\bar{v}\gamma^j\big] + \text{c.c.} \\ + C_A^2 \operatorname{Tr}\big(U_p U_p^\dagger \sigma_j U_n U_n^\dagger \sigma_k\big)\cdot \operatorname{Tr}\big[(\alpha_A-\gamma_5)u\bar{u}(\alpha_A+\gamma_5)\gamma^j v\bar{v}\gamma^k\big] \quad \text{(C.3)}$$

(note that we have suppressed here the labels e and ν at the corresponding spinors, but this cannot lead to any confusion). The spin summation indicated above is carried out in several steps. Using (C.2) and the relation

$$\sum_{\text{spin}} U_p U_p^\dagger = 2M\cdot \mathbb{1}$$

(with $\mathbb{1}$ denoting the 2×2 unit matrix, cf. (1.39)), as well as the identities (A.67), one gets

$$\sum_{\text{spin } p,e,\bar{\nu}} |\mathcal{M}|^2 = 4M^2 C_V^2 \operatorname{Tr}\Big(\frac{1+\sigma_3}{2}\Big) \cdot \operatorname{Tr}\big[(1-\alpha_V\gamma_5)(\not{p}_e + m_e) \times (1+\alpha_V\gamma_5)\gamma_0 \not{p}_{\bar{\nu}}\gamma_0\big] + 4M^2 C_A C_V \operatorname{Tr}\Big(\frac{1+\sigma_3}{2}\sigma_j\Big) \cdot \operatorname{Tr}\big[(\alpha_A - \gamma_5)(\not{p}_e + m_e)(1+\alpha_V\gamma_5)\gamma_0\not{p}_{\bar{\nu}}\gamma^j\big] + \text{c.c.} + 4M^2 C_A^2 \operatorname{Tr}\Big(\sigma_j \frac{1+\sigma_3}{2}\sigma_k\Big) \cdot \operatorname{Tr}\big[(\alpha_A - \gamma_5)(\not{p}_e + m_e) \times (\alpha_A + \gamma_5)\gamma^j \not{p}_{\bar{\nu}}\gamma^k\big] \tag{C.4}$$

Next, with the help of standard identities for the Dirac and Pauli matrices (see Appendix A, in particular (A.52)), Eq. (C.4) is recast as

$$\begin{aligned}\sum_{\text{spin } p,e,\bar{\nu}} |\mathcal{M}|^2 &= 4M^2 C_V^2 \cdot \operatorname{Tr}\big[(1+\alpha_V^2 - 2\alpha_V\gamma_5)\not{p}_e\gamma_0\not{p}_{\bar{\nu}}\gamma_0\big] \\ &+ 4M^2 C_A C_V \cdot \operatorname{Tr}\big[\big(\alpha_A + \alpha_V - (1+\alpha_A\alpha_V)\gamma_5\big)\not{p}_e\gamma_0\not{p}_{\bar{\nu}}\gamma^3\big] + \text{c.c.} \\ &+ 4M^2 C_A^2 \cdot \operatorname{Tr}\big[(1+\alpha_A^2 - 2\alpha_A\gamma_5)\not{p}_e\gamma^j\not{p}_{\bar{\nu}}\gamma^j\big] \\ &- 4M^2 C_A^2 \cdot i\epsilon_{jk3}\operatorname{Tr}\big[(1+\alpha_A^2 - 2\alpha_A\gamma_5)\not{p}_e\gamma^j\not{p}_{\bar{\nu}}\gamma^k\big]\end{aligned} \tag{C.5}$$

This expression can be simplified considerably just on the basis of symmetry arguments. Indeed, in the first line on the right-hand side of (C.5), the term involving γ_5 vanishes because of antisymmetry of the Levi-Civita symbol $\epsilon_{\mu\nu\rho\sigma}$; the same is true for the third line. In the second line, the term with γ_5 gives effectively zero, since the corresponding trace is purely imaginary and thus it gets cancelled when combined with its c.c. counterpart. Finally, in the fourth line, the term without γ_5 does not contribute, as the trace in question is symmetric under $j \leftrightarrow k$ and ϵ_{jk3} is antisymmetric. Thus, (C.5) is reduced to

$$\begin{aligned}\sum_{\text{spin } p,e,\bar{\nu}} |\mathcal{M}|^2 &= 4M^2 C_V^2(1+\alpha_V^2) \cdot \operatorname{Tr}(\not{p}_e\gamma_0\not{p}_{\bar{\nu}}\gamma_0) \\ &+ 8M^2 C_A C_V(\alpha_A + \alpha_V) \cdot \operatorname{Tr}(\not{p}_e\gamma_0\not{p}_{\bar{\nu}}\gamma^3) \\ &+ 4M^2 C_A^2(1+\alpha_A^2) \cdot \operatorname{Tr}(\not{p}_e\gamma^j\not{p}_{\bar{\nu}}\gamma^j) \\ &+ 16iM^2 C_A^2\alpha_A \cdot \operatorname{Tr}(\not{p}_e\gamma_1\not{p}_{\bar{\nu}}\gamma_2\gamma_5)\end{aligned} \tag{C.6}$$

Evaluating the traces in (C.6), one gets, after some simple manipulations,

$$\begin{aligned}\sum_{\text{spin } p,e,\bar{\nu}} |\mathcal{M}|^2 = {} & 16M^2C_V^2(1+\alpha_V^2)(E_eE_{\bar{\nu}} + \vec{p}_e \cdot \vec{p}_{\bar{\nu}}) \\ & + 32M^2C_AC_V(\alpha_A + \alpha_V)(E_e p_{\bar{\nu}}^3 + E_{\bar{\nu}} p_e^3)) \\ & + 16M^2C_A^2(1+\alpha_A^2)(3E_eE_{\bar{\nu}} - \vec{p}_e \cdot \vec{p}_{\bar{\nu}}) \\ & + 64M^2C_A^2\alpha_A(E_e p_{\bar{\nu}}^3 - E_{\bar{\nu}} p_e^3) \end{aligned} \tag{C.7}$$

Now, components of the vectors $\vec{p}_e$ and $\vec{p}_{\bar{\nu}}$ can be parametrized in terms of spherical angles as

$$\begin{aligned}\vec{p}_e &= \big(|\vec{p}_e| \sin\vartheta_e \cos\varphi_e, \quad |\vec{p}_e| \sin\vartheta_e \sin\varphi_e, \quad |\vec{p}_e| \cos\vartheta_e\big) \\ \vec{p}_{\bar{\nu}} &= \big(|\vec{p}_{\bar{\nu}}| \sin\vartheta_{\bar{\nu}} \cos\varphi_{\bar{\nu}}, \quad |\vec{p}_{\bar{\nu}}| \sin\vartheta_{\bar{\nu}} \sin\varphi_{\bar{\nu}}, \quad |\vec{p}_{\bar{\nu}}| \cos\vartheta_{\bar{\nu}}\big)\end{aligned} \tag{C.8}$$

Then the scalar product $\vec{p}_e \cdot \vec{p}_{\bar{\nu}}$ is expressed as

$$\vec{p}_e \cdot \vec{p}_{\bar{\nu}} = |\vec{p}_e|.|\vec{p}_{\bar{\nu}}| \sin\vartheta_e \sin\vartheta_{\bar{\nu}} \cos(\varphi_e - \varphi_{\bar{\nu}}) + |\vec{p}_e|.|\vec{p}_{\bar{\nu}}| \cos\vartheta_e \cos\vartheta_{\bar{\nu}} \tag{C.9}$$

We are interested in the angular distribution of the electron with respect to the direction of neutron polarization, which means that the relevant variable is ϑ_e. To obtain the quantity in question, one has to integrate the expression (C.7) over directions of $\vec{p}_{\bar{\nu}}$. Obviously, for a fixed $\vec{p}_e$, one has

$$\begin{aligned}\int_0^{2\pi} \cos(\varphi_e - \varphi_{\bar{\nu}}) d\varphi_{\bar{\nu}} &= 0 \\ \int_0^{\pi} \cos\vartheta_{\bar{\nu}} \sin\vartheta_{\bar{\nu}} d\vartheta_{\bar{\nu}} &= 0\end{aligned} \tag{C.10}$$

(recall that the element of the relevant solid angle is $d\Omega_{\bar{\nu}} = \sin\vartheta_{\bar{\nu}} d\vartheta_{\bar{\nu}} d\varphi_{\bar{\nu}}$). Thus, relations (C.8), (C.9) and (C.10) make it clear that an integration over the directions of antineutrino momentum eliminates from (C.7) all terms involving the scalar product $\vec{p}_e \cdot \vec{p}_{\bar{\nu}}$ or the $p_{\bar{\nu}}^3$. As a result, one obtains

$$\begin{aligned}\int \frac{d\Omega_{\bar{\nu}}}{4\pi} \sum_{\text{spin } p,e,\bar{\nu}} |\mathcal{M}|^2 = {} & 16M^2E_eE_{\bar{\nu}}\big[C_V^2(1+\alpha_V^2) + 3C_A^2(1+\alpha_A^2) \\ & + \big(2C_AC_V(\alpha_A + \alpha_V) - 4C_A^2\alpha_A\big)\beta\cos\vartheta_e\big]\end{aligned} \tag{C.11}$$

with $\beta = |\vec{p}_e|/E_e$. This is precisely the formula (1.113) of Chapter 1 (there we have set, conventionally, $E = E_e$).

Appendix D

Massive Vector Bosons

Relativistic theory of free massive particles with spin 1 is based on the Proca equation

$$\partial_\mu F^{\mu\nu} + m^2 B^\nu = 0 \tag{D.1}$$

where

$$F^{\mu\nu} = \partial^\mu B^\nu - \partial^\nu B^\mu \tag{D.2}$$

and $B^\mu = B^\mu(x)$ is a four-vector under Lorentz transformations of space–time coordinates. Similarly, as any other relativistic wave equation, (D.1) can either be used as one-particle equation of relativistic quantum mechanics, or it is treated as the equation of motion of a classical vector field that is subsequently quantized in terms of spin-1 particles with non-zero mass.

Before discussing solutions and other properties of Eq. (D.1), the following important comment is in order here. Substituting (D.2) into (D.1), one gets

$$(\Box + m^2)B^\nu - \partial^\nu(\partial \cdot B) = 0 \tag{D.3}$$

where we have denoted $\partial \cdot B \equiv \partial_\mu B^\mu$. Acting on (D.3) with ∂_ν, it is easy to see that one is ultimately left with $m^2 \partial \cdot B = 0$ and, since $m \neq 0$, this yields

$$\partial_\mu B^\mu = 0 \tag{D.4}$$

In other words, a "Lorenz condition" follows directly from the equation of motion (D.1); obviously, the crucial point in this respect is that $m \neq 0$ (remember that for Maxwell equations, the Lorenz condition has to be added by hand). Thus, looking back at (D.3), it becomes clear that instead of (D.1), one can write a pair of equations

$$(\Box + m^2)B^\mu = 0, \qquad \partial_\mu B^\mu = 0 \tag{D.5}$$

(i.e. one has the Klein–Gordon equation for each component B^μ, supplemented with the Lorenz condition). Now, since a passage in the reverse direction, i.e. from (D.5) to (D.1), is quite obvious, one can conclude that Eq. (D.1) is *equivalent* to (D.5). The condition (D.4) involves only the first time derivative and represents, in fact, a constraint on the components of the four-vector B^μ: only three of them are thus independent, corresponding to the three internal degrees of freedom of a massive spin-1 particle.

Let us now describe briefly the plane-wave solutions of Eq. (D.5). For such a solution, we use an Ansatz

$$B_\mu(x) = \varepsilon_\mu(k)\mathrm{e}^{-ikx} \tag{D.6}$$

with $k = (k^0, \vec{k})$. $\varepsilon_\mu(k)$ is called, in analogy with an electromagnetic plane wave, a "polarization vector"; its components are, in general, complex. We omit here the usual normalization factor, as this is inessential for the present purpose. Note that $\varepsilon_\mu(k)$ plays a similar role as $u(k)$ in a plane-wave solution of the Dirac equation. Obviously, another independent solution of Eq. (D.5) is obtained by complex conjugation of (D.6). Substituting (D.6) into (D.5), the Klein–Gordon equation yields immediately the mass-shell condition for k, i.e.

$$k^2 = (k_0)^2 - \vec{k}^2 = m^2 \tag{D.7}$$

(we shall assume, conventionally, that $k_0 > 0$) and the Lorenz condition turns into the requirement of transversality of the polarization vector in the four-dimensional momentum space:

$$k^\mu \varepsilon_\mu(k) = 0 \tag{D.8}$$

It is not difficult to realize that for a given k, there are three independent four-vectors $\varepsilon(k)$ satisfying the condition (D.8) and, moreover, they are *space-like*. The argument goes as follows. Due to the four-vector character of k and $\varepsilon(k)$, the scalar product $k \cdot \varepsilon(k)$ has the same value in any Lorentz frame. In particular, one can pass to the rest system of k, in which $k = k^{(0)} = (m, \vec{0})$; from (D.8), it is then obvious that the time component of any polarization vector must vanish in such a system. This means, generally, that $\varepsilon(k)$ is of space-like character. It is also clear that there are just three such vectors — these correspond to the three linearly independent spatial directions in the rest frame. Note also that the normalization of an $\varepsilon(k)$ is conventionally fixed by

$$\varepsilon(k) \cdot \varepsilon^*(k) = -1 \tag{D.9}$$

(here, we take into account that the $\varepsilon(k)$ may be complex).

The polarization vectors in question are labelled, for a given k, as $\varepsilon^\mu(k,\lambda)$, with $\lambda = 1,2,3$. A particularly useful triad can be defined in the following manner. $\varepsilon(k,1)$ and $\varepsilon(k,2)$ are taken in the form

$$\begin{aligned} \varepsilon^\mu(k,1) &= \big(0,\ \vec{\varepsilon}^{(1)}(\vec{k})\big) \\ \varepsilon^\mu(k,2) &= \big(0,\ \vec{\varepsilon}^{(2)}(\vec{k})\big) \end{aligned} \tag{D.10}$$

where $\vec{\varepsilon}^{(\lambda)}$, $\lambda = 1,2$ are two linearly independent vectors lying in the plane perpendicular to $\vec{k}$ (it means that $\vec{k}\cdot\vec{\varepsilon}^{(1)} = 0$, $\vec{k}\cdot\vec{\varepsilon}^{(2)} = 0$). As for $\varepsilon(k,3)$, this is chosen to have its spatial part directed along the $\vec{k}$. Thus, it can be written as

$$\varepsilon^\mu(k,3) = \left(\varepsilon^0,\ \alpha\frac{\vec{k}}{|\vec{k}|}\right) \tag{D.11}$$

with $\alpha > 0$. The parameters ε^0 and α are determined uniquely by making use of the conditions (D.8) and (D.9) and one gets

$$\varepsilon^\mu(k,3) = \left(\frac{|\vec{k}|}{m},\ \frac{k_0}{m}\frac{\vec{k}}{|\vec{k}|}\right) \tag{D.12}$$

where $k_0 = \sqrt{\vec{k}^2 + m^2}$.

In usual terminology, $\varepsilon(k,3)$ is called **longitudinal polarization** vector, while $\varepsilon(k,1)$ and $\varepsilon(k,2)$ correspond to two independent **transverse polarizations**. For practical purposes, it is convenient to introduce a specific symbol for longitudinal polarization: thus, we will usually denote $\varepsilon(k,3)$ as $\varepsilon_L(k)$. As regards the transverse polarizations shown in (D.10), $\vec{\varepsilon}^{(1)}$ and $\vec{\varepsilon}^{(2)}$ can be chosen, e.g., as two real (and mutually orthogonal) vectors; in such a case, we speak of "linear polarizations". Next, one can also form complex vectors

$$\vec{\varepsilon}_\pm = \frac{1}{\sqrt{2}}(\vec{\varepsilon}^{(1)} \pm i\vec{\varepsilon}^{(2)}) \tag{D.13}$$

corresponding to "circular polarizations". Needless to say, such a terminology is based on a straightforward analogy with electromagnetic plane waves. Note that the considered polarization vectors obviously satisfy orthonormality relations

$$\varepsilon(k,\lambda)\cdot\varepsilon^*(k,\lambda') = -\delta_{\lambda\lambda'} \tag{D.14}$$

In the context of relativistic quantum mechanics of spin-1 bosons, it is important to note that the plane waves specified above describe the states with definite helicities (and fixed energy–momentum): in particular, the circular transverse polarizations

$$\varepsilon(k, \pm) = (0,\ \vec{\varepsilon}_{\pm}) \tag{D.15}$$

correspond to helicities ± 1 (right-handed and left-handed motion, respectively) and the longitudinally polarized plane wave carries the helicity zero. For more details concerning this issue, see, e.g., Appendix H in [Hor]. Thus, a "canonical" set of polarization vectors can be taken as consisting of $\varepsilon(k, \pm)$ given by (D.13) and $\varepsilon_L(k)$,

$$\varepsilon_L^\mu(k) = \left(\frac{|\vec{k}|}{m},\ \frac{k_0}{m} \frac{\vec{k}}{|\vec{k}|} \right) \tag{D.16}$$

An astute reader may observe that the longitudinal polarization vector of spin-1 boson coincides with the spin four-vector describing helicity of a spin-$\frac{1}{2}$ fermion. Of course, this is not surprising as the relevant requirements are formally the same in both cases; however, the physical meaning of the two quantities is different: as we noted before, $\varepsilon(k)$ plays the role of a one-particle wave function in momentum space.

In practical calculations, one needs some further particular properties of the polarization vectors $\varepsilon(k, \lambda)$. First, from (D.16), one can infer quite easily that in the high-energy limit, components of $\varepsilon_L(k)$ behave essentially as the four-momentum k itself; in explicit terms, the relevant statement reads

$$\varepsilon_L^\mu(k) = \frac{k^\mu}{m} + O\left(\frac{m}{|\vec{k}|}\right) \quad \text{for } |\vec{k}| \gg m \tag{D.17}$$

(of course, the remainder in (D.17) could also be written as $O(m/k_0)$). On the other hand, components of a transverse polarization vector $\varepsilon_T(k)$ cannot grow indefinitely: there is an obvious bound $|\varepsilon_T^\mu(k)| \le 1$ set by the euclidean norm of $\vec{\varepsilon}^{(\lambda)}$, $\lambda = 1, 2$ in (D.10). Another important formula is the "completeness relation"

$$\sum_{\lambda=1}^{3} \varepsilon_\mu(k, \lambda) \varepsilon_\nu^*(k, \lambda) = -g_{\mu\nu} + \frac{1}{m^2} k_\mu k_\nu \tag{D.18}$$

One should note that this is an analogue of the identities (A.67) for Dirac spinors. A straightforward proof of Eq. (D.18) goes as follows. For a given k satisfying $k^2 = m^2$, one considers the unit time-like vector

$$\varepsilon^\mu(k, 0) = \frac{1}{m} k^\mu \tag{D.19}$$

together with the space-like polarization vectors $\varepsilon(k,\lambda)$ described above. The $\varepsilon(k,\lambda)$, $\lambda = 0,1,2,3$ obviously satisfy an orthonormality relation

$$\varepsilon(k,\lambda)\cdot\varepsilon^*(k,\lambda') = g_{\lambda\lambda'} \tag{D.20}$$

and form a basis in the four-dimensional space endowed with the usual metric. The latter statement means that

$$\varepsilon_\mu(k,0)\varepsilon^*_\nu(k,0) - \sum_{\lambda=1}^{3}\varepsilon_\mu(k,\lambda)\varepsilon^*_\nu(k,\lambda) = g_{\mu\nu} \tag{D.21}$$

(this can be verified easily by multiplying both sides of (D.21) with $\varepsilon^\nu(k,\lambda')$, taking consecutively $\lambda' = 0,1,2,3$ and utilizing (D.20)). From (D.21) then immediately follows the result (D.18) for the polarization sum in question.

For reader's convenience, let us also add that there is an independent and frequently used argument for (D.18), which can be formulated in the following way. Since $\varepsilon(k,\lambda)$, $\lambda = 1,2,3$ are supposed to be four-vectors, the polarization sum on the left-hand side of (D.18) should be a second rank Lorentz tensor depending on the four-momentum k. Thus, on general grounds, one can write

$$\sum_{\lambda=1}^{3}\varepsilon_\mu(k,\lambda)\varepsilon^*_\nu(k,\lambda) = Ag_{\mu\nu} + Bk_\mu k_\nu \tag{D.22}$$

where the coefficients A and B may only depend on k^2; however, one has $k^2 = m^2$ and, therefore, A and B are simply constants. Now, multiplying (D.22) with k^μ and utilizing (D.8), one gets the constraint

$$A + Bm^2 = 0 \tag{D.23}$$

Further, one can raise, e.g., the index ν in both sides of (D.22) and take then the corresponding trace; this yields

$$4A + Bm^2 = -3 \tag{D.24}$$

Solving (D.23) and (D.24), one obtains

$$A = -1, \quad B = \frac{1}{m^2} \tag{D.25}$$

and the result (D.18) is thus recovered.

Let us now discuss quantization of the free massive vector field. For simplicity, we shall consider the case of a real (hermitian) field. The relevant Lagrangian density can be written as

$$\mathscr{L} = -\frac{1}{4}F_{\mu\nu}F^{\mu\nu} + \frac{1}{2}m^2 B_\mu B^\mu \tag{D.26}$$

(here and in what follows, we usually write simply B_μ instead of $B_\mu(x)$, etc.). It is not difficult to verify that (D.26) yields (D.1) as the corresponding equation of motion. Indeed, from (D.26), one gets

$$\frac{\delta \mathscr{L}}{\delta(\partial_\mu B_\nu)} = -F^{\mu\nu} \tag{D.27}$$

and

$$\frac{\delta \mathscr{L}}{\delta B_\nu} = m^2 B^\nu \tag{D.28}$$

Using now these results in the Euler–Lagrange equation

$$\partial_\mu \frac{\delta \mathscr{L}}{\delta(\partial_\mu B_\nu)} - \frac{\delta \mathscr{L}}{\delta B_\nu} = 0 \tag{D.29}$$

one recovers immediately Eq. (D.1).

According to our previous analysis, only three of the four field components B_μ are to be taken as independent, since they are constrained by

$$\partial^\mu B_\mu = 0 \tag{D.30}$$

(cf. the remark following Eq. (D.5)). For the purpose of canonical quantization, we take B^j, $j = 1, 2, 3$ as the relevant independent variables ("generalized coordinates") and B_0 is understood as a solution of the constraint (D.30). The corresponding canonically conjugate momenta are defined in the usual way:

$$\pi_j \equiv \frac{\delta \mathscr{L}}{\delta(\partial_0 B_j)} \tag{D.31}$$

(the reader should not be confused by the seemingly non-covariant position of the indices – π_j is simply a convenient notation for canonical momentum associated with B_j). Using (D.27), the definition (D.31) yields

$$\pi_j = F_{0j} = \partial_0 B_j - \partial_j B_0 \tag{D.32}$$

(note that (D.27) also makes it clear that the canonical momentum conjugate to B_0 would be identically zero).

Let us see how the constraint (D.30) can be solved, i.e. whether and how B_0 can be expressed in terms of our canonical variables. To this end, it is convenient to utilize directly the original form (D.1) of the equation of motion. One thus gets

$$B_0 = -\frac{1}{m^2}\partial_j F_{0j} \tag{D.33}$$

(of course, the symbol ∂_j ($= -\partial^j$) stands for $\partial/\partial x^j$ as usual). Taking now into account (D.32), Eq. (D.33) is recast as

$$B_0 = -\frac{1}{m^2}\partial_j \pi_j \tag{D.34}$$

and this, of course, is a crucial result since the last expression involves only derivatives of canonical momenta with respect to the space coordinates.

For canonical quantization, one postulates the equal-time (E.T.) commutation relations

$$\begin{aligned} [B_j(x),\, B_k(y)]_{\text{E.T.}} &= 0 \\ [\pi_j(x),\, \pi_k(y)]_{\text{E.T.}} &= 0 \\ [B_j(x),\, \pi_k(y)]_{\text{E.T.}} &= i\delta_{jk}\delta^3(\vec{x}-\vec{y}) \end{aligned} \tag{D.35}$$

Any component $B_\mu(x)$ is a solution of (D.5) and thus can be written in terms of a plane-wave expansion

$$\begin{aligned} B_\mu(x) = \sum_{\lambda=1}^{3} \int \frac{d^3k}{(2\pi)^{3/2}(2k_0)^{1/2}} \big[&a(k,\lambda)\varepsilon_\mu(k,\lambda)\mathrm{e}^{-ikx} \\ &+ a^+(k,\lambda)\varepsilon_\mu^*(k,\lambda)\mathrm{e}^{ikx}\big] \end{aligned} \tag{D.36}$$

where $k_0 = \sqrt{\vec{k}^2 + m^2}$ and $\varepsilon_\mu(k,\lambda)$ are polarization vectors described above. We assume that the field operator B_μ is hermitian, so that $a^+(k,\lambda)$ is hermitian conjugate of $a(k,\lambda)$, i.e. $a^+(k,\lambda) = a^\dagger(k,\lambda)$. Of course, $a(k,\lambda)$ and $a^+(k,\lambda)$ are to be identified with annihilation and creation operators corresponding to particles (vector bosons) with definite energy–momentum and spin (polarization).

For convenience, we also introduce the linear combinations

$$\begin{aligned} a_\mu(k) &\equiv \sum_{\lambda=1}^{3} a(k,\lambda)\varepsilon_\mu(k,\lambda) \\ a_\mu^+(k) \equiv a_\mu^\dagger(k) &= \sum_{\lambda=1}^{3} a^+(k,\lambda)\varepsilon_\mu^*(k,\lambda) \end{aligned} \tag{D.37}$$

$a_\mu(k)$ and $a_\mu^+(k)$ can be calculated from (D.36) and expressed in terms of $B_\mu(x)$ and time derivatives $\dot{B}_\mu(x) = \partial_0 B_\mu(x)$ (for the relevant technique, see, e.g., [BjD]). Employing canonical commutation relations (D.35) and the result (D.34) (as well as Eq. (D.30)), one can evaluate all possible commutators of B_μ and $\dot{B}_\nu$ for $\mu, \nu = 0, 1, 2, 3$ and thus one is also able to determine all commutators involving the momentum-space operators $a_\mu(k)$ and $a^+_\nu(k')$. The calculation is rather tedious, but the result is rewarding and easy to remember:

$$
\begin{aligned}
[a_\mu(k),\, a_\nu(k')] &= 0 \\
[a^+_\mu(k),\, a^+_\nu(k')] &= 0 \\
[a_\mu(k),\, a^+_\nu(k')] &= \left(-g_{\mu\nu} + \frac{1}{m^2} k_\mu k_\nu\right) \delta^3(\vec{k} - \vec{k}')
\end{aligned}
\tag{D.38}
$$

(let us stress again that the four-momenta labelling the operators a, a^+ are on the mass shell, i.e. $k_0 = \sqrt{\vec{k}^2 + m^2}$). Next, making use of the orthonormality properties of the polarization vectors $\varepsilon(k, \lambda)$, one can solve Eq. (D.37) and turn subsequently the relations (D.38) into an algebra of the operators $a(k, \lambda)$ and $a^+(k, \lambda)$. The calculation is straightforward and the result is, as expected,

$$
\begin{aligned}
[a(k, \lambda),\, a(k', \lambda')] &= 0 \\
[a^+(k, \lambda),\, a^+(k', \lambda')] &= 0 \\
[a(k, \lambda),\, a^+(k', \lambda')] &= \delta_{\lambda\lambda'} \delta^3(\vec{k} - \vec{k}')
\end{aligned}
\tag{D.39}
$$

Of course, for the interpretation of $a(k, \lambda)$ and $a^+(k, \lambda)$ as annihilation and creation operators, one has to calculate the relevant physical quantities (energy, momentum, etc.) for the considered quantized field. Here, we take the connection of the operators a, a^+ with one-particle vector boson states for granted; the main purpose of the preceding discussion was to emphasize that massive vector field is a constrained system that can be canonically quantized in a straightforward way — by solving explicitly the constraint in terms of canonical variables (see Eq. (D.34)).

Now, we are going to briefly discuss the Feynman propagator. One can start with the definition

$$
i\mathcal{D}_{\mu\nu}(x - y) = \langle 0 | T\big(B_\mu(x) B_\nu(y)\big) | 0 \rangle \tag{D.40}
$$

where the time-ordered operator product (or simply T-product) in (D.40) is conventionally defined by means of the Heaviside step function; let us recall that such a definition reads, in general,

$$T\big(A(x)B(y)\big) = \theta(x_0 - y_0)A(x)B(y) + \theta(y_0 - x_0)B(y)A(x) \qquad \text{(D.41)}$$

if one considers two bosonic operators A and B depending on space–time coordinates. Using in (D.40) the decomposition (D.36), one obtains (after a somewhat tedious calculation) the result for $\mathcal{D}_{\mu\nu}$ in the usual form of Fourier integral:

$$\mathcal{D}_{\mu\nu}(x-y) = \int \frac{d^4q}{(2\pi)^4}\left(\frac{P_{\mu\nu}(q)}{q^2 - m^2 + i\varepsilon} - \frac{1}{m^2} g_{0\mu} g_{0\nu}\right) e^{iq(x-y)} \qquad \text{(D.42)}$$

where

$$P_{\mu\nu}(q) = -g_{\mu\nu} + \frac{1}{m^2} q_\mu q_\nu \qquad \text{(D.43)}$$

and the $+i\varepsilon$ prescription has the usual meaning as in any other Feynman propagator.

A remarkable feature of the expression (D.42) is that, apart from the "normal" covariant term involving the tensor $P_{\mu\nu}(q)$, there is a non-covariant contribution proportional to $g_{0\mu}g_{0\nu}$; obviously, this has contact character, since the integration of the exponential factor yields four-dimensional delta function. Note that the appearance of such a term is related to the non-covariant nature of the conventional T-product (D.41). Thus, (D.42) can be written as

$$\mathcal{D}_{\mu\nu}(x-y) = \mathcal{D}^{\text{covar.}}_{\mu\nu}(x-y) - \frac{1}{m^2} g_{0\mu} g_{0\nu} \delta^4(x-y) \qquad \text{(D.44)}$$

where

$$\mathcal{D}^{\text{covar.}}_{\mu\nu}(x-y) = \int \frac{d^4q}{(2\pi)^4} D^{\text{covar.}}_{\mu\nu}(q) e^{iq(x-y)}$$

with

$$D^{\text{covar.}}_{\mu\nu}(q) = \frac{-g_{\mu\nu} + m^{-2} q_\mu q_\nu}{q^2 - m^2 + i\varepsilon} \qquad \text{(D.45)}$$

The presence of the non-covariant term in the propagator (D.44) seems to be a disturbing feature of the theory of massive vector bosons. Nevertheless, in Feynman diagram calculations within common field theory models, one does employ the familiar form (D.45), simply omitting the non-covariant terms. A basic reason for that is, briefly, the following. To develop the perturbation expansion in the usual Dirac picture, one passes from an interaction Lagrangian $\mathscr{L}_{\text{int}}$ to a corresponding Hamiltonian

$\mathscr{H}_{\text{int}}$. It turns out that $\mathscr{H}_{\text{int}}$ differs from $-\mathscr{L}_{\text{int}}$ by an additional term, which cancels exactly the contribution of the contact non-covariant term in the propagator (D.44). A technical discussion of this issue would go beyond the scope of this appendix; we have mentioned it here in order to make the reader aware of subtleties and possible pitfalls of the canonical operator quantization of the massive vector field. For a detailed exposition, see, e.g., [Chg].

It is useful to know that there is another independent way to arrive at the covariant form (D.45). It is based on the observation that the propagator in question can also be understood as the (causal) Green's function of the Proca equation (D.1); as we shall see, such an approach is well suited for practical calculations. Let us show how this works. To find the covariant propagator function $\mathcal{D}_{\mu\nu}(x)$, one has to solve the equation

$$(\Box + m^2)\mathcal{D}^{\mu}{}_{\nu}(x) - \partial^{\mu}\big(\partial_{\lambda}\mathcal{D}^{\lambda}{}_{\nu}(x)\big) = g^{\mu}{}_{\nu}\delta^4(x) \tag{D.46}$$

(let us recall again that $g^{\mu}{}_{\nu} = \delta^{\mu}_{\nu}$). Performing Fourier transformation, i.e. defining a function $D_{\mu\nu}(q)$ through

$$\mathcal{D}_{\mu\nu}(x) = \int \frac{d^4q}{(2\pi)^4} D_{\mu\nu}(q)\mathrm{e}^{iqx} \tag{D.47}$$

one gets from (D.46) the system of linear algebraic equations

$$(-q^2 + m^2)D^{\mu}_{\nu}(q) + q^{\mu}q_{\lambda}D^{\lambda}_{\nu}(q) = g^{\mu}_{\nu} \tag{D.48}$$

that can be written compactly as

$$L^{\mu}_{\lambda}D^{\lambda}_{\nu} = g^{\mu}_{\nu} \tag{D.49}$$

with

$$L^{\mu}_{\lambda} = (-q^2 + m^2)g^{\mu}_{\lambda} + q^{\mu}q_{\lambda} \tag{D.50}$$

The $D^{\mu\nu}(q)$ is a second rank tensor depending on a four-vector q and, therefore, its most general form reads

$$D^{\mu\nu}(q) = D_T(q^2)P^{\mu\nu}_T(q) + D_L(q^2)P^{\mu\nu}_L(q) \tag{D.51}$$

where

$$\begin{aligned} P^{\mu\nu}_T &= g^{\mu\nu} - \frac{q^{\mu}q^{\nu}}{q^2} \\ P^{\mu\nu}_L &= \frac{q^{\mu}q^{\nu}}{q^2} \end{aligned} \tag{D.52}$$

Denoting as P_T and P_L the 4×4 matrices whose elements coincide with the mixed components of tensors (D.52), one finds easily that

$$P_T^2 = P_T, \quad P_L^2 = P_L, \quad P_T P_L = P_L P_T = 0 \tag{D.53}$$

Thus, the matrices P_T and P_L represent orthogonal projectors (this is the main advantage of the form (D.51) over a parametrization in terms of the basis made simply of $g_{\mu\nu}$ and $q_\mu q_\nu$). The matrix L defined in (D.50) can be recast, accordingly, as

$$L = (-q^2 + m^2)P_T + m^2 P_L \tag{D.54}$$

The matrix equation (D.49) can now be solved easily by utilizing the relations (D.53); taking into account that the unit matrix on the right-hand side of (D.49) can be decomposed as $P_T + P_L$, one readily gets

$$D_T = \frac{1}{-q^2 + m^2}, \quad D_L = \frac{1}{m^2} \tag{D.55}$$

for $q^2 \neq m^2$. This is the desired answer; substituting (D.55) into (D.51), one recovers the result (D.45).

Of course, within the approach described above, one has to make the replacement $m^2 \to m^2 - i\varepsilon$ in the propagator denominator by hand (relying on the general knowledge of properties of causal Green functions). In this context, one should keep in mind that there are infinitely many Green functions, which are all solutions of the original Eq. (D.46) (to any particular solution of the inhomogeneous equation (D.46), one may add an arbitrary solution of the corresponding homogeneous equation); by removing the singularity at $q^2 = m^2$ in a specific way (e.g., through the $i\varepsilon$ prescription), the ambiguity is fixed. In any case, the nice feature of the tensor method explained above is that it provides a very efficient tool for finding the algebraic form of the propagator in momentum space; such an approach can be used conveniently in many other situations.

In closing this appendix, let us add that most of the previous results can be generalized almost without changes to the case of a complex (non-hermitian) vector field. Denoting, for convenience, the four components of a field such as W_μ^-, a corresponding free Lagrangian can be written as

$$\mathscr{L} = -\frac{1}{2}(\partial_\mu W_\nu^- - \partial_\nu W_\mu^-)(\partial^\mu W^{+\nu} - \partial^\nu W^{+\mu}) + m^2 W_\mu^- W^{+\mu} \tag{D.56}$$

where $W_\mu^+ = (W_\mu^-)^*$ in a classical theory or $W_\mu^+ = (W_\mu^-)^\dagger$ in the quantum case. W_μ^- and W_μ^+ are treated as independent dynamical variables (for this reason, the coefficients in (D.56) differ from those in (D.26)). The plane-wave expansion of a quantized field $W_\mu^\pm$ is written as

$$W_\mu^-(x) = \sum_{\lambda=1}^{3} \int \frac{d^3k}{(2\pi)^{3/2}(2k_0)^{1/2}} \left[b(k,\lambda)\varepsilon_\mu(k,\lambda)\mathrm{e}^{-ikx} + d^+(k,\lambda)\varepsilon_\mu^*(k,\lambda)\mathrm{e}^{ikx}\right]$$

$$W_\mu^+(x) = \sum_{\lambda=1}^{3} \int \frac{d^3k}{(2\pi)^{3/2}(2k_0)^{1/2}} \left[b^+(k,\lambda)\varepsilon_\mu^*(k,\lambda)\mathrm{e}^{ikx} + d(k,\lambda)\varepsilon_\mu(k,\lambda)\mathrm{e}^{-ikx}\right] \tag{D.57}$$

where b and b^+ are the annihilation and creation operators of particles (conventionally taken to be the W^- bosons), and d and d^+ play an analogous role for the antiparticles (W^+). The other symbols have the same meaning as in (D.36). The algebra of creation and annihilation operators now reads

$$\begin{aligned} [b(k,\lambda),\, b(k',\lambda')] &= [d(k,\lambda),\, d(k',\lambda')] = 0 \\ [b(k,\lambda),\, b^+(k',\lambda')] &= [d(k,\lambda),\, d^+(k',\lambda')] = \delta_{\lambda\lambda'}\delta^3(\vec{k}-\vec{k}') \\ [b(k,\lambda),\, d(k',\lambda')] &= 0, \quad [b(k,\lambda),\, d^+(k',\lambda')] = 0 \end{aligned} \tag{D.58}$$

where we have omitted commutators that follow from (D.58) by hermitian conjugation. Note that from the representation (D.57), one can infer the following rule for external lines corresponding to vector bosons: an incoming line always contributes a factor of $\varepsilon_\mu(k,\lambda)$ and the outgoing line a factor of $\varepsilon_\mu^*(k,\lambda)$, independent of whether the line in question represents a particle or antiparticle. The Feynman propagator can be defined through the time-ordered product of $W_\mu^-(x)$ and $W_\nu^+(y)$; the formula (D.45) remains unchanged.

Appendix E

Basics of the ABJ Anomaly

In this appendix, we derive the basic formula for the Adler–Bell–Jackiw (ABJ) axial anomaly employed in Section 7.9. The anomaly has many facets and there are many different ways how to derive it; accordingly, the relevant literature is vast. Our discussion is aimed at an uninitiated reader, and for this purpose, we adopt here a traditional elementary approach, which nevertheless provides substantial insight into the nature and origin of the axial anomaly.

Let us start with the VVA triangle graph (see Fig. E.1), which represents a correlation function of two vector currents and one axial-vector current made of a single fermion (Dirac) field.[1]

A formal expression for the VVA amplitude can be written as

$$T_{\alpha\mu\nu}(k,p;m) = \int \frac{d^4l}{(2\pi)^4} \mathrm{Tr}\left(\frac{1}{\not{l}-\not{k}-m}\gamma_\mu \frac{1}{\not{l}-m}\gamma_\nu \frac{1}{\not{l}+\not{p}-m}\gamma_\alpha\gamma_5\right) + \left[(k,\mu) \leftrightarrow (p,\nu)\right] \tag{E.1}$$

where m stands for the mass of the fermion circulating in the loop. Another relevant quantity, closely related to (E.1), is

$$T_{\mu\nu}(k,p;m) = \int \frac{d^4l}{(2\pi)^4} \mathrm{Tr}\left(\frac{1}{\not{l}-\not{k}-m}\gamma_\mu \frac{1}{\not{l}-m}\gamma_\nu \frac{1}{\not{l}+\not{p}-m}\gamma_5\right) + \left[(k,\mu) \leftrightarrow (p,\nu)\right] \tag{E.2}$$

Note that (E.2) corresponds to a triangle loop obtained from the original VVA graph by replacing the axial-vector vertex with a pseudoscalar one

[1]More precisely, the quantity in question is a Fourier transform of the vacuum expectation value of the above-mentioned three currents. To keep our discussion as general as possible, we do not impose any particular restrictions on the external four-momenta k and p.

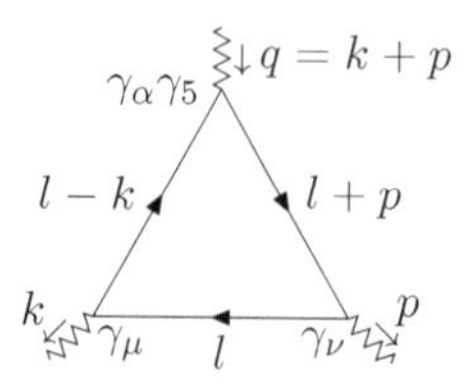

Figure E.1. *VVA triangle graph*: a closed fermionic loop with two vector (V) vertices and one axial-vector (A) vertex. External lines attached to the vertices merely symbolize the incoming and outgoing momenta. Additional contribution of the crossed graph with $(k, \mu) \leftrightarrow (p, \nu)$ has to be included in the full VVA amplitude.

(i.e. by $\gamma_\alpha\gamma_5 \to \gamma_5$). Thus, the quantity (E.2) can be naturally called a VVP amplitude. When speaking of the expression for $T_{\alpha\mu\nu}$, we stress the adjective *formal*: the integral in (E.1) has in fact an ultraviolet (UV) divergence and its proper definition requires a special care. We shall discuss this issue later on, and now let us focus on the $T_{\mu\nu}$.

At first sight, the degree of divergence of the integral in (E.2) would seem to be the same as that of (E.1). However, it turns out that — for purely algebraic reasons — the integral (E.2) is perfectly convergent! To see this, let us recast the expression (E.2) in the usual manner as

$$T_{\mu\nu}(k, p; m) = \int \frac{d^4 l}{(2\pi)^4} \frac{\mathrm{Tr}\big[(\not{l} - \not{k} + m)\gamma_\mu(\not{l} + m)\gamma_\nu(\not{l} + \not{p} + m)\gamma_5\big]}{[(l-k)^2 - m^2](l^2 - m^2)[(l+p)^2 - m^2]} \tag{E.3}$$

Working out the trace in (E.3), one finds out that this is simplified drastically, and the result is

$$\begin{aligned} &\mathrm{Tr}\big[(\not{l} - \not{k} + m)\gamma_\mu(\not{l} + m)\gamma_\nu(\not{l} + \not{p} + m)\gamma_5\big] \\ &\quad = -m\mathrm{Tr}(\gamma_\mu\gamma_\nu\not{k}\not{p}\gamma_5) = -4im\epsilon_{\mu\nu\rho\sigma}k^\rho p^\sigma \end{aligned} \tag{E.4}$$

(the reader is recommended to verify this independently, utilizing the familiar properties of traces of the Dirac matrices, summarized in Appendix A). Thus, we see that the l-dependence of the integrand in (E.2) is entirely due to its denominator and, consequently, the whole integrand behaves as l^{-6} for $l \to \infty$; this means that the integral (E.2) is actually even more convergent than necessary.[2]

[2]Note that in (hyper)spherical coordinates, one can write d^4l schematically as $l^3 dl d\Omega$ (with $d\Omega$ denoting the angular part); in this way, (E.2) is eventually reduced to a radial integral involving, asymptotically, $l^{-6} \cdot l^3 dl = l^{-3} dl$.

Now, taking into account (E.4), it is not difficult to realize that the contribution of the crossed term in (E.3) is the same as that of the direct one. Thus, we have

$$T_{\mu\nu}(k,p;m) = -8im\epsilon_{\mu\nu\rho\sigma}k^{\rho}p^{\sigma}\int\frac{d^4l}{(2\pi)^4} \times \frac{1}{[(l-k)^2-m^2][(l+p)^2-m^2](l^2-m^2)} \tag{E.5}$$

For the purpose of later discussion, we are going to recast the last expression in terms of an integral over Feynman parameters. This is done as follows. First, one introduces an integral representation of the integrand in (E.5) by means of the general formula

$$\frac{1}{ABC} = 2\int_0^1 dx\int_0^{1-x} dy\frac{1}{\left[Ax+By+C(1-x-y)\right]^3} \tag{E.6}$$

Then, after some simple manipulations, the expression (E.5) is rewritten as

$$\begin{aligned}T_{\mu\nu}&(k,p;m)\\ &= -16im\epsilon_{\mu\nu\rho\sigma}k^{\rho}p^{\sigma}\int_0^1 dx\int_0^{1-x} dy\int\frac{d^4l}{(2\pi)^4}\\ &\times\frac{1}{\left[(l-xk+yp)^2+x(1-x)k^2+y(1-y)p^2+2xyk\cdot p-m^2\right]^3}\end{aligned} \tag{E.7}$$

As a next step, one performs the shift $l-xk+yp\to l$ in the loop-momentum integral; (E.7) thus becomes

$$T_{\mu\nu}(k,p;m) = -16im\epsilon_{\mu\nu\rho\sigma}k^{\rho}p^{\sigma}\int_0^1 dx\int_0^{1-x} dy \times\int\frac{d^4l}{(2\pi)^4}\frac{1}{\left[l^2-C(x,y;k,p,m^2)\right]^3} \tag{E.8}$$

where we have denoted

$$C(x,y;k,p,m^2) = m^2-x(1-x)k^2-y(1-y)p^2-2xyk\cdot p \tag{E.9}$$

Now, integration over the loop momentum can be carried out by means of the general formula

$$\int\frac{d^nl}{(2\pi)^n}\frac{(l^2)^r}{(l^2-C+i\varepsilon)^s} = \frac{i}{(4\pi)^{\frac{n}{2}}}(-1)^{r-s}C^{r+\frac{n}{2}-s}\frac{\Gamma(r+\frac{n}{2})\Gamma(s-r-\frac{n}{2})}{\Gamma(\frac{n}{2})\Gamma(s)} \tag{E.10}$$

valid in n dimensions for values of r and s such that the integral converges (C is an essentially arbitrary real parameter and we may suppose, for convenience, that $C > 0$; note that we have also retrieved the $i\varepsilon$ term omitted in (E.8) for brevity). Using (E.10) in (E.8), one gets the desired Feynman-parametric representation of the VVP amplitude in question:

$$T_{\mu\nu}(k,p;m) = -\frac{1}{2\pi^2}\epsilon_{\mu\nu\rho\sigma}k^\rho p^\sigma \int_0^1 dx \int_0^{1-x} dy \, \frac{m}{C(x,y;k,p,m^2)} \tag{E.11}$$

Let us now examine the relevant Ward identities for the VVA amplitude. The preliminary discussion that follows is heuristic and "naive" (i.e. non-rigorous) in the sense that we ignore temporarily the divergent nature of the considered integrals; a regularization of the UV divergences will be taken into account in the second step of our investigation.

We start with an evaluation of the quantity $k^\mu T_{\alpha\mu\nu}$, which corresponds to the four-divergence of one of the vector currents involved in the VVA triangle graph. Using the formal representation (E.1), one has

$$\begin{aligned} k^\mu T_{\alpha\mu\nu}(k,p;m) &= \int \frac{d^4l}{(2\pi)^4}\mathrm{Tr}\left(\frac{1}{\not{l}-\not{k}-m}\not{k}\frac{1}{\not{l}-m}\gamma_\nu\frac{1}{\not{l}+\not{p}-m}\gamma_\alpha\gamma_5\right) \\ &\quad + \int \frac{d^4l}{(2\pi)^4}\mathrm{Tr}\left(\frac{1}{\not{l}-\not{p}-m}\gamma_\nu\frac{1}{\not{l}-m}\not{k}\frac{1}{\not{l}+\not{k}-m}\gamma_\alpha\gamma_5\right) \end{aligned} \tag{E.12}$$

To simplify the expression (E.12), one employs the following simple algebraic trick. In the first integral, the $\not{k}$ is recast as

$$\not{k} = (\not{l}-m) - (\not{l}-\not{k}-m) \tag{E.13}$$

and, similarly, in the second integral, one writes

$$\not{k} = (\not{l}+\not{k}-m) - (\not{l}-m) \tag{E.14}$$

These substitutions result in a partial cancellation of propagator denominators and (E.12) is rewritten as a sum of four integrals, namely

$$\begin{aligned} k^\mu T_{\alpha\mu\nu}(k,p;m) &= \int \frac{d^4l}{(2\pi)^4}\mathrm{Tr}\left(\frac{1}{\not{l}-\not{k}-m}\gamma_\nu\frac{1}{\not{l}+\not{p}-m}\gamma_\alpha\gamma_5\right) \\ &\quad - \int \frac{d^4l}{(2\pi)^4}\mathrm{Tr}\left(\frac{1}{\not{l}-m}\gamma_\nu\frac{1}{\not{l}+\not{p}-m}\gamma_\alpha\gamma_5\right) \end{aligned}$$

$$+\int \frac{d^4l}{(2\pi)^4} \mathrm{Tr}\left(\frac{1}{\not l - \not p - m}\gamma_\nu \frac{1}{\not l - m}\gamma_\alpha\gamma_5\right)$$

$$-\int \frac{d^4l}{(2\pi)^4} \mathrm{Tr}\left(\frac{1}{\not l - \not p - m}\gamma_\nu \frac{1}{\not l + \not k - m}\gamma_\alpha\gamma_5\right) \tag{E.15}$$

Now, it is easy to see that the first and fourth integrals in (E.15) mutually cancel — this becomes clear when one performs the shift $l \to l + k - p$ in the first integral. As for the second and third integrals, these can be shown to vanish (separately) on symmetry grounds: it is not difficult to realize that each of them would be a second rank *pseudotensor*, depending on a single four-vector p. However, one obviously cannot construct such an object because of full antisymmetry of the Levi-Civita pseudotensor (that would have to be involved in a corresponding expression). Thus, we arrive at the identity

$$k^\mu T_{\alpha\mu\nu}(k, p; m) = 0 \tag{E.16}$$

In view of the symmetry of the VVA amplitude (E.1) under $(k, \mu) \leftrightarrow (p, \nu)$, Eq. (E.16) also immediately implies

$$p^\nu T_{\alpha\mu\nu}(k, p; m) = 0 \tag{E.17}$$

The identities (E.16) and (E.17) are in fact anticipated results, since the vector current (made of a single free Dirac field) is conserved.

In the same manner, we can calculate the quantity $q^\alpha T_{\alpha\mu\nu}$ that expresses four-divergence of the axial-vector current within the VVA triangle graph. Using in (E.1) the trace cyclicity, it is convenient to start with

$$q^\alpha T_{\alpha\mu\nu}(k, p; m) = \int \frac{d^4l}{(2\pi)^4} \mathrm{Tr}\left(\gamma_\mu \frac{1}{\not l - m}\gamma_\nu \frac{1}{\not l + \not p - m}\not q\gamma_5 \frac{1}{\not l - \not k - m}\right)$$

$$+\int \frac{d^4l}{(2\pi)^4} \mathrm{Tr}\left(\gamma_\nu \frac{1}{\not l - m}\gamma_\mu \frac{1}{\not l + \not k - m}\not q\gamma_5 \frac{1}{\not l + \not p - m}\right) \tag{E.18}$$

Taking into account that $q = k + p$, $\not q\gamma_5$ in the first integral can be recast, for obvious reasons, as

$$\not q\gamma_5 = (\not l + \not p - m)\gamma_5 + \gamma_5(\not l - \not k - m) + 2m\gamma_5 \tag{E.19}$$

and, similarly, in the second integral, one writes

$$\not{q}\gamma_5 = (\not{l} + \not{k} - m)\gamma_5 + \gamma_5(\not{l} - \not{p} - m) + 2m\gamma_5 \tag{E.20}$$

Then, using the symmetry argument explained above and remembering the definition (E.2), one arrives at the result

$$q^\alpha T_{\alpha\mu\nu}(k,p;m) = 2mT_{\mu\nu}(k,p;m) \tag{E.21}$$

Again, this looks like an expected result, since it corresponds to a "partial conservation" of the axial-vector current made of a massive Dirac field.

The relations (E.16), (E.17) and (E.21) represent "naive" (or "canonical") Ward identities (WI) for the VVA amplitude. The corresponding nomenclature sounds quite naturally: the equations (E.16) and (E.17) are called **vector WI**, while Eq. (E.21) is the **axial WI**. As we stressed earlier, in deriving them, we have entirely ignored all possible complications that could be due to the UV divergences in the considered loop-momentum integrals. Now, we are going to make up for this flaw. In particular, we will regularize the formal expression for the contribution of the VVA diagram by means of the Pauli–Villars (PV) method (see, e.g., [ItZ]). This consists in subtracting from (E.1) the contribution of an analogous loop, in which the original fermion mass is replaced by an auxiliary regulator mass M (of course, the subtraction is made at the level of the corresponding integrand). Since the integral in (E.1) is only linearly divergent, one such PV subtraction is sufficient. In explicit terms, the PV-regularized VVA amplitude reads

$$\begin{aligned} T^{\text{reg.}}_{\alpha\mu\nu}(k,p;M) = \int \frac{d^4l}{(2\pi)^4} \Bigg\{ &\text{Tr}\left(\frac{1}{\not{l} - \not{k} - m}\gamma_\mu \frac{1}{\not{l} - m}\gamma_\nu \frac{1}{\not{l} + \not{p} - m}\gamma_\alpha\gamma_5\right) \\ &- \text{Tr}\left(\frac{1}{\not{l} - \not{k} - M}\gamma_\mu \frac{1}{\not{l} - M}\gamma_\nu \frac{1}{\not{l} + \not{p} - M}\gamma_\alpha\gamma_5\right)\Bigg\} \\ &+ \big[(k,\mu) \leftrightarrow (p,\nu)\big] \end{aligned} \tag{E.22}$$

The salient feature of this regularization procedure is that it preserves automatically the vector WI (note that precisely the same effect occurs in the familiar example of the vacuum polarization graph in spinor QED).[3]

For a general Feynman graph, one cannot simply remove the UV cutoff by performing the limit $M \to \infty$ in the regulated expression (before

[3]The point is that within such a scheme, the internal fermion lines entering the vector vertex carry the same mass M and the vector current conservation is thus maintained.

doing that, the quantity in question has to be renormalized properly). However, the convergence properties of the VVA triangle graph are subtle and rather amusing. In particular, although the integral in (E.1) is certainly UV divergent in a strict mathematical sense, it turns out that the limit $M \to \infty$ for the PV-regularized expression (E.22) does exist! This statement is non-trivial and will not be proved here; the interested reader can find a very detailed treatment of the convergence properties of the VVA diagram, e.g., in a paper by the present author and O.I. Zavialov [91]. An upshot of all this is as follows. A "renormalized" contribution of the VVA graph can be defined in a straightforward way as

$$T^{\text{ren.}}_{\alpha\mu\nu}(k,p;m) = \lim_{M\to\infty} T^{\text{reg.}}_{\alpha\mu\nu}(k,p;m,M) \tag{E.23}$$

Moreover, since vector WI hold for any value of the regularization parameter M, the $T^{\text{ren.}}_{\alpha\mu\nu}$ must obviously satisfy them as well, i.e. one has

$$k^\mu T^{\text{ren.}}_{\alpha\mu\nu}(k,p;m) = 0, \quad p^\nu T^{\text{ren.}}_{\alpha\mu\nu}(k,p;m) = 0 \tag{E.24}$$

Let us now focus on the axial WI. All manipulations that led to the naive identity (E.21) are now legal for regularized quantities. Then, keeping in mind the simple structure of the definition (E.22), it is easy to realize that the "intermediate" identity

$$q^\alpha T^{\text{reg.}}_{\alpha\mu\nu}(k,p;m,M) = 2mT_{\mu\nu}(k,p;m) - 2MT_{\mu\nu}(k,p;M) \tag{E.25}$$

must be valid for any finite value of M. In view of (E.23), the limit $M \to \infty$ can be performed in (E.25) and one thus obtains

$$q^\alpha T^{\text{ren.}}_{\alpha\mu\nu}(k,p;m) = 2mT_{\mu\nu}(k,p;m) - \lim_{M\to\infty} 2MT_{\mu\nu}(k,p;M) \tag{E.26}$$

Now, we come to the crucial point of our discussion. The appearance of the second term on the right-hand side of Eq. (E.26) indicates a possible deviation from the naive identity (E.21); it only remains to be seen whether such an extra term is indeed non-vanishing. Using our previous result for the $T_{\mu\nu}$, one finds easily that the answer is *yes*; from (E.11) and (E.9), one readily gets

$$\begin{aligned} \lim_{M\to\infty} 2MT_{\mu\nu}(k,p;M) &= -\frac{1}{\pi^2}\epsilon_{\mu\nu\rho\sigma}k^\rho p^\sigma \lim_{M\to\infty}\int_0^1 dx \int_0^{1-x} dy \\ &\quad \times \frac{M^2}{M^2 - x(1-x)k^2 - y(1-y)p^2 - 2xyk\cdot p} \\ &= -\frac{1}{2\pi^2}\epsilon_{\mu\nu\rho\sigma}k^\rho p^\sigma \end{aligned} \tag{E.27}$$

Thus, we arrive at an identity

$$q^{\alpha} T^{\text{ren.}}_{\alpha\mu\nu}(k, p; m) = 2m T_{\mu\nu}(k, p; m) + \frac{1}{2\pi^2} \epsilon_{\mu\nu\rho\sigma} k^{\rho} p^{\sigma} \tag{E.28}$$

which is precisely Eq. (7.191) quoted in the main text.

The second term on the right-hand side of (E.28) is the celebrated **ABJ axial anomaly** [82, 83]. Accordingly, the relation (E.28) is often called the **anomalous axial Ward identity**. Clearly, the labels "anomaly" and "anomalous" are of historical origin: they reflect the fact that the discovery of the ABJ anomaly was indeed a kind of surprise, taking into account that in many other situations, naive results (in the above sense) often prove to be correct, i.e. they are recovered when an appropriate regularization is included. On the other hand, our preceding discussion should have made it clear that, as a matter of fact, there is nothing anomalous about the anomaly: it emerges as a result of a proper definition of the VVA amplitude in question (whereas the naive Ward identities are derived by sloppy manipulations with ill-defined quantities). The mechanism, by which the anomaly is generated, becomes quite transparent within our approach. If one wants to maintain the vector current conservation, the corresponding pair of internal fermion lines must carry the same regulator mass M and this must be so for the two neighbouring vector vertices. Due to the extremely simple topology of the triangle graph, M thus automatically appears in both internal lines entering the axial-vector vertex and, consequently, the axial WI gets modified.

For completeness, let us add that the evaluation of the anomaly can be generalized in such a way that one need not rely on a particular regularization procedure; we adopted here the PV method since it is instructive and transparent for a first reading. The essence of the anomaly phenomenon can be described succinctly as follows. There is no consistent way of defining the contribution of the VVA graph, such that the naive vector and axial WI would hold simultaneously; in particular, when the vector WI are imposed, the axial WI inevitably picks up the extra term shown in (E.28). As we noted before, the ABJ anomaly has many interesting aspects and the relevant literature is rich. The reader seeking a deeper knowledge of the subject can find an appropriate introduction, e.g., in the review article [84] or in the comprehensive monograph [Ber]; needless to say, these sources contain many other relevant references.

References

[1] J. D. Wells, *Studies in history and philosophy in modern physics*, **62** (2018) 36.
[2] E. Fermi, *Z. Phys.* **88** (1934) 161.
[3] G. Gamow and E. Teller, *Phys. Rev.* **49** (1936) 895.
[4] T. D. Lee and C. N. Yang, *Phys. Rev.* **104** (1956) 254.
[5] C. S. Wu, E. Ambler, R. W. Hayward, D. D. Hoppes and R. P. Hudson, *Phys. Rev.* **105** (1957) 1413.
[6] R. L. Workman *et al.* (Particle Data Group), *PTEP* **2022**(8) (2022) 083C01.
[7] M. Goldhaber, L. Grodzins and A. W. Sunyar, *Phys. Rev.* **109** (1958) 1015.
[8] H. C. Andersen, *Favourite Fairy Tales*, Transl. M. R. James (Faber Fanfares, London, 1978), p. 157.
[9] D. R. Lide, ed. *CRC Handbook of Chemistry and Physics*, 77th Edition (CRC Press, Boca Raton, 1997).
[10] T. D. Lee and C. N. Yang, *Phys. Rev.* **105** (1957) 1671.
[11] L. D. Landau, *Nucl. Phys.* **3** (1957) 127; A. Salam, *Nuovo Cim.* **5** (1957) 299.
[12] R. P. Feynman and M. Gell-Mann, *Phys. Rev.* **109** (1958) 193.
[13] E. C. G. Sudarshan and R. E. Marshak, *Phys. Rev.* **109** (1958) 1860.
[14] G. Danby, J. M. Gaillard, K. Goulianos, L. M. Lederman, N. Mistry, M. Schwartz and J. Steinberger, *Phys. Rev. Lett.* **9** (1962) 36.
[15] F. Scheck, *Phys. Rept.* **44** (1978) 187.
[16] L. Michel, *Proc. Phys. Soc. A* **63**, 514 (1950); C. Bouchiat and L. Michel, *Phys. Rev.* **106** (1957) 170.
[17] J. Tiomno and J. A. Wheeler, *Rev. Mod. Phys.* **21** (1949) 153.
[18] F. Reines, H. S. Gurr and H. W. Sobel, *Phys. Rev. Lett.* **37** (1976) 315.
[19] R. C. Allen *et al.*, *Phys. Rev. Lett.* **55** (1985) 2401.
[20] N. Cabibbo, *Phys. Rev. Lett.* **10** (1963) 531.
[21] S. C. Adler *et al.* (E787 Collaboration), *Phys. Rev. Lett.* **79** (1997) 2204 [arXiv:hep-ex/9708031].
[22] M. Gell-Mann, *Phys. Lett.* **8** (1964) 214.
[23] S. Gershtein and Ya. Zeldovich, *Sov. Phys. JETP* **2** (1956) 576.
[24] C. Jarlskog, In *Proc. 1974 CERN School of Physics*, CERN Report 74-22, p. 1.
[25] M. Gell-Mann, *Phys. Rev.* **111** (1958) 362.

[26] Y. K. Lee, L. W. Mo and C. S. Wu, *Phys. Rev. Lett.* **10** (1963) 253.
[27] J. H. Christenson, J. W. Cronin, V. L. Fitch and R. Turlay, *Phys. Rev. Lett.* **13** (1964) 138.
[28] F. Reines and C. L. Cowan, *Phys. Rev.* **113** (1959) 273.
[29] T. D. Lee and C. N. Yang, *Phys. Rev. Lett.* **4** (1960) 307; B. L. Ioffe, L. B. Okun and A. P. Rudik, *Sov. Phys. JETP Lett.* **20** (1965) 128; see also T. Appelquist and J. D. Bjorken, *Phys. Rev. D* **4** (1971) 3726.
[30] T. Kinoshita and D. R. Yennie, In *Quantum Electrodynamics*, Ed. T. Kinoshita (World Scientific, Singapore, 1990), p. 1.
[31] J. M. Cornwall, D. N. Levin and G. Tiktopoulos, *Phys. Rev. D* **10** (1974) 1145 [Erratum-*ibid.* D **11** (1975) 972].
[32] M. Gell-Mann, M. L. Goldberger, N. M. Kroll and F. E. Low, *Phys. Rev.* **179** (1969) 1518.
[33] K. J. Kim and Y.-S. Tsai, *Phys. Rev. D* **7** (1973) 3710.
[34] K. Hagiwara, R. D. Peccei, D. Zeppenfeld and K. Hikasa, *Nucl. Phys. B* **282** (1987) 253.
[35] G. 't Hooft, *Nucl. Phys. B* **33** (1971) 173.
[36] C. H. Llewellyn Smith, *Phys. Lett. B* **46** (1973) 233.
[37] S. D. Joglekar, *Annals Phys.* **83** (1974) 427.
[38] R. Kleiss, In *Proc. 1989 Trieste Summer School in High Energy Physics and Cosmology.* The ICTP Series in Theoretical Physics, Vol. 6 (World Scientific, Singapore, 1990), p. 404.
[39] S. L. Glashow, *Nucl. Phys.* **22** (1961) 579.
[40] S. Weinberg, *Phys. Rev. Lett.* **19** (1967) 1264.
[41] A. Salam, In *Elementary Particle Physics, Proc. Nobel Symposium*, No. 8 (ed. N. Svartholm, Almqvist & Wiksell, Stockholm, 1968), p. 367.
[42] C. N. Yang and R. L. Mills, *Phys. Rev.* **96** (1954) 191.
[43] J. Goldstone, *Nuovo Cim.* **19** (1961) 154.
[44] M. Baker and S. L. Glashow, *Phys. Rev.* **128** (1962) 2462.
[45] J. Goldstone, A. Salam and S. Weinberg, *Phys. Rev.* **127** (1962) 965.
[46] Y. Nambu, *Phys. Rev. Lett.* **4** (1960) 380.
[47] P. W. Higgs, *Phys. Rev. Lett.* **13** (1964) 508; *Phys. Rev.* **145** (1966) 1156.
[48] G. 't Hooft and M. J. Veltman, *Nucl. Phys. B* **44** (1972) 189; *Nucl. Phys. B* **50** (1972) 318.
[49] G. 't Hooft, *Nucl. Phys. B* **35** (1971) 167.
[50] C. E. Vayonakis, *Lett. Nuovo Cim.* **17** (1976) 383; M. S. Chanowitz and M. K. Gaillard, *Nucl. Phys. B* **261** (1985) 379. G. J. Gounaris, R. Kögerler and H. Neufeld, *Phys. Rev. D* **34** (1986) 3257.
[51] P. W. Anderson, *Phys. Rev.* **130** (1963) 439.
[52] F. Englert and R. Brout, *Phys. Rev. Lett.* **13** (1964) 321.
[53] P. W. Higgs, *Phys. Lett.* **12** (1964) 132.
[54] G. S. Guralnik, C. R. Hagen and T. W. Kibble, *Phys. Rev. Lett.* **13** (1964) 585.
[55] T. W. Kibble, *Phys. Rev.* **155** (1967) 1554.
[56] S. Weinberg, *Phys. Rev. D* **7** (1973) 1068.

[57] B. W. Lee, C. Quigg and H. B. Thacker, *Phys. Rev. D* **16** (1977) 1519.
[58] L. Durand and J. L. Lopez, *Phys. Rev. D* **40** (1989) 207.
[59] G. C. Branco, P. M. Ferreira, L. Lavoura, M. N. Rebelo, M. Sher and J. P. Silva, *Phys. Rept.* **516** (2012), 1–102 [arXiv:1106.0034 [hep-ph]].
[60] S. Kanemura, T. Kubota and E. Takasugi, *Phys. Lett. B* **313** (1993), 155–160 [arXiv:hep-ph/9303263 [hep-ph]].
[61] A. G. Akeroyd, A. Arhrib and E. M. Naimi, *Phys. Lett. B* **490** (2000), 119–124 [arXiv:hep-ph/0006035 [hep-ph]].
[62] J. Hořejší and M. Kladiva, *Eur. Phys. J. C* **46** (2006), 81–91 [arXiv:hep-ph/0510154 [hep-ph]].
[63] G. Aad *et al.* [ATLAS], *Phys. Lett. B* **716** (2012), 1–29 [arXiv:1207.7214 [hep-ex]].
[64] S. Chatrchyan *et al.* [CMS], *Phys. Lett. B* **716** (2012), 30–61 [arXiv:1207.7235 [hep-ex]].
[65] M. Veltman, *The Higgs System*, in *Perspectives on Higgs Physics*, Ed. G. Kane (World Scientific, Singapore, 1993), p. 1.
[66] P. Sikivie, L. Susskind, M. B. Voloshin and V. I. Zakharov, *Nucl. Phys. B* **173** (1980) 189.
[67] M. S. Chanowitz, M. Golden and H. Georgi, *Phys. Rev. D* **36** (1987) 1490.
[68] M. S. Chanowitz and M. Golden, *Phys. Lett. B* **165** (1985) 105.
[69] H. Georgi and S. L. Glashow, *Phys. Rev. Lett.* **28** (1972) 1494.
[70] S. L. Glashow, J. Iliopoulos and L. Maiani, *Phys. Rev. D* **2** (1970) 1285.
[71] B. Aubert *et al.*, *Phys. Rev. Lett.* **33** (1974) 1404; J. E. Augustin *et al.*, *Phys. Rev. Lett.* **33** (1974) 1406.
[72] M. Kobayashi and T. Maskawa, *Prog. Theor. Phys.* **49** (1973) 652.
[73] S. Weinberg, *Phys. Rev. Lett.* **37** (1976) 657.
[74] M. L. Perl *et al.*, *Phys. Rev. Lett.* **35** (1975) 1489.
[75] K. Kodama *et al.* (DONUT Collaboration), *Phys. Lett. B* **504** (2001) 218 [arXiv:hep-ex/0012035].
[76] S. W. Herb *et al.*, *Phys. Rev. Lett.* **39** (1977) 252.
[77] F. Abe *et al.* (CDF Collaboration), *Phys. Rev. Lett.* **73** (1994) 2662 [Erratum-*ibid.* **74** (1995) 1891]; S. Abachi *et al.* (D0 Collaboration), *Phys. Rev. Lett.* **74** (1995) 2632 [arXiv:hep-ex/9503003].
[78] K. Fujikawa, B. W. Lee and A. I. Sanda, *Phys. Rev. D* **6** (1972) 2923.
[79] B. W. Lee and J. Zinn-Justin, *Phys. Rev. D* **5** (1972) 3121; 3137; 3155.
[80] Y.-P. Yao and C.-P. Yuan, *Phys. Rev. D* **38** (1988) 2237; J. Bagger and C. Schmidt, *Phys. Rev. D* **41** (1990) 264; H.-J. He, Y.-P. Kuang and X. Li, *Phys. Rev. D* **49** (1994) 4842.
[81] J. Hořejší, *Czech. J. Phys.* **47** (1997) 951 [arXiv:hep-ph/9603321].
[82] J. S. Bell, *Nucl. Phys. B* **60** (1973) 427.
[83] S. L. Adler, *Phys. Rev.* **177** (1969) 2426. J. S. Bell and R. Jackiw, *Nuovo Cim. A* **60** (1969) 47.
[84] J. Hořejší, *Czech. J. Phys.* **42** (1992) 241; 345.
[85] C. Bouchiat, J. Iliopoulos and Ph. Meyer, *Phys. Lett. B* **38** (1972) 519.
[86] C. P. Korthals Altes and M. Perrottet, *Phys. Lett. B* **39** (1972) 546.

[87] D. J. Gross and R. Jackiw, *Phys. Rev. D* **6** (1972) 477.
[88] A. Djouadi, V. Driesen and C. Junger, *Phys. Rev. D* **54** (1996), 759–769 [arXiv:hep-ph/9602341 [hep-ph]].
[89] J. J. Lopez-Villarejo and J. A. M. Vermaseren, *Phys. Lett. B* **675** (2009), 356–359 [arXiv:0812.3750 [hep-ph]].
[90] P. Vilain *et al. Phys. Lett. B* **335** (1994) 246.
[91] J. Hořejší and O. I. Zavialov, *Czech. J. Phys.* **39** (1989) 478.

Bibliography

[AbL] E. S. Abers and B. W. Lee, Gauge theories, *Phys. Rep.* **9C** (1973) 1.

[Adv] B. Maglich, ed. Discovery of parity violation in weak interactions, In *Adventures in Experimental Physics*, Vol. 3 (World Science Education, Princeton, 1972), p. 93.

[Alt] G. Altarelli, Collider physics within the standard model: A Primer, In *Lecture Notes in Physics*, Vol. 937, (Springer Open, 2017).

[Bai] D. Bailin, *Weak Interactions* (Adam Hilger Ltd., Bristol, 1982).

[BaL] D. Bailin and A. Love: *Introduction to Gauge Field Theory* (Institute of Physics Publishing, Bristol, 1993).

[Ber] R. A. Bertlmann, *Anomalies in Quantum Field Theory* (Oxford University Press, Oxford, 1996).

[Bil] S. M. Bilenky, C. Giunti and W. Grimus, Phenomenology of neutrino oscillations, *Prog. Part. Nucl. Phys.* **43** (1999) 1. [arXiv:hep-ph/9812360].

[Bra] G. C. Branco, L. Lavoura and J. P. Silva, *CP Violation* (Oxford University Press, Oxford, 1999).

[Brn] J. Bernstein, Spontaneous symmetry breaking, gauge theories, the Higgs mechanism and all that, *Rev. Mod. Phys.* **46** (1974) 7.

[BjD] J. D. Bjorken and S. D. Drell, *Relativistic Quantum Mechanics* (McGraw-Hill, New York, 1964); *Relativistic Quantum Fields* (McGraw-Hill, New York, 1965).

[CaG] R. N. Cahn and G. Goldhaber, *The Experimental Foundations of Particle Physics* (Cambridge University Press, Cambridge, 1991).

[Cah] R. N. Cahn, The eighteen arbitrary parameters of the standard model in your everyday life, *Rev. Mod. Phys.* **68** (1996) 951.

[Chg] S. J. Chang, *Introduction to Quantum Field Theory* (World Scientific, Singapore, 1990).

[ChL] Ta-Pei Cheng and Ling-Fong Li, *Gauge Theory of Elementary Particle Physics* (Oxford University Press, Oxford, 2000).

[CoB] E. D. Commins and P. H. Bucksbaum, *Weak Interactions of Leptons and Quarks* (Cambridge University Press, Cambridge, 1983).

[Col] S. Coleman, *Aspects of Symmetry* (Cambridge University Press, Cambridge, 1985).

[Dob] A. Dobado, A. Gómez-Nicola, A. L. Maroto and J. R. Peláez, *Effective Lagrangians for the Standard Model* (Springer-Verlag, Berlin Heidelberg, 1997).

[Don] J. F. Donoghue, E. Golowich and B. R. Holstein, *Dynamics of the Standard Model* (Cambridge University Press, Cambridge, 1992).

[FaR] Fayyazuddin and Riazuddin, *A Modern Introduction to Particle Physics* (World Scientific, Singapore, 1992).

[GeN] M. Gell-Mann and Y. Ne'eman, *The Eightfold Way* (Perseus Publishing, Cambridge, Massachusetts, 2000).

[Geo] H. Georgi, *Weak Interactions and Modern Particle Theory* (Addison-Wesley, Redwood City, 1984).

[Gre] W. Greiner and B. Müller, *Gauge Theory of Weak Interactions* (Springer-Verlag, Berlin Heidelberg, 1996).

[Gun] J. F. Gunion, H. E. Haber, G. Kane and S. Dawson, *The Higgs Hunter's Guide* (Perseus Publishing, Cambridge, 1990).

[HaM] F. Halzen and A. D. Martin, *Quark and Leptons: An Introductory Course in Modern Particle Physics* (John Wiley & Sons, New York, 1984).

[Hua] K. Huang, *Quarks, Leptons and Gauge Fields* (World Scientific, Singapore, 1992).

[Hor] J. Hořejší, *Introduction to Electroweak Unification: Standard Model from Tree Unitarity* (World Scientific, Singapore, 1994).

[ItZ] C. Itzykson and J.-B. Zuber, *Quantum Field Theory* (McGraw-Hill, New York, 1980).

[Jac] J. D. Jackson: *The Physics of Elementary Particles* (Princeton University Press, Princeton, 1958).

[Kay] B. Kayser, *The Physics of Massive Neutrinos* (World Scientific, Singapore, 1989).

[LaL] V. B. Berestetskii, E. M. Lifshitz and L. P. Pitaevskii, In *Quantum Electrodynamics*, Landau and Lifshitz Course of Theoretical Physics, Vol. 4 (Butterworth-Heinemann, Oxford, 1999).

[Lan] P. Langacker, *The Standard Model and Beyond* (2nd Edition, CRC Press, Boca Raton, 2017).

[Mar] R. E. Marshak, *Conceptual Foundations of Modern Particle Physics* (World Scientific, Singapore, 1993).

[Mes] A. Messiah, *Quantum Mechanics* (Dover Publications, Inc., Mineola, New York, 1999).

[MRR] R. E. Marshak, Riazuddin and C. P. Ryan, *Theory of Weak Interactions in Particle Physics* (Wiley-Interscience, New York, 1969).

[Nak] N. Nakanishi and I. Ojima, *Covariant Operator Formalism of Gauge Theories and Quantum Gravity* (World Scientific, Singapore, 1990).

[Pal] P. B. Pal, *An Introductory Course of Particle Physics* (CRC Press, Boca Raton, 2015).

[Pas] E. A. Paschos, *Electroweak Theory* (Cambridge University Press, Cambridge, 2007).

[PeS] M. Peskin and D. V. Schroeder, *An Introduction to Quantum Field Theory* (Addison-Wesley, Reading, MA, 1995).

[Pok] S. Pokorski, *Gauge Field Theories*, 2nd Edition (Cambridge University Press, Cambridge, 2000).

[Rai] L. O'Raifeartaigh, *Group Structure of Gauge Theories* (Cambridge University Press, Cambridge, 1986).

[Ren] P. Renton, *Electroweak Interactions* (Cambridge University Press, Cambridge, 1990).

[Ryd] L. H. Ryder, *Quantum Field Theory* (Cambridge University Press, Cambridge, 1996).

[Sak] J. J. Sakurai, *Modern Quantum Mechanics* (Addison-Wesley, Reading, MA, 1994).

[Tay] J. C. Taylor, *Gauge Theories of Weak Interactions* (Cambridge University Press, Cambridge, 1976).

[Tel] V. L. Telegdi, *Mind Over Matter: The Intellectual Content of Experimental Physics*, CERN Report 90–09 (1990).

[Ven] W. Venus, A LEP summary, Plenary talk at *Europhysics Conference on High Energy Physics*, Budapest, 2001, published in the JHEP Proceedings, hep2001/284.

[Vog] F. Boehm and P. Vogel, *Physics of Massive Neutrinos* (Cambridge University Press, Cambridge, 1992).

[Wat] P. Watkins, *Story of the W and Z* (Cambridge University Press, Cambridge, 1986).

[Wei] S. Weinberg, *The Quantum Theory of Fields*, Vol. II (Cambridge University Press, Cambridge, 1996).

Index

Printed in the United States
by Baker & Taylor Publisher Services